Asbjørn Rolstadås (Ed.)

Computer-Aided Production Management

With 169 Figures

Springer-Verlag Berlin Heidelberg New York
London Paris Tokyo

Professor Dr. Asbjørn Rolstadås
Production Engineering Laboratory
NTH-SINTEF
Richard Birkelandsv. 2b
N-7034 Trondheim-NTH, Norway

ISBN-13:978-3-642-73320-8   e-ISBN-13:978-3-642-73318-5
DOI: 10.1007/978-3-642-73318-5

Library of Congress Cataloging-in-Publication Data. Computer-aided production management / A. Rolstadås (ed.).   p.   cm. Includes index. ISBN-13:978-3-642-73320-8 1. Production management — Data processing. 2. Computer integrated manufacturing systems. 3. Flexible manufacturing systems. I. Rolstadås, A. (Asbjørn), 1944—. TS155.6.C646 1988 658.5'0028'5-dc19   88-6627   CIP

This work is subject to copyright. All rights are reserved, whether the whole or part of the material is concerned, specifically the rights of translation, reprinting, re-use of illustrations, recitation, broadcasting, reproduction on microfilms or in other ways, and storage in data banks. Duplication of this publication or parts thereof is only permitted under the provisions of the German Copyright Law of September 9, 1965, in its version of June 24, 1985, and a copyright fee must always be paid. Violations fall under the prosecution act of the German Copyright Law.

© 1988 IFIP International Federation for Information Processing,
16 place Longemalle, CH-1204 Geneva, Switzerland
Softcover reprint of the hardcover 1st edition 1988

The use of registered names, trademarks, etc. in this publication does not imply, even in the absence of a specific statement, that such names are exempt from the relevant protective laws and regulations and therefore free for general use.

Typesetting: K+V Fotosatz GmbH, Beerfelden;

2145/3140-543210 — Printed on acid free paper

Preface

The purpose of this book is to discuss the state of the art and future trends in the field of computerized production management systems. It is composed of a number of independent papers, each presented in a chapter. Some of the widely recognized experts in the field around the world have been asked to contribute. I owe each of them my sincere gratitude for their kind cooperation. I am also grateful to Peter Falster and Jim Browne for their kind support in helping me to review topics to be covered and to select the authors.

This book is a result of the professional work done in the International Federation of Information Processing Technical Committee IFIP TC5 "Computer Applications in Technology" and especially in the Working Group WG5.7 "Computer-Aided Production Management". This group was established in 1978 with the aim of promoting and encouraging the advancement of the field of computer systems for the production management of manufacturing, offshore, construction, electronic and similar and related industries. The scope of the work includes, but is not limited to, the following topics:

1) design and implementation of new production planning and control systems taking into account new technology and management philosophy;
2) CAPM in a CIM environment including interfaces to CAD and CAM;
3) project management and cost engineering;
4) knowledge engineering in CAPM;
5) CAPM for Flexible Manufacturing Systems (FMS) and Flexible Assembly Systems (FAS);
6) methods and concepts in CAPM;
7) economic and social implications of CAPM.

The Working Group has the following specific tasks in furthering the area of its scope:
a) maintaining liaison with the other appropriate national and international organizations and IFIP's TCs and WGs working in related fields;
b) conducting working conferences and symposia as deemed appropriate in furthering its scope;
c) stimulating and sponsoring research investigations and social studies into the various aspects of the topics of its scope.

This book is one of the ways in which the Working Group 5.7 can meet its objectives.

The book is divided into six parts. Each part contains a number of papers. Part I covers stages of development in production management and is intended as an introduction. Recently various PM concepts or philosophies have received a great deal of attention, and Part II contains papers on the most interesting ones, like MRP, JIT, OPT, period batch control, and all-embracing production control. The fundamental techniques that are used in production management are discussed in Part III. These cover graph theory, simulation, operations research, and artificial intelligence.

Parts II and III form the theoretical foundation of PM. Part IV addresses computerization of PM systems. It contains papers on databases, user interfaces, systems analysis, fourth generation languages, etc. In Part V some important aspects are further discussed. Scheduling, planning in FMS, forecasting, stock control, and integration of PM into CIM are discussed. In the last part, Part VI, some industrial applications are discussed. Different industries are covered, such as small companies, the car industry, the aircraft industry, job shops, the electro-mechanical industry, and the electronic industry.

IFIP WG5.7 has been working with a Glossary of Production Management that has been published. A short summary of this is presented as an appendix.

The viewpoints in this book are those of the authors as individuals and not necessarily those of their companies/organizations.

It is my sincere hope that this book will be of interest to production planners and computer experts as well as to academicians and industrial people.

Trondheim, Norway, February 1988   Asbjørn Rolstadås

Contents

Part I
Stages of Development in Production Management

Chapter 1. Production Management Systems
    Asbjørn Rolstadås 3

Part II
Production Management Philosophies

Chapter 2. MRP/MRP II
    John Harhen 23
Chapter 3. Just-in-Time Production – A New Formulation and Algorithm of the Flow Shop Problem
    Hajime Yamashina 37
Chapter 4. The Drum-Buffer-Rope (DBR) Approach to Logistics
    Oded Cohen 51
Chapter 5. Period Batch Control
    John L. Burbidge 71
Chapter 6. All-Embracing Production Control
    Gideon Halevi 77

Part III
Fundamental Techniques

Chapter 7. Graph Theoretical Approaches
    Peter Falster 97
Chapter 8. Simulation and Simulation Models
    Jim Browne 123
Chapter 9. Operations Research Models and Techniques
    Wing S. Chow, Sunderesh Heragu, and Andrew Kusiak 135
Chapter 10. Artificial Intelligence Approach to Production Planning
    Andrew Kusiak 149

Part IV
The Computerized Production Management System

Chapter 11. Databases
    Johan C. Wortmann 169
Chapter 12. User Interface
    Eero Eloranta 181
Chapter 13. Systems Analysis Techniques
    Guy Doumeingts 201
Chapter 14. Fourth Generation Languages
    Jarle Aaram 225
Chapter 15. Design of a Generalized Job Shop Control System and PM Packages
    Harinder Jagdev 233
Chapter 16. Validation of Job Shop Control Software — A Case Study
    Harinder Jagdev 253

Part V
Some Important Aspects of Production Management Functions

Chapter 17. Production Scheduling
    John R. King 267
Chapter 18. Production Planning and Scheduling in Flexible Manufacturing Systems
    Kathryn E. Stecke 281
Chapter 19. Forecasting and Stock Control
    Birger Rapp 289
Chapter 20. Integration of PM into CIM
    Gideon Halevi 303

Part VI
Industrial Applications

Chapter 21. Multi-Product Batch Production on a Single Machine — A Problem Revisited
    Samuel Eilon 319
Chapter 22. Production Control in Small Companies
    Kai Mertins 345
Chapter 23. Production Control in the Car Industry
    Wolfgang D. Thurow 355
Chapter 24. Production Control in the Aircraft Industry
    Bernd Hirsch and Gustav Humbert 363
Chapter 25. Job Shop Production Control
    Oddmund Oterhals 375
Chapter 26. Production Control in the Electromechanical Industry
    Siegfried Augustin 385
Chapter 27. Production Control in the Electronics Industry
    Ichiro Inoue 393

Appendix

A Drafted PM Glossary
John L. Burbidge 399

Address List of Authors 403

ns)
Part I

Stages of Development
in Production Management

Part I

Chapter 1
Production Management Systems*

Asbjørn Rolstadås

Asbjørn Rolstadås is Professor in Production Engineering at the University of Trondheim, Norwegian Institute of Technology, Norway. He has a masters degree (1968) and a Dr. ing. degree (1972) from the same university. He is also connected to SINTEF — the largest research organization in Norway — where he is responsible for research within production management. He has previously done research on automation (numerical control) of manufacturing processes and technological planning. His current research comprises project management and production planning and control, in special computer applications in these fields. Professor Rolstadås is the author of several books and many papers. He is a member of IFIP TC 5 and WG 5.7.

1 Introduction

Production management is defined as the planning and control of all activities necessary to produce a set of products. For a typical metalworking job shop type of company, production management thus comprises important functions such as:

- Product development and technological planning
- Production planning and control
- Materials planning and control
- Quality assurance
- Cost engineering.

The product development and the technological planning belong to the product cycle and will result in all technical specifications necessary for production. The remainder of the above-mentioned functions are all mainly concerned with the production cycle, i.e. the daily production. However, quality assurance and cost engineering also apply in the product cycle. In addition, quality assurance will transfer feedback between the two cycles. This is illustrated in Fig. 1, where the arrows indicate information flow. It can be seen how the product cycle pro-

---

* This paper was first published in IFIP Congress 86 proceedings.

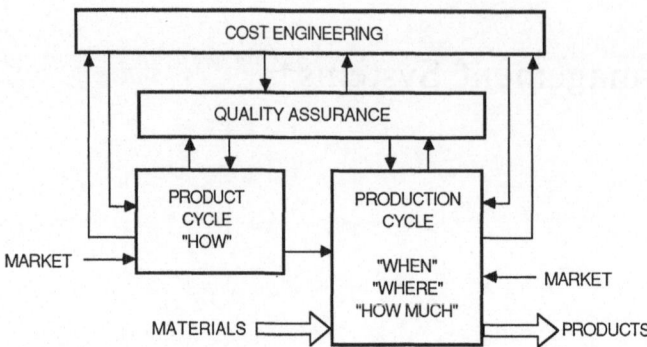

Fig. 1. Connectance between production management functions

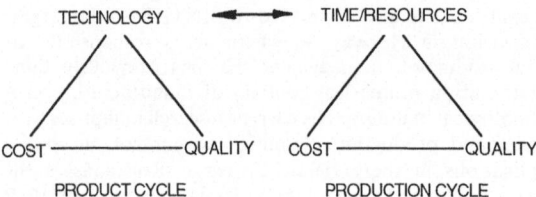

Fig. 2. Product and production triangles

vides information for the production cycle. This, together with market information, controls the materials flow.

Quality assurance will have feedback loops on several levels. The lowest levels are within each of the cycles (product and production). The level above will transfer feedback between the two of them. The same applies to cost engineering. However, in this case there is also a feedback loop connected to the quality assurance function.

The different functions interconnect as shown in Fig. 2. In, for instance, the production cycle, the triangle indicates that increased quality involves increased costs and increased time/resources. Production and materials planning and control are usually associated with time/resources, and this term is used in Fig. 2.

An overall management of production must take advantage of the connectance shown in Fig. 2 to be successful. In practice, however, these functions are usually the responsibility of different organizational functions and the mutual interaction is low. Also, these organizational functions may often have their own computer systems tailored to serve their own purpose and with very little communication with the others to serve needs they might have.

In this paper, the main focus is put on production and materials planning and control, but some thoughts are given to integration with other functions.

2 The Structure of Production Planning and Control Systems

A production planning and control (PPC) system will have an environment consisting of:

- Suppliers
- Customers
- Users
- Systems operation.

It could be said that the suppliers and customers are *affected* by the PPC systems whilst the users are *serviced* by it. Systems operation is responsible for providing a running system in accordance with approved requirements and specifications.

Table 1 lists the common functions of production planning and control, split under planning and control. Master scheduling is the long-term planning. Requirements planning, scheduling and loading, and dispatching constitute the detailed operational planning. Sequencing is the shop floor planning. Purchasing controls the materials flow into the company. Order processing is customer order handling.

The control functions are divided in four to control

- The size of the inventory
- The progress of the jobs
- The load on the machines
- The flow of materials.

Figure 3 shows a typical hardware/software configuration for a computerized production planning and control system. The hardware will consist of a computer with some mass storage possibilities and a number of terminals. The software will usually utilize a database management system and some basic software for handling the user dialogue on the terminal. On top of this are the user application programs.

The computer facilities may be dedicated or time sharing depending upon need and data-processing volume as well as type of hardware involved. Process-

Table 1. Production planning and control functions

Planning	Control
● Master scheduling	● Inventory control
● Requirements planning	● Progress control
● Scheduling and loading	● Load control
● Dispatching	● Materials control
● Sequencing	
● Purchasing	
● Order processing	

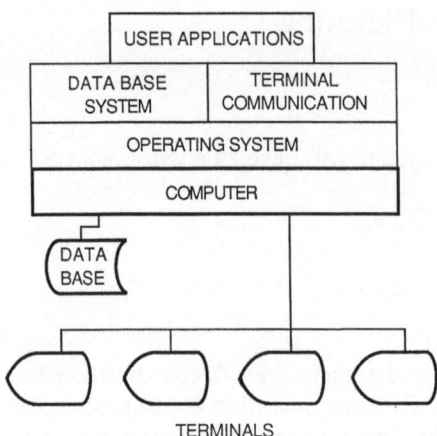

Fig. 3. Typical configuration for a computerized PPC system

Table 2. Application areas of production management (after Eloranta, Räisänen, and Sulonen [1])

Function	Manual	Batch	Real time
Order processing	65	32	16
Master scheduling	62	47	4
Inventory control	37	66	22
MRP (net)	43	37	4
Factory data collection	70	47	6

ing mode may be batch, real time, or a mix. Some of the planning functions require heavy processing capacity and are perhaps best served by batch processing. However, all control functions as well as easy planning functions and processing changes in plans ought to be done in real time.

Several surveys of the application of DP to PPC have been published. An extract from one of the more recent [1] is shown in Table 2. It is interesting to note that manual systems are still widely used and that most EDP systems are still based on batch processing. There is, however, good reason to believe that this picture will change in the next few years, since a new generation of systems is now growing.

Implementation of PPC systems in companies can be carried out in four different ways as shown in Fig. 4:

- Centralized
- Centralized/decentralized
- Decentralized
- Integrated.

The historical development of the solutions has been in the sequence shown above. Centralized systems constitute the classical approach. This has a centralized database, but terminals may well be distributed to the various user departments.

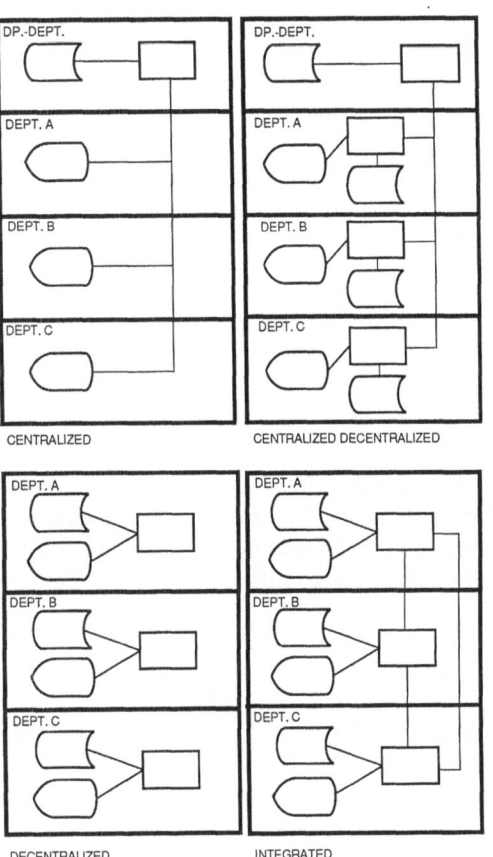

Fig. 4. Implementations of PPC systems

The next step is still a centralized solution since it consists of a centralized database. However, it has taken advantage of decentralized computer power and data storage. Thus some processing is done locally and some data are stored locally. These local data may be extracts of the main centralized database and they may be the day's updatings to be transferred later. The local data should be regarded only as workfiles, still leaving the central database as the main source.

The third stage is a fully decentralized solution. All data and systems are here decentralized. The figure only shows the systems responsibility, i.e. the responsibility for programs and data is decentralized. The computing power may still be provided from a central mainframe, but this is of no interest in this connection.

The third stage may be found in industry, but the first two are most common. The pure decentralized solutions have a disadvantage in lack of communication possibilities. However, this will gradually develop, which means that the systems will be decentralized and integrated.

In Fig. 4, integration is shown through the various departments of the company, but only focusing on PPC. In the future, integration with the cost control system, the technological planning system, and the machine tool operation will develop.

3 Production Management Fundamentals

Production of mechanical parts and products can be classified according to the following types:

- Single-piece production
- Batch production
    - Order production
    - Stock production
- Mass production.

In principle it is the lot size that will vary from one piece to an infinite number. Here batch production is considered. This can be of two types, dependent on how the order is released. Either the company produces from forecasts to stock or it starts production based on order receipt. In any case, the planning decisions to be taken are:

- How much to order
- When to order

assuming that the machines to be used are fixed (which in practice, of course, is not completely realistic).

Any planning system will aim at fixing those two decisions for all components and subassemblies. A PPC system can be considered as an information system containing several variables. Professor John L. Burbidge [2] has classified these variables in four categories:

- System design parameters
- Regulatory parameters
- Input parameters
- Output parameters.

This classification is shown in the production model in Fig. 5.

The values of the system design parameters (S) represent a choice between two alternatives. The regulatory parameters (P) are input variables to which a manager can assign an arbitrary value at will. The input variables (I) are imposed by the environment, and the manager of the production system is supposed to have no control over them.

The output variables can only be altered indirectly by the manager by changing regulatory or system design parameters. Professor Burbidge has identified more than 200 variables and he has defined the connectance between the variables. The idea is to consider effects on the total system based on changes in one isolated variable. There are two main types of connectance:

- Limitation
- Induction.

1. Production Management Systems

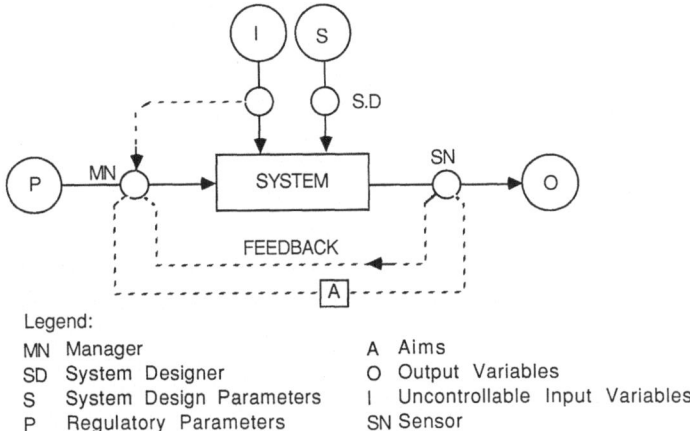

Legend:
MN Manager   A   Aims
SD System Designer   O   Output Variables
S   System Design Parameters   I   Uncontrollable Input Variables
P   Regulatory Parameters   SN Sensor

Fig. 5. Model of a production system

Limitation-changes in the value of a system variable may affect a connected parameter in the following ways: They may fix the value of the parameter or change the limits to the possible range of values that the parameter may have. Changes in the value of a system variable induce changes in the value of some other connected variable or variables.

Models like Burbidge's connectance model help us to understand the mechanisms involved in production control and are tools to ease the design of PPC systems.

PPC system exist in many variants. Almost every company seems to have its own version. However, there are some "basic" operations in PPC. To identify and specify these "base operations" is a future task of great importance. Some work in the field has, however, been done by Production Engineering Laboratory NTH-SINTEF, Norway and the Electrical Power Engineering Department, Technical University of Denmark.

The idea is that such base operations shall serve as building blocks for a complete system. They must therefore map the fundamental tasks of PPC.

In principle there would be at least four types of base operations:

- Transaction oriented
- Operation oriented
- Database management
- User communication.

The last two types are concerned with management of all data to and from the database and the design of the user dialogue over the terminal. Such functions can normally be covered by basic software supplied by the computer manufacturer.

The first two types are used to design the system transactions or, more precisely, the transaction programs. The transaction level is concerned with major PPC tasks whilst the operation level is concerned with more invariant detailed

Table 3. Some possible base operations in a production planning system

Base Operations		Function	Input	Output
Transaction level	Operation level			
Forecast	• Forecast model • Maintain model •	To forecast demand of a product over a period	Sales figures	Sales plan
Master scheduling	• Computer load profile • Load unlimited • Load limited •	To compute load profile and load on departments	Sales plan	Production plan
Requirements planning	• Net requirement • Gross requirement • Lot size • Time phasing	To compute requirements of all parts	Production plan	Requirement plan
Order	• Order point • Periodic order •	To release production or purchasing order	Inventory or stock level	Orders
Schedule	• Include operations • Operation time • Load unlimited • Load limited	Detailed scheduling and loading	Orders or requirement plan	Schedule
Dispatch	• • •	Dispatching and sequencing orders	Schedule	Orders

tasks. Examples are given in Table 3. These examples only concern planning functions. Similar functions will apply during control.

It seems that no one has yet been able to properly identify and describe the fundamental principles touched upon above. This is a clear hindrance for further development in the field and calls for considerable research.

4 Some Development Trends

The amount of data to be processed in production planning and control is very large. It is therefore quite natural that computers were applied early in this field. In fact the efficiency in a planning system is very heavily dependent upon how fast planning functions can be carried out or reviewed. The application of data processing has undergone marked development since the late 1950s. The development can be recognized as three generations:

1. Integrated batch systems (1950s)
2. Interactive real-time systems (1970s)
3. User-adapted systems (1980s).

In the first generation, batch processing in a computing centre was the dominant type. The systems were built integrated to collect all data in one common databank organized in files such as product file, product structure file, machine tool file, operations file, etc.

Separate program modules to cover functions such as forecasting, requirements planning, scheduling and loading, stock control, purchasing, etc. were developed and executed in sequence, operating on the same database.

The second generation was characterized by real-time processing. Terminals were brought onto the shop floor, and for the first time the users came in direct contact with the DP planning system.

Real-time systems offered the possibilities of fully updated data, and were soon better accepted amongst the users than the batch systems. An important change in principle was that decision-making was left with the user. Due to the direct contact with the user and the heavy influence on his working environment, the labour organizations became interested in influencing the system's design. User participation is today an accepted way of developing real-time systems.

The third generation is characterized by user adaptation. Systems to be developed in the 80s require a simple way of adaptation to users' requirements. The user participation on systems development to a large extent prohibits the use of standard system packages in the traditional sense. Most companies have therefore developed their own systems. The solution to this is very high-level languages which enable problem formulation in a problem-oriented language.

The 1990s will call for a new type of production and materials planning and control system. The development towards this fourth generation of management information systems has already started. It is influenced by four types of trend-setting factors, as follows.

- New production principles. In this group is found the application of group technology to form product-oriented materials flow based on cellular manufacturing. This calls for new production management fundamentals.
- Changed information requirement. More user-friendly human-machine communication is one important issue. Another is use of consequence analysis in planning.
- New system design criteria. There is a tendency towards decentralized systems and systems responsibility. Prototyping and user participation in systems development is an important issue. A stronger integration with the process control of the machine tools and handling devices can also be foreseen.
- New tools. A number of new tools is already available which will enable development of a completely new type of system. Among these are computer graphics, text processing, relational databases, and very high level languages.

5 Systems Development

The life cycle time of a computerized PPC system will consist of the following phases:

1. Prestudy
2. Specification
3. Design
4. Programming und implementing
5. Maintenance.

During the prestudy phase the problem is defined and aims and scope are determined. In the specification phase all functional requirements are developed, and user dialogue is specified. The design phase will transfer the functional specifications to system specifications, i.e. database structure, program sections, subroutines, etc. The programming and implementation phase will develop, test, and implement the programs and will provide documentation and train users. During maintenance it is important that user experience is collected and fed back to system improvements.

There are numerous systems development techniques available to formalize the process described in the preceding. Some of these are:

- JSP — Jackson Structured Programming
- SADT — Structured Analysis and Design Technique
- IDEFO — Developed from SADT for use in the US Air Force project "Integrated Computer Aided Manufacturing"
- ISAC — Information System Work and Analysis of Changes
- GRAI — Developed at the University of Bordeaux, France, especially for use in production management.

It is beyond the purpose of this paper to discuss all these methods. The idea is to demonstrate the variety available. However, a few remarks are given below concerning the GRAI method.

As mentioned, this method is one of the very few developed especially for production management. The main authors are G. Doumeingts and D. Breuil. The method has been applied with success in a number of French companies. A brief description has been published [3].

System development techniques can be characterized by:

- Decomposition principle
- Graphic tools
- Conceptual models.

The GRAI method uses a decomposition principle which is a combination of top-down and bottom-up. Decomposition is done by isolating decision centres in the company by means of a form shown in Fig. 6. The decision centres may be found on different levels in the organization.

After the identification of decision centres, a top-down analysis is run to establish their structure and their connectance to other functions. During the bottom-up analysis the elements of each decision centre are determined. Such elements are:

- Decisions made
- Variables affected
- Decision constraints
- etc.

1. Production Management Systems

FUNCTIONS DECISION LEVELS HORIZON PERIOD	EXTERNAL INFORMATION	TO DESIGN	TO PURCHASE	TO PLAN	TO MANUFACTURE	TO INSPECT	TO DELIVER
H P							
H P					DECI- SION CENTER		
H P							

Fig. 6. Construction of decision centers in GRAI method

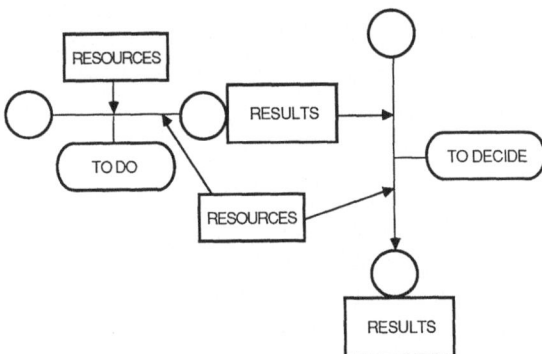

Fig. 7. Example of GRAI nets

The results are presented as "GRAI nets" as shown by the example in Fig. 7.

A problem in designing systems is to take full advantage of the user's experience and requirements. Even if the system is fully defined through the user and system specifications, it is difficult for the end-users to fully understand the properties of the system until they have tried it in real life. Experience shows that after implementation many user viewpoints are created, calling for a changed design.

To overcome this problem, a system prototype can be developed. A prototyping tool called "base operations" was developed by Schmidt, Theilgaard, Oterhals, and Jordanger in a joint Norwegian/Danish research project. A set of basic routines for handling:

- Database management
- User communication
- PPC functions

was programmed in APL. These functions represent a base or a framework which can be easily modified according to users' wishes. The prototype functions like the real system, but does not use a full-scale database, and the database and the programs are not optimized. The method has been applied in practice and shows that a prototype can be developed in a few weeks. Prototyping can also be applied to decide on specifications for standard packages that a company intends to implement.

Program development for PPC systems is time-consuming and expensive. In addition the requirement for maintenance is large, i.e. the systems seem to need continuous modifications. To overcome this problem, very high-level languages or fourth generation tools have been developed. During the last 2 or 3 years a large number of such tools have come on the market, many of them applicable on a personal computer.

For management information systems the two oldest and perhaps best known systems are:

− Admins
− Mims.

However, a number of other systems could be mentioned, such as:

− Basis
− CSP
− Focus
− Knowledgman
− Mapper
− Mimer.

Again a discussion of all these systems falls beyond the purpose of this text. The point is that the necessary tools for development of computer systems for PPC exist in a large variety today. Thus the problems of production management are shifted away from problems of providing the computerized solution back to problems of finding the proper PPC techniques.

6 User Communication

Modern PPC systems use general display terminals for user communication, almost without exception. Of course, paper listings are still used, but the "immediate" dialogue is handled over a display terminal. Mostly the displays are of alphanumeric type. However, graphic colour displays are to an increasing degree being used, especially in connection with systems based on personal computers.

Looking a few years ahead, there is reason to believe that graphic colour displays will be dominant since they provide possibilities of presenting data in a more "readable" format than computer listing.

A management information system can use computer graphics in several ways as illustrated in Fig. 8.

Fixed dialogue means that the user has access to information through a set of preprogrammed transactions for storing, updating, and retrieval. This is the normal situation in production management applications (stock control, schedul-

1. Production Management Systems

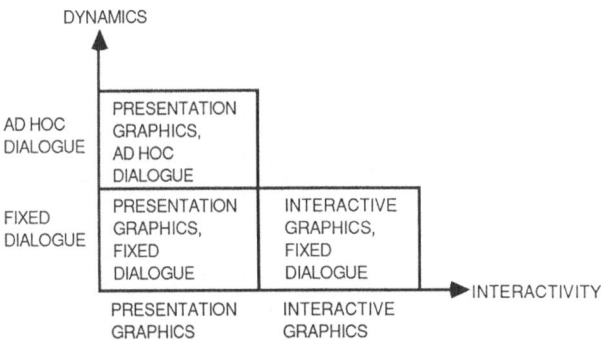

Fig. 8. Classes of graphic systems

ing, ...). *Ad hoc* dialogue means that the user spontaneously defines his information needs in a non-procedural way, – hence, allowing access to selections/combinations of information not foreseen by the preprogrammed transactions. *Ad hoc* dialogue should be possible only for information retrieval, as the information contained in the system could otherwise be changed in an uncontrolled way.

Presentation graphics means that information is retrieved from the database and presented to the user as diagrams. Interactive graphics means that the user can change a picture presented to him and update stored information through such changes.

User communication is extremely important in real-time systems since the enduser is directly confronted with the terminal. At the same time the question of what information should be presented is very important. Report generators help the user to compile his own report the way he needs it. Computer graphics

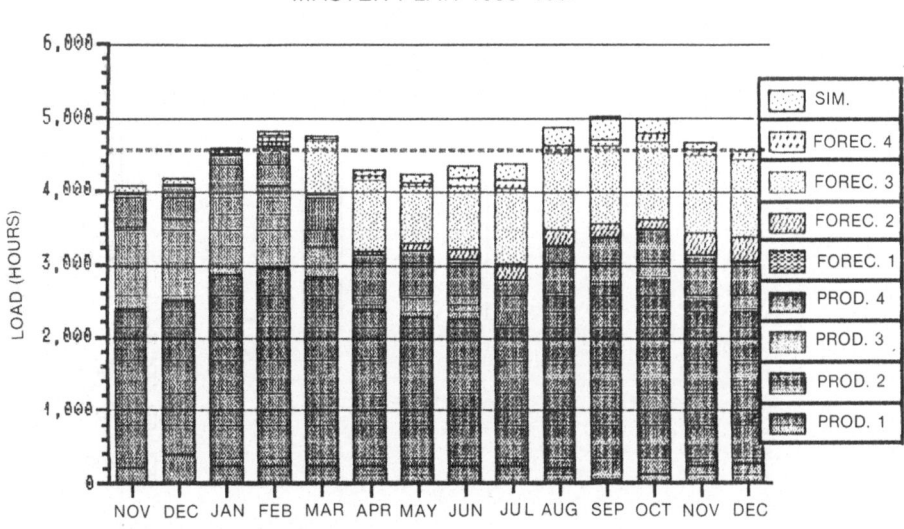

Fig. 9. Example of a graphical output of a master schedule

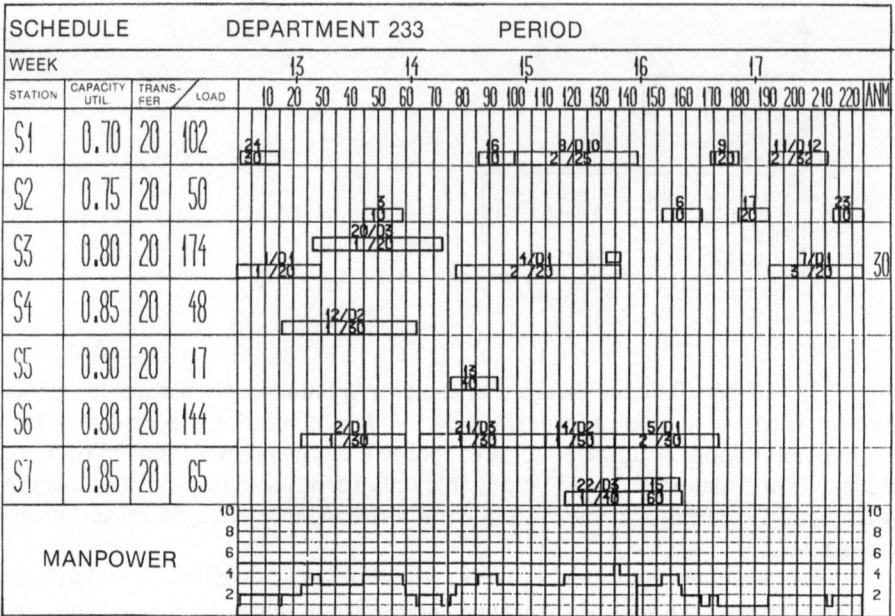

Fig. 10. Example of graphical scheduling

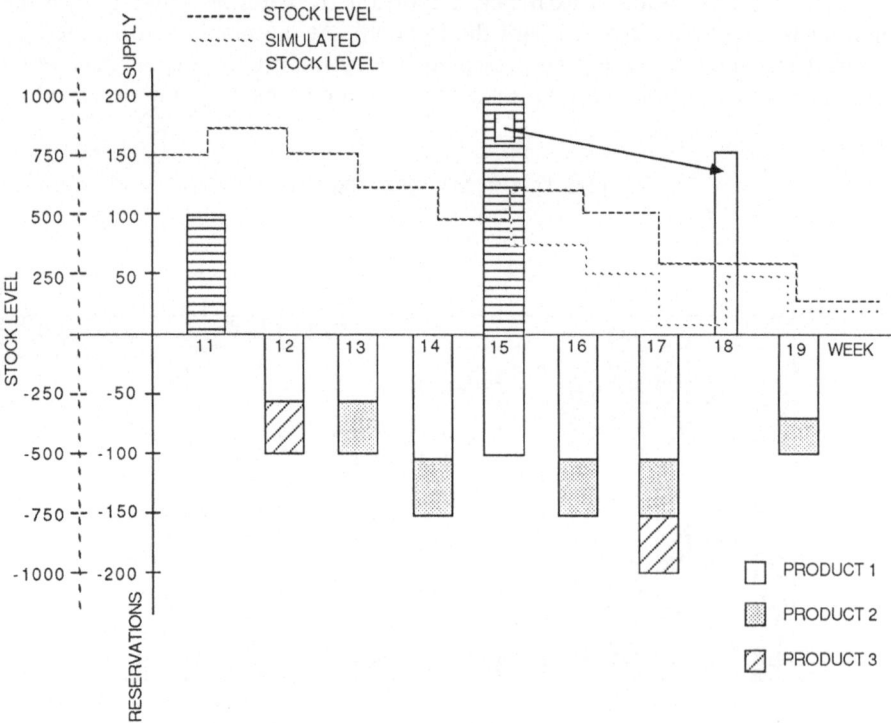

Fig. 11. Example of application of computer graphics to stock control

enables presentation of data in the form of curves, bar charts, histograms, etc., which is much easier to interpret than tables of text and numbers. Figure 9 shows an example of a master plan presented as a histogram. Total load in hours is given for each month. The histogram shows the load for each of the products 1, 2, 3, and 4. Both orders and forecasts are recognized in addition to simulated load. Master scheduling is an area well suited to business graphics. (Colours make it easier to distinguish the different load patterns.) Another well-suited area is scheduling interactively using a Gantt-form. The display is then actually used as a sketch-pad. An example is given in Fig. 10.

There have been tremendous efforts in creating automatic scheduling programs, but success has been very limited. This is because scheduling is not a closed mathematical problem – no priority rules are formal. Thus effective scheduling requires the intuition and creativeness of human thought. The idea of interactive, graphic scheduling is to combine the strengths of the computer with these qualities.

Other examples of application of computer graphics are stock control and precedence networks. Figure 11 shows an example of stock control. Graphical representations of planned supplies and reservations, with resulting levels, gives better insight. By different colours, different sources of reservations and supplies can be identified. Consequences of moving supplies/reservations in time, or of changing volumes, are easily shown (thin bar, dotted line).

7 Some New Production Management Principles

The production environment is changing in order to obtain more efficient production. A higher degree of automation is being obtained. Changed production layout based on more rational materials flow is developing. Product-oriented layout seems to have advantages over process-oriented. Plans need more frequent updating, calling for consequence analysis of decisions.

In total, these factors call for completely new planning principles which in the next step will require new systems. This will probably be the prospect for the late 1980s and early 1990s.

In this paper only a few factors are briefly mentioned. They are:

- Cellular manufacturing
- Simulation
- Just-in-time production
- Optimized production.

Cellular manufacturing is based on the complete manufacturing of a family of parts in a group of machines. The whole group of machines will be occupied while such a lot (containing a family) is being manufactured. This makes today's production planning theory obsolete. Requirements planning explodes products into single parts, and groups them into optimal lots. Here the grouping needs to be done to families. Furthermore, the theory of lot sizing, balancing setup and inventory costs, will not be valid. For these and other reasons the whole production planning process from requirements planning to further details will have to be redesigned. Only limited work has been done so far, mainly on the scheduling of cells and FMS systems.

Simulation is a tool extremely powerful in PPC. Simulation can serve two purposes:

- System design
- Decision support.

By system design is understood the design of a production system. Layout problems provide one good example. The change of production layout in a workshop is a rather comprehensive task. Before physical rearrangement of equipment, it is worthwhile to explore all consequences. Simulation models provide an appropriate tool for this task. By defining routing, priority rules, operation times, etc., a set of orders can be simulated with respect to load on machine tools, queues, inventories, etc.

To be able to perform a simulation, a model must be designed. This model must map the problem that is to be investigated. For example, it would be possible to simulate a production plan or a change in a plan and to outline all associated consequences. This would require a model containing the relevant machine tools, all orders to be run, and all process routings. A large number of simulation languages is available, and they offer good support in establishing the model. Once the model is established, the order can be dispatched and the load on machine tools and the flow of production can be simulated. In this way simulation can aid the taking of decisions concerning:

- Schedule
- Change in capacity
- Use of overtime
- Priority rules of sequencing
- Delivery times
- Stock levels
- Use of subcontractors
- Labour allocation
- Use of alternate routings
- etc.

If the model is properly designed, there is practically no end to the possibilities of decision support. However, the costs of establishing a model may be comprehensive, since the quality of the simulation results are dependent on the quality of the model.

In recent years Japanese companies, especially in the car manufacturing industry, have had great success with a new production philosophy – "Just-in-time" production. The philosophy is very simple: produce the required items just-in-time for the moment they are needed, neither sooner nor later. The ultimate goal is to reduce production costs by eliminating unnecessary inventories. Important factors in this context are:

- Reduction of lead time
- Correct quality
- Smooth production
- Flow manufacturing
- Balance of flow, not capacity.

Perhaps the most important of these issues is reduction of production lead time. Lead time will in principle be minimum if the lot size is 1 and the setup time is 0. Achievement of correct quality involves aiming at zero defects. Quality circles seem to be an important tool to achieve this.

Smooth production requires flexible capacity, which in many European countries is unrealistic. Flow type layout makes control and manufacture easier.

The balance of flow and not capacity means that the flow of material should exactly meet the requirement. Full utilization of capacity is not important, since it will result in unnecessary inventories.

Another new PPC philosophy is optimized production technology (OPT). OPT is an analytical technique aiming at "making money". Three criteria are used:

- Throughput
- Inventories
- Operating expenses.

OPT is an available software product designed to achieve increased throughput with decreased inventories and minimum operating expenses. The key is optimized schedules. OPT focuses on bottlenecks in the production. An hour lost at a bottleneck is an hour lost for the total system. The setup times at a bottleneck should be minimized since this will directly increase production capacity. The batch size on a bottleneck machine should thus be large. OPT claims that inventories are a function of the amount required to keep the bottlenecks busy. Another important aspect of OPT is the recognition of two lot sizes: the transfer batch and the manufacturing batch.

The principles of OPT and just-in-time production seem very promising. These techniques, together with changed production flow, will enable more rational production in the future. Simulation is a necessary tool to explore the consequences of changed layout or new PPC principles.

8 References

1. Eloranta, E., Räisänen, J., Sulonen, R., Computer Applications in Production and Engineering, pp. 40–53, Warman, E.A. (ed.). North-Holland, Amsterdam, 1983
2. Burbidge, J. L., Production Management Systems, pp. 3–16, Hübner, H. (ed.). North-Holland, Amsterdam, 1984
3. Maloubier, H., Breuil, D., Doumeingts, G., Garard, J., Advances in Production Management Systems, pp. 127–142, Doumeingts and Carter (eds.). North-Holland, Amsterdam, 1984

Part II
Production Management Philosophies

Chapter 2
MRP / MRP II

John Harhen

John Harhen is a senior software engineer with Digital Equipment Corporation, in Marlboro, Massachusetts, USA. He holds a B.E. and M. Enge. Sci. in Industrial Engineering from University College Galway, Ireland. He is currently pursuing a Ph.D. programme at Univ. Mass. and his research interests relate to the application of knowledge-based systems technology to manufacturing. He has participated on the ESPRIT project on manufacturing control systems for computer-integrated manufacturing. He has also participated in the implementation and education of MRP II at Digital's Clonmel plant, an implementation which is now rated as "Class A". He is a co-author of the book "Production Management Strategies: a CIM perspective". He is a member of the Institute of Engineers of Ireland, and is certified as a fellow in production and inventory management by APICS.

1 Introduction

MRP has been quite likely the most widely implemented large-scale Production Management System since the early 1970s, with several thousand[1] MRP type systems implemented. My aim here is to review the state of the art of MRP technology. This discussion addresses both the following:

- *Materials Requirements Planning (MRP)*, which generates a schedule of manufacturing and purchase orders to meet a given demand
- *Manufacturing Resource Planning (MRP II)*, which is an extension of MRP to support the integrated management of many of the functions of the manufacturing enterprise.

---

[1] While it is difficult to assess the total number of MRP implementations, it is reasonable to suggest that it is of the order of several thousands. The 1982 survey by Anderson et al. [1] of APICS members in 2 of the 14 APICS regions in the USA indicated 433 companies using MRP in these two regions alone. Promotional material from the more popular independent software companies selling MRP packages often claims of the order of 200 to 600 implementations each. A recent study [2] of firms employing 100 of more employees in South Carolina indicated that 31% of these firms were users of MRP. A 1981 publication by Wight [3, p.75] presents without support a claim that 8000 firms in the USA were using some form of MRP by mid-1981.

Three perspectives are adopted in reviewing the state of the art:

1. The state of practice of MRP/MRP II
2. Current research and development related to MRP/MRP II
3. The state of the philosophical debate underlying MRP/MRP II.

Before embarking on this discussion, a little history is presented, followed by a brief review of the production management procedures embedded within MRP and MRP II systems.

2 History of MRP

MRP originated in the early 1960s in the USA as a computerized approach to the planning of materials acquisition and production. The definitive textbook on the technique is undoubtedly Joseph Orlicky's [4] 1975 publication.

Undoubtedly the technique had been manually practised in aggregate form prior to the Second World War in several locations in Europe. However, what Orlicky realized was that a computer enabled the application of the technique at a detailed level that would make it effective in managing manufacturing inventories.

These early computerized applications of MRP were built around a Bill of Materials (BOM) processor, which converted a discrete plan of production for a parent item into a discrete plan of production or purchasing for component items. This was done by exploding the requirements for the top-level product, through the Bill of Materials, to generate component demand and then comparing the projected gross demand with available inventory and open orders, over the planning time horizon and at each level in the BOM. These systems were implemented on large mainframe computers and run in centralized materials departments for large companies. As time passed, the installations of the technique became more widespread and various operational functions were added to extend the range of task that these software systems supported. In particular, these extensions included Master Production Scheduling (MPS), Rough-Cut Capacity Planning (RCCP), Capacity Requirements Planning (CRP), Production Activity Control (PAC)[2], and Purchasing. The combination of the planning (MPS, MRP, CRP) and execution modules (PAC and Purchasing), with the potential for feedback from the execution cycle to the planning cycle, was termed *Closed Loop MRP*. With the addition of certain financial modules, as well as the extension of Master Production Scheduling to deal with Master planning and the support of Business Planning in financial terms, it was realized that the resultant system offered an integrated approach to the management of manufacturing resources. This extended MRP was labelled *Manufacturing Resource Planning* or MRP II. Since 1980 the number of MRP installations has continued to increase

---

[2] Production Activity Control is the term favoured by the American Production and Inventory Control Society to cover activities traditionally described by Shop Floor Control.

as MRP applications became available at lower cost on minicomputers and microcomputers.

MRP's popularity stems from the "MRP crusade" which was launched by the American Production and Inventory Control Society (APICS) in the early 1970s. The focus of this campaign was to convince people that MRP was the solution, as it represented an integrated communication and decision support system that supports the management of the total manufacturing business. It was emphasized that to succeed, MRP implementation programmes required management commitment and total work-force education. The role of optimizing techniques drawn from Operations Research and management science was frowned upon. The real problems, it was declared, were problems of discipline, education, understanding, and communication. This message was promoted by APICS and a stream of almost evangelical consultants, and finally echoed by a computer industry eager to expand the range of applications it could offer.

One of the significant reasons why MRP was the technique that was adopted, was that it made use of the computers ability to centrally store and provide access to the large body of information that seemed necessary to run a company. It helped coordinate the activities of various functions in the manufacturing firm such as engineering, production and materials. Thus, the attraction of MRP II, lay not only in its role as decision making support, but more importantly in its integrative role within the manufacturing organization.

Today, there is some concern as to how systems of the style of MRP can be integrated into a Computer Integrated Manufacturing (CIM) environment and also concerns about its adequacy as compared to alternate philosophies such as Kanban/Just In Time (JIT) and proprietary techniques such as Optimized Production Technology (OPT) [6]. Questions are being raised as to the effectiveness of MRP, as managers see the extent of the effort required to implement such systems and then the frequent failures to realize the promised benefits.

3 The Technique of Materials Requirements Planning

The starting point for MRP is the recognition that products to be manufactured or assembled can be represented by a Bill of Material. A Bill of Material (BOM) describes the parent/child relationship between an assembly and its component parts or raw material – this is illustrated in Fig. 1. As can be seen, the bills of material may have an arbitrary number of levels and have typically purchased items at the bottom levels in the hierarchy. An implicit assumption is that there is an adequate part numbering system in the company to differentiate all parts and components at various stages of manufacturing, where planning intervention may be required. The MRP system is based very simply on the fact that the BOM relationship allows one derive the demand for component material based on the demand for the parent item. MRP was thus proposed as a technique for managing dependent component demand, by transmitting the independent demand for top level products and spares, through the component hierarchy as represented by the BOM.

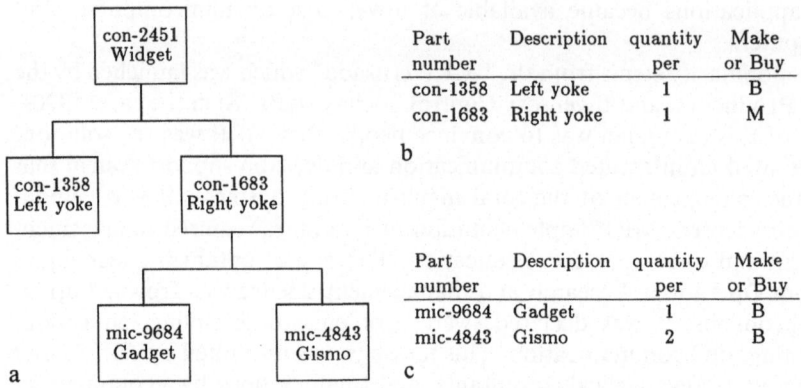

Fig. 1. a Bill of Material hierarchy. **b** BOM for "Widget". **c** BOM for "Right Yoke"

Widget	1	2	3	4	5	6
Production plan	100	100	100	100	100	100

Right yoke	1	2	3	4	5	6
Gross requirements	100	100	100	100	100	100
Scheduled receipt		250				
On hand 225	125	275	175	75	-25	-125
Net requirement					25	100
Planned Order release	250					

Gismo	1	2	3	4	5	6
Gross requirements	500					

action message to "schedule a new order"

Fig. 2. Illustration of MRP technique

The MRP calculation procedure is extremely simple. In MRP, time is assumed to be discrete. Typically, it is represented as a series of one-week intervals, though systems which operate on daily planning periods are readily available.

Consider for illustration how the demand for the "Right yoke" and "Gismo" in Fig. 1 might be derived from the "Widget". The following are the inputs to the planning process. The lead time for building a "Right yoke" is 4 weeks for a lot size of 250. The current inventory on hand of the "Right yokes" is 225 and there is an open manufacturing order due for completion in week 2. There is a production plan to build 100 widgets every week, and this gives rise to a gross requirement each week for 100 "Right yokes".

The application of the MRP technique to the problem is illustrated in Fig. 2. The time horizon in the simple example is 6 weeks[3]. This simple example illustrates the technique of MRP through the following procedures:

---

[3] In real MRP II systems, time horizons should extend beyond the longest cumulative lead time for a product. Anderson et al. [1] found in a 1981 study that the average MPS was 40 weeks.

1. Netting off the gross requirement against projected inventory and taking into account any open orders scheduled for receipt, thus yielding net requirements
2. Conversion of the net requirement to a planned order quantity using a lot size
3. Placing a planned order in the appropriate period by backward scheduling from the required date by the lead time to fulfil the order for that component
4. Generating appropriate action and exception messages to guide the user's attention
5. Explosion of parent item planned production to gross requirements for a component.

This procedure starts out at the top-level production plan and works down level by level and component by component of the BOM, until all parts are planned. Demand for a component can derive from any of the products in which it is used, as well as independent spares parts orders.

Much of what remains to be known about the mechanics of the technique of MRP really comprises implementation details, though important from the perspective of operating an MRP system. Some fall within the domain of software engineering, some within the domain of decision science. None really changes the fundamental procedure. In any MRP system, the decision-making procedures are no more complex than the basic arithmetic which I have just illustrated.

These residual features involve such concerns as the following.

- *Net change or regenerative*, which describe alternative approaches to the recalculation of an existing plan on the basis of changes in the input to that plan. Net change MRP systems enable continuous replanning, whereas regenerative technique are typically applied in weekly replanning cycles.
- *Bucketed or bucketless*, i.e. the time representation used in the system. Bucketed systems limit the time horizon that may be considered and the granularity of timing that may be ascribed to an order. Bucketless systems enable daily visibility to an order's date of requirement.
- *Low-level coding*, which helps determine the sequence in which processing of part requirements is done.
- *Firm planned order*, which describes how a planner may force the MRP system to plan in a particular way, thus overriding lot size or lead time rules.
- *Pegging*, which allows the user to identify the sources of demand for a component.
- *Calculation of lot sizes*, though they were never emphasized as being at the core of the MRP paradigm [6].
- *Database design*, the decision of how to store the various data used in the system.

4 Manufacturing Resource Planning: MRP II

Manufacturing Resource Planning represents an extension of the functionality of the MRP system to deal with many other functions beyond materials planning,

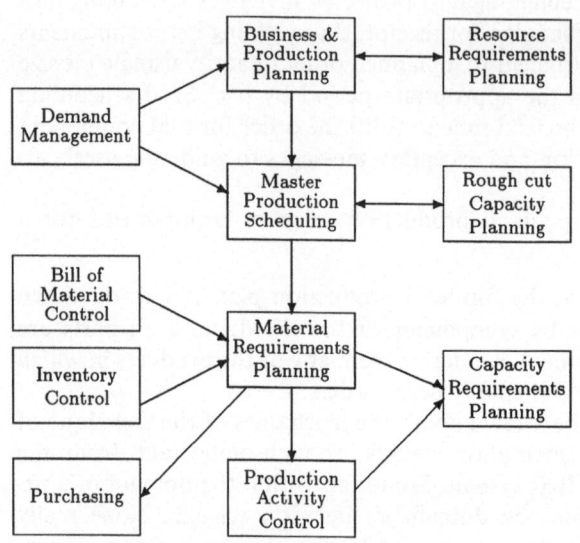

Fig. 3. Manufacturing Resource Planning: MRP II

inventory control, and BOM control. These extensions were natural and not very complicated, as for instance the addition of transaction-processing software to help the purchasing, inventory, and financial functions of the firm. In supporting the extension of decision-making support, similar and quite reasonable assumptions are made and similar procedures are applied as those of MRP. In this way, MRP was extended to support Master Planning, Rough-Cut Capacity Planning, Production Activity Control, and Capacity Requirements Planning.

The modular structure of a typical MRP II system is shown in Fig. 3. A brief description of the various modules is given below.

Master Planning and Rough-Cut Capacity Planning are two techniques that are employed in parallel. MRP systems vary somewhat in the support they give to these functions. The functions are well treated by Berry et al. [7].

Typically, the master schedule allows for a system-generated forecast, a manually entered forecast, a schedule of actual customer orders received, and a very simple algorithm for combining the above into a working forecast of demand. The user can specify a master production schedule. Each order is treated as a firm planned manufacturing order. There is a netting process very similar to MRP, as the forecast demand is netted with the MPS and current inventory to generate a projection of inventory on-hand and available to promise. The differnce with MRP is that demand will only propagate from the scheduled MPS, and not from the projected requirements. This means that the MPS will not influence manufacturing or purchasing orders without the intervention of the master planner.

In Rough-Cut Capacity Planning (RCCP), a Bill of Resource is attached to each of the master scheduled products. This bill of resource describes what capacity of various key facilities and/or people is required to produce one unit of the item. The provision is made for using lead time offsets. No consideration is given to component inventories and the rough-cut capacity plan is driven

2. MRP/MRP II

solely by exploding the MPS against the Bill of Resource. In this way, MPS and the rough-cut capacity requirements plan can be developed interactively.

Master planning systems often allow for planning at multiple levels, and similar techniques to MPS and RCCP can be applied at the more aggregate business planning and production planning and the associated resource requirements plan.

MRP output can be used for Capacity Requirements Planning. This is done by exploding the manufacturing orders (planned and actual) through the routing file in the Production Activity Control system. This generates a detailed profile of what capacity is required in each work-centre. Required capacity is then compared with available capacity, and overload/underload conditions are identified. CRP generates a more detailed capacity profile than that generated by RCCP. However, CRP is only performed after each MRP run, and these runs are typically performed only once a week. Therefore, CRP does not facilitate interactive planning, and it is used primarily as a verification tool.

At the Production Activity Control level, functions have been added that describe the process routing for fulfilling a manufacturing order (i.e. the sequence of steps a part goes through in its manufacturing process), as well as work-centre and standard time information. Backward scheduling from the MRP due date is used to determine when operations should start, by taking into account the work-centre lead time as well as the setup time, run time, and lot size for an order. Work-in-progress (WIP) tracking facilitates the prioritizing of manufacturing orders through a daily dispatch list. Capacity control, in the form of input−output control, is provided in some systems.

This limited summary of the production management functions of MRP II systems is sufficient to introduce a review of the state of the practice, research, and art of MRP systems.

5 The Effectiveness of MRP II Systems

There is a significant divergence between what is available in MRP software packages and what has turned out to be useful. Therefore, to assess MRP/MRP II, and to understand what is the state of the art, is not a question of understanding the range of functions and options that are embedded in MRP II software. If that were the question, then the answer would be that state-of-the-art MRP is represented by a bucketless and net change MRP II system, which puts appropriate emphasis on master planning and supports the other functions (Capacity, PAC, etc.) adequately.

The question of what is the state of the art of MRP II, relates primarily to understanding the effectiveness of MRP systems for the companies that use them.

The pragmatics of operating an MRP system, though covered by Orlicky, find their loudest expression in the works of Ollie Wight [3]. Wight proposed a classification scheme that rates how well companies operate their MRP system.

The scheme is quite simple and involves a series of 25 questions which relate to the technical capability of the MRP software package, the accuracy of supporting data, the amount of education that has been provided to the employers, and the results achieved by using the system. MRP system use is rated between Class "A", which represents excellence, and Class "D" which represents a situation where the only people using the system are those in the MIS department.

Among the criteria that measure effective use of MRP are the following:

- MRP should use planning periods no larger than a week.
- The frequency of replanning should be weekly or more frequent.
- If people are effectively using the system to plan, then the shortage list should have been eliminated.
- Delivery performance is 95% or better for vendors, the manufacturing shop, and the MPS.
- Performance in at least two of the following three business goals has improved: inventory, productivity, and customer service.

The various surveys taken through the years indicate several problems with MRP system implementations.

- Only a very small percentage of users of MRP consider themselves to be successfully operating their MRP systems. Many systems are "installed" as opposed to "implemented", i.e. the formal system is not the real system. The 1982 APICS study by Anderson et al. [1, 8] showed that only 9.5%[4] of MRP users considered themselves to be *"Class A" MRP II* users, while 61.3% considered themselves to be *"Class C"* or *"Class D" MRP II* users.
- Master Production Scheduling (MPS) is not computerized by MRP users as often as might be expected. The Anderson et al. study found that 52.2%[5] of MRP users had computerized their MPS.
- Capacity Requirements Planning also has a relatively low utilization by MRP users. The Anderson study found that 37.7%[6] of MRP users had computerized their Capacity Planning System.
- In relatively few cases is computerized Production Activity Control implemented. Anderson et al. found that 30.5%[7] had done so.

On the other hand, important performance improvements stemming from MRP II were reported. For instance in the Anderson et al. study, the following improvements were identified.

- Average inventory performance improved from 3.2 to 4.3 turns.
- Average delivery lead times fell from 71.4 to 58.9 days.

---

[4] The LaForge et al. study [2], though more recent (1986), had a much smaller sample size. They found the percentage of *"Class A" MRP II* users was 25%. The authors did not draw as a conclusion that the percentage of firms with "Class A" MRP II had increased, and chose instead to say that the results supported the general pattern of MRP performance found by Anderson et al.

[5] The LaForge study [2] found the percentage of MRP users with computerized MPS was 61%.

[6] LaForge [2] reported 42%.

[7] LaForge [2] reported 52%.

- Average delivery performance rose from 61.4% to 76.6%.
- Average number of expediters fell from 10.1 to 6.5.

The findings were also supported by LaForge et al. What emerges from these surveys is that MRP II has definitely brought benefits to these manufacturing companies and that the costs associated with the implementation were on average well repaid. "Class A" MRP II represents in many ways the state of the art of MRP II. However, many firms have still not achieved it.

6 Current Research and Development

The main thrust of current development of MRP is the continued application of software engineering techniques to improve the MRP II system, in terms of its user interface, its inter-connectivity with other systems, data management, and the provision of low-cost delivery systems. None of these developments has changed the basic MRP procedure.

At the user interface level, recent developments have concerned user access to, and presentation of, information. Linking spreadsheets to MRP financial information is a developed capability in several MRP systems. The layering of decision support tools such as fourth generation languages on top of the MRP II database is also becoming commonplace. This is facilitated by the fact that today many MRP systems implement their data storage in database management systems as opposed to simpler file systems. Some user interface research has explored the provision of significant graphical display in MRP system output, particularly where concerned with capacity management and WIP management. While technically feasible, this had up till now been restricted by the availability of low-cost terminals with adequate graphics capability. Fortunately this cost barrier seems to be falling.

The problem of interconnecting MRP II to other manufacturing information systems is currently attracting a great deal of research attention. It is part of the general CIM problem, and today at any manufacturing conference one will frequently find papers presenting case studies of how an MRP II system was linked to a CAD system, an Automatic Storage and Retrieval System (AS/RS), or a Flexible Manufacturing System (FMS). The appropriateness of various approaches to constructing these interconnections is not fully understood today, partly because the application of modular design in the large-scale, multi-vendor environment that is CIM, is not itself understood. There are significant open questions around how control systems which drive intelligent manufacturing systems, such as Flexible Manufacturing Systems (FMS), will link with MRP II. Similar problems exist with Bill of Materials information, which appears in a company in many forms, in both design engineering and manufacturing engineering systems.

Today, MRP systems can be interconnected with such systems by customized interfaces, but this *"interconnection"* does not represent an *"integration"* in accordance with an understood architecture. Initial research in this direction is being focused on the developments of appropriate architectures and standards to cope with the problem of a large-scale, multi-vendor, multi-application environment. This is evidenced by activities around the higher levels of the Manufacturing Automation Protocol (MAP) specification [9], the work of the National Bureau of Standards in the USA [10], and various efforts within the European Community's ESPRIT programme, project 477 for instance [11]. One of the potential consequences of these efforts may be that MRP II will have to become more modular than is currently the case. Certain existing modules, such as Bills of Materials and Production Activity Control, may move outside the MRP II environment, and have a separate existence at the interface of other environments such as Computer-Aided Process Planning (CAPP), equipment control, materials handling control, with MRP II.

Another development related to interconnection of MRP II systems is Electronic Data Interchange (EDI). Emerging standards activity in the area of EDI will play an important role in facilitating the interconnection of the MRP system of one company with those of its suppliers. This has two purposes: to shorten the purchasing cycle, and to transfer appropriate information for longer-range materials and capacity planning to suppliers. This is currently being practised to a limited extent, and the pressure to move to "Just-in-Time" manufacturing will accelerate this process.

On the factory floor, automatic data collection systems have been linked with the materials tracking systems of MRP II systems, thus providing real-time access to WIP information. The interfaces are available off the shelf from some MRP II system suppliers.

MRP II packages have recently become available at relatively low cost that run on low-end minicomputers and microcomputers. These may represent viable approaches for the small firm. Nevertheless the dominant research and development effort in MRP II has been towards enhancing the functionality of systems to cope with the needs of large-scale manufacturing enterprises. Little has been written about the suitability and acceptance of MRP as an appropriate approach for the very small firm.

Other research has focused on issues such as MRP/JIT/OPT comparisons [12], with limited use of simulation models to test theories. Hall presented an extensive comparison of MRP and Kanban. In particular, he documents an ongoing effort to apply both the MRP and Kanban concepts in conjunction: "Synchro-MRP" [13]. Synchro cards similar to Kanban cards are used to control the flow of material in the repetitive flow areas in the factory. The synchro cards are generated from the mature portion of the Master Production Schedule/Final Assembly Schedule. These synchro cards can, in the same way as Kanban cards, be used to control the amount of WIP in the factory, and so identify flow problems in the process. The supporting MRP system generates two types of manufacturing orders: a typical job shop order, and a flow order. Parts are closed against the flow order as completed, but parts are not produced or conveyed in this order size, and are instead produced according to a more detailed schedule. Synchro MRP requires the same rigorous engineering of the process as is the case with Kanban.

Attention has also focused on how to approach the "Just-in-Time" ideal by using MRP II [14]. Approaches include use of blow-through BOMs, recognition of operation overlap in shop routings, use of a daily rather than weekly schedule, and devoting specific attention to lot sizing and lead time rules. The emerging view is that MRP provides the planning framework within which the JIT execution system will operate.

7 The Philosophy Underlying MRP/MRP II

To understand what is the state of the art in MRP also requires some consideration of the state of the philosophy of MRP users. Two major influences have emerged to put pressure on the MRP II paradigm. There is the decision science concern of how MRP II compares with emerging popular alternative philosophies such as "Just-in-Time" and OPT. There is also the software engineering concern of where the limits of an MRP II system should be and how MRP integrates with proliferating factory automation and the controllers that supervise it. At already stated, the road to CIM seems to involve some redefinition of what functions remain within or move outside the MRP II system, particularly in the areas of BOM management and PAC.

An essential and core proposal of the MRP paradigm is that the production management system should use a very simple technique, so that people may understand what decisions are being made by the computer and what human interventions are appropriate. Once people understand, and are motivated, then they can be expected to assume responsibility for ensuring that the large amount of raw data that the system processes is accurate and that the recommendations that the system makes are valid.

Therefore, since finite loading algorithms are necessarily heuristic and probably difficult for the layman to understand, they have been frowned upon by the MRP community. MRP II is seen in the role of an infinite-loading/decision support tool, wherein the users develop the schedule particularly using the "what-if" support provided by the MPS system. MRP II is also a hierarchical scheduling system, as scheduling decisions are made at three levels of aggregation: the MPS level, the MRP level, and the PAC level. This prompts the question of whether decision support/infinite loading should be the paradigm that is used across each of the levels.

Hierarchical scheduling systems seem to be the way of the future. To the extent that manufacturing processes become more highly engineered, automated, and predictable, and the people who operate these processes become more highly skilled, then it seems less reasonable to adopt infinite-loading/decision support strategy for scheduling manufacturing, particularly at PAC/MRP level, regardless of the "keep the responsibility with the people" argument. On the other hand, to the extent that manufacturing processes remain people-intensive and less predictable, then finite loading is likely a dubious proposition. Moreover, it is unlikely, in any case, that at the master planning level any approach other than a decision support will ever be acceptable. The claim that MRP II will be made obsolete by OPT [5] with its partitioned forward-finite/

backward-infinite heuristic seems unlikely to be fulfilled. Nevertheless, it seems likely that both finite and infinite paradigms will survive, and in the future we will continue to see emerging hybrid approaches.

The debate around JIT seems to have reached some tentative understanding around some of the issues. MRP users can and have learned a great deal from Kanban, about the evils of work in progress. This has stimulated great attention to cycle times throughout the whole manufacturing process. It seems that in very repetitive manufacturing, MRP performance is inferior to a well-engineered process using Kanban. However, in non-repetitive situations, such as job shop and small-batch production, full Kanban is difficult to implement and MRP remains a very workable solution. In between these situations, there is much room for hybrid applications of both techniques.

In comparing scheduling philosophy, it is important to distinguish between the technique in itself and the technique as applied in a real system. The choice of a techniques may not be overly important as compared with the need to improve the management and engineering process that supports the application of the technique. The point was well made recently by Galvin [15, p. 92], when he stated that "Apparently the techniques employed are not such a dominant factor as we have been lead to believe. The cohesive power of a successful system would appear to be due to the readiness and concerted efforts of all functional groups to achieve a common plan. If you haven't got a common plan, then you haven't got a system."

8 Conclusions

This paper has attempted to discuss what is the state of the art of MRP/MRP II. Practice, research and philosophy have all been considered. In conclusion, MRP/MRP II is a viable approach to Production Management, with a proven track record. While MRP II will continue to be widely applied in its present form, it may nevertheless be subject in the future to radical modularization so as to re-emerge in new hybrid production management environments.

9 References

1. Anderson, J., Schroeder, R., Tupy, S., and White, E., Materials requirements planning systems: the state of the art. Production and Inventory Management, Q.4 1982, pp. 51–67. American Production and Inventory Control Society, Falls Church, VA
2. LaForge, R., and Sturr, V., MRP practices in a random sample of manufacturing firms. Production and Inventory Management, Q.3 1986, pp. 129–137. American Production and Inventory Control Society, Falls Church, VA
3. Wight, O., MRP II – Unlocking America's Productivity Potential. CBI Publishing, Boston, MA, 1981

4. Orlicky, J., MRP Materials Requirements Planning, The New Way of Life in Production and Inventory Management. McGraw-Hill, New York, 1975
5. Fox, R., Build your own OPT. In APICS 28th Annual International Conference Proceedings, pp. 568–572. American Production and Inventory Control Society, Falls Church, VA, 1985
6. Orlicky, F., A note on exercises in sterility. Production and Inventory Management, Q.3 1975, pp. 90–91. American Production and Inventory Control Society, Washington, DC
7. Berry, W., Vollman, T., and Whybark, D., Master Production Scheduling, Principles and Practice. American Production and Inventory Control Society, Washington, DC, 1979
8. Anderson, J., Schroeder, R., Tupy, S., and White, E., A study of MRP benefits and costs. Journal of Operations Management, Vol. 2, No. 1, October 1981, pp. 1–9. American Production and Inventory Control Society, Washington, DC
9. Kosmalski, D., Manufacturing Automation Protocol Specification. General Motors Corporation, Warren, MI, 1984
10. Simpson, J., Hocken, R., and Albus, J., The Automated Manufacturing Research Facility of the National Bureau of Standards. Journal of Manufacturing Systems, Vol. 1, No. 1, 1984, pp. 17–32
11. Actis Dato, M., Ehret, O., and Barta, G., Control systems for integrated manufacturing: the CAM solution. In ESPRIT '85: Status Report of Ongoing Work, Commision of the European Communities (ed.), North-Holland, Amsterdam, 1986
12. Browne, J., Joyce, R., and Shivnan, J., Review of Materials Requirements Planning, Just-in-Time and Optimised Production Technology, report to ESPRIT project 477. Department of Industrial Engineering, University College, Galway, Ireland, 1985
13. Hall, R., Driving the Productivity Machine: Production Planning and Control in Japan. American Production and Inventory Control Society, Falls Church, VA, 1981
14. Sonnerburg, R., The MRP and Just-in-Time marriage. In APICS 26th Annual International Conference Proceedings, pp. 683–687. American Production and Inventory Control Society, Falls Church, VA, 1985
15. Galvin, P., Visions and realities: MRP as system. Production and Inventory Management, Q.3 1986, pp. 91–95. American Production and Inventory Control Society, Falls Church, VA

Chapter 3
Just-in-Time Production — A New Formulation and Algorithm of the Flow Shop Problem

Hajime Yamashina

Dr. Hajime Yamashina is a Professor in Production Engineering at the Department of Precision Mechanics, Faculty of Engineering, Kyoto University, Japan. He received B. S., M. S., and D. ing. from Kyoto University. His research interests cover production and inventory control, scheduling and sequencing, quality control, computer aided manufacturing, FMS, FA, etc.

1 Introduction

After the first oil crisis, Japan went into the low economic growth period and the market started to show saturation as shown in Fig. 1. Accordingly, customers' needs have become diversified, and forecasting customers' demands has consequently become quite difficult. Figure 2 schematically illustrates the market situations in the past, before the first oil crisis, and after the oil crisis, giving the example of the mug market. This example simply tells of today's market circumstances of diversified customers' demands.

What mainly count as competitive factors in the market for a manufacturing company are prices, qualities including services and delivery dates. In what order these items count heavily depends on the product. It is normally the case that when a customer wants to buy a certain product from the company, he wants to place his order with specifications at the latest possible day, and that once he has made a decision, he wants to have it immediately. If he has to wait for a long time until he gets it, most probably he will lose his wish to buy it. If he finds a similar product made by another company, if the price and quality of it are almost the same, and if that product can be delivered to him much faster than that from the first company, then he may buy it. Therefore, delivery dates are surely a very important factor for the company to be competitive.

To meet such a customer's demand, the company can take either of the following two solutions.

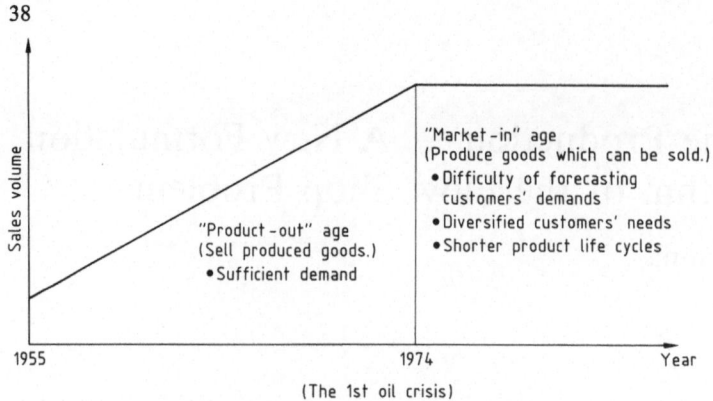

Fig. 1. Market situations before and after the 1st oil crisis

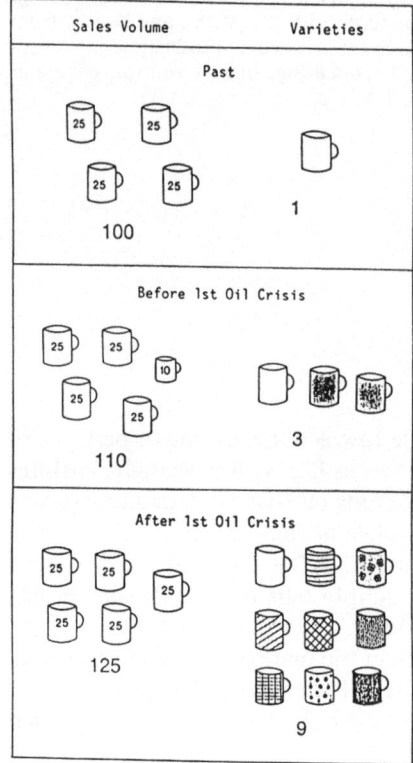

Fig. 2. Economic circumstances

(a) Keep a very high stock of finished goods. Then, whatever demand comes, there is no danger of losing the customer, since the company can meet the demand immediately. This solution has a very attractive characteristic for the company since it enables the company to have stable production.

In fact, in the West there are many companies which employed or are still employing this solution. But, in such companies it is often the case that they don't have the very items customers want, although they have a high stock of

3. Just-in-Time Production

finished goods, and that the customers have to wait for a long time until they get them.

The drawbacks of this solution are as follows.

- Tied up capital in the finished stock can be dangerously high.
- Some of the finished stock may go to dead stock because the total demands are limited as shown in Fig. 2 and because the product life cycles are getting shorter and shorter.
- Retailers, wholesalers, and manufacturers have in many cases only limited space for keeping inventory, thus it is very difficult for them to keep all the stock.

Because of these reasons, this seems not to be the right solution.

(b) To have very short throughput times to replenish the stock of finished goods.
One function of inventory is to act as a buffer between sales and production. In other words, it separates the sales function and the production function and enables each to function independently. The mass production system was possible owing to this buffering function of inventory as shown in Fig. 3(a). But the present mature and diversified market has made this solution infeasible as discussed in (*a*) above. Few stock can be accepted between sales and production. In principle, with few stock, production must confront directly the diversified and unforeseen market as shown in Fig. 3(b). This means that merely minimizing production cost is not a sufficient criterion to operate the production system. A new element, flexibility, has become a crucial concern for operating the production system. If the production system cannot cope with the market needs quickly enough, obviously shortage of products and excessive inventory take place. This will be followed by the decrease of sales and the loss of the company's reputation. Thus, it has become very important to be able to provide customers with the products meeting their demands at the right time with reasonable prices – in other words, to implement just-in-time production.

To meet the demand of just-in-time production, the Japanese have found as a solution shortening throughput times. If throughput times are short enough, any demand can be met whatever it is, and redundant stock can be eliminated.

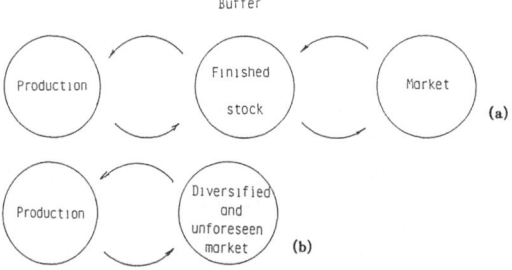

Fig. 3. Production must confront directly the diversified and unforeseen market with few stock between sales and production: (**a**) the mass production system; (**b**) the just-in-time production system

The three key elements of shortening throughput times are (1) product-oriented layout, (2) short setup times, and (3) small lot production [1]. In product-oriented layout, a number of processes are arranged in series so that the flow of products is unidirectional. The drawbacks of product-oriented layout are as follows.

- It often requires more machines. If the machines are expensive, it may not be economically justified.
- It often requires multi-purpose workers in order to improve labour productivity. They might not be available because of relations with the labour union.

To compensate for the drawbacks, to secure flexibility, and to facilitate workload balancing, product-oriented layout often takes a flow type configuration, each stage consisting of several identical machines. It is required that such a flow shop be operated as to produce only the necessary products at the necessary time in the necessary quantity with possibly minimal flow times. If the shop cannot be operated properly, its efficiency will decrease significantly because of the loss of setup times and the interference losses caused by starving and blocking phenomena, with the consequence that it becomes impossible to achieve planned production requirements in a given period. In order to reduce such losses, various countermeasures can be taken:

- Introduction of more machines to the bottleneck stage
- Increase of buffer capacity between stages
- Sequencing jobs at each stage.

Investment in a new machine for the bottleneck stage will increase the cost of the shop. The increase of buffer capacity between stages will increase flow times and make just-in-time production difficult. Thus, the possibility of improving efficiency has been investigated by scheduling jobs at each stage properly.

The following sections present a completely new formulation and effective algorithm of the flow shop problem for just-in-time production, assuming that setup times are short enough to enable small-lot production.

2 Problem Statement

For products requiring the same processes, a series of machines are laid out in tandem according to the processes. Assume the following.

(a) The flow shop consists of m stages which are numbered 1, 2, ..., m, a loading stage in front of the m stages, and an unloading stage at the end of the m stages. Stage p ($1 \leq p \leq m$) consists of m_p identical machines which are numbered $M_{1p}, M_{2p}, \ldots, M_{m_p p}$.

Figure 4 shows an example of a 2-stage flow shop, each stage consisting of 3 identical machines. This example is used below to illustrate this solution procedure.

3. Just-in-Time Production

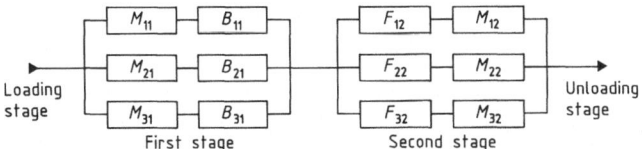

Fig. 4. An example of a 2-stage flow shop, each stage consisting of 3 identical machines

Table 1. Processing times and lot sizes for example

i	1	2	3	4	5	6	7	8	9
C_{i1}	494	562	891	2665	1703	1154	210	858	469
C_{i2}	2396	945	485	690	1532	351	110	1981	471
l_i	3	3	3	3	3	3	3	3	3

(b) There is a set of n jobs, $N = \{1, 2, \ldots, n\}$, to be processed by the shop. The lot size of job i ($1 \le i \le n$) is l_i. The first processed product of a lot is called the initial product of the lot.

(c) Each job requires all the m stages in the numbered order. The operations of job i are correspondingly numbered $O_{i1}, O_{i2}, \ldots, O_{im}$. Operation O_{ip} can be processed by any machine of the pth stage, and requires c_{ip} units of time. To facilitate later computations, "time" is considered to be an integer quantity.

For the example illustrated in Fig. 4, assume that there are nine jobs to be processed by the shop and that their processing times and lot sizes are as shown in Table 1.

(d) Setup times for the operations are sequence-dependent. Let $s_{ij,p}$ denote the setup time required for job j after job i is completed at stage p. Setup times $s_{ij,p}$ constitute an $n \times n$ matrix $S_p = (s_{ij,p})$. Suffix p is omitted unless necessary.

(e) Associated with a machine, there is a buffer in front of the machine (called front buffer) and a buffer at the back of the machine (called back buffer), except for the machines at the first stage and the last stage. The machines of the first stage have only back buffers, while the machines of the last stage have only front buffers. The front buffers of stage p are numbered $F_{1p}, F_{2p}, \ldots, F_{m_p p}$; the back buffers, $B_{1p}, B_{2p}, \ldots, B_{m_p p}$. The processed products of a batch stored tentatively in a back buffer of stage p ($p \le m-1$) are fed into one of the front buffers of stage ($p+1$). They are fed into the same front buffer without intervention by any other products. Thus, if the front buffer is still engaged to receive the rest of products of the batch, a processed product stored tentatively in another back buffer of stage p cannot be fed to the front buffer.

In order to keep the necessary buffer area as small as possible for minimizing flow times, and to minimize the control function of product transportation timing between two stages, the earliest finished initial product among the initial products of different batches kept in the back buffers of stage p will be fed to the disengaged front buffer of stage ($p+1$), or one of the disengaged front buffers if there are more than one disengaged front buffer as shown in Fig. 5. Thus, only in the case that there are more than one such disengaged buffer, a scheduling problem

Table 2. Setup times for the example

		1	2	3	4	5	6	7	8	9
	1		2695	1833	147	477	217	6213	2785	8154
	2	208		2505	325	15875	227	5377	9331	4562
	3	1625	1841		833	5904	327	1020	7598	4149
	4	162	9251	1325		337	4146	1445	433	2105
$S_1 =$	5	8400	1311	842	416		587	1205	4119	1223
	6	208	1018	701	9992	4021		438	1786	3141
	7	5838	3925	1426	4001	3253	9259		626	925
	8	3268	9512	1012	492	4526	523	528		518
	9	5925	1861	5213	1992	3177	3541	3019	642	

		1	2	3	4	5	6	7	8	9
	1		1661	2108	745	4799	9826	6723	977	1532
	2	3361		264	252	620	11	13976	49	7194
	3	3293	259		898	496	1426	156	7085	729
	4	11630	319	1181		706	739	3591	150	1433
$S_2 =$	5	4229	1011	626	822		425	102	5293	7893
	6	4084	21	15	4	1669		22	1854	2068
	7	2489	15792	13379	2375	1316	13916		201	2490
	8	95	54	2386	172	5786	314	524		2305
	9	6319	8083	6195	1299	4911	2165	3102	406	

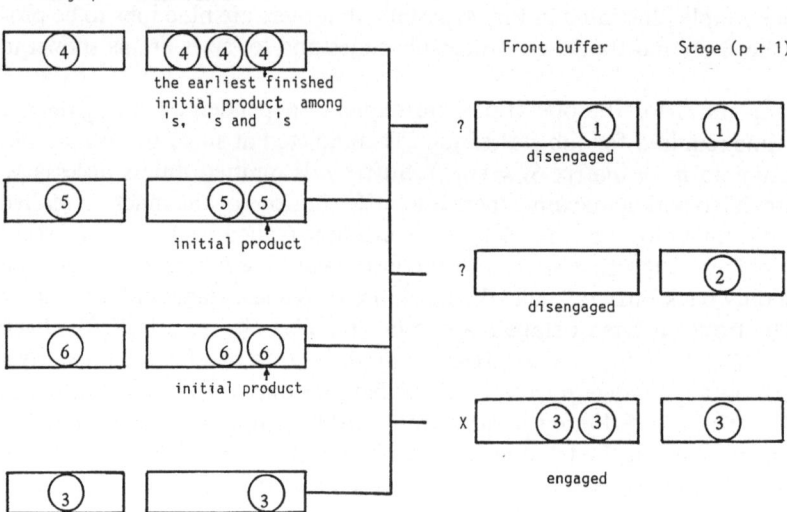

Fig. 5. How the initial products of different batches kept in the back buffers of stage p will be fed to the disengaged front buffer of stage $(p+1)$

occurs in regard to which front buffer the earliest finished initial product should be fed.

(f) Transport times are negligible. A schedule is called a complete schedule and denoted by a large letter A with an appropriate subscript if it includes all the

3. Just-in-Time Production

n jobs. Otherwise, a schedule is called a subschedule and denoted by a small letter a with an appropriate subscript.

Let A_p denote a complete schedule of stage p as follows:

$$A_p = \begin{matrix} M_1 \\ M_2 \\ \vdots \\ M_{m_{pp}} \end{matrix} \begin{bmatrix} 1[1], s_{1[1],1[2]}, 1[2], \ldots \\ 2[1], s_{2[1],2[2]}, 2[2], \ldots \\ \vdots \\ m_p[1], s_{m_p[1]m_p[2]}, m_p[2], \ldots \end{bmatrix} \qquad (1)$$

where $k[q]$ is the job processed at the qth position in sequence on the kth machine.

The completion time of job i at stage p under subschedule a_p is denoted by $t[a_p]_{i,p}$. The notation $[a_p]$ is omitted unless necessary.

If $W_p(a_p)$ is defined in the following way:

$$W_p(A_p) = \max_{1 \leq i \leq n} (t[A_p]_{i,p}) \qquad (2)$$

then the problem is to find the *optimal overall schedule* $A = (A_1, A_2, \ldots, A_m)$ which minimizes the make span of stage m, $W_m(A_m)$.

3 Algorithm

The algorithm commences with a possible value of $W_m(A_m)$, w_m, and calculates a possible value of $W_{m-1}(A_{m-1})$, w_{m-1}, and so on in the following way.

From the workload and the setup times needed at stage p,

$$w_p \geq \frac{\sum_{i=1}^{n} c_{ip} l_i + \sum_{k=1}^{m_p} \sum_{q=1}^{} s_{k[1]k[q+1]}}{m_p}. \qquad (3)$$

In the inequality (3), the value of $\sum_{i=1}^{n} c_{ip} l_i$ is fixed.

For the example,

$$\sum_{i=1}^{n} c_{i1} l_i = 27018, \quad \sum_{i=1}^{n} c_{i2} l_i = 26883.$$

A lower bound of $\sum_{k=1}^{m_p} \sum_{q=1}^{} s_{k[q]k[q+1]}$ can be obtained by either solving the m_p-travelling salesmen problem or by subtracting an amount equal to the smallest element in each row or column until each column and row has at least one zero.

For the example,

$$\sum_{k=1}^{3} \sum_{q=1}^{3} s_{k[q]k[q+1], 1} = 2490, \quad \sum_{k=1}^{3} \sum_{q=1}^{3} s_{k[q]k[q+1], 2} = 1739$$

by solving the 3-travelling salesmen problems for the first and second stages, respectively. Thus, $w_1 \geq 9836$, $w_2 \geq 9541$.

On the other hand, from the flow shop assumption,

$$w_p + \min_{1 \leq i \leq n} [c_{i(p+1)}] \leq w_{p+1} \quad (1 \leq p \leq m-1). \tag{4}$$

For the example,

$$w_1 + \min_{1 \leq i \leq 9} c_{i2} = 9836 + 110 = 9946 \leq w_2.$$

A predetermined lower bound of make span of stage m can be established in this way. It is denoted by w'_m. Also, an upper limit of make span of stage m can be obtained for which it is certain that a feasible overall schedule can be obtained, by applying an appropriate heuristic method. It is denoted by w''_m.

For the example, $w'_2 = 9946$. For an upper limit, for instance, $w''_2 = 13246$.

The basic idea of the algorithm is to ascertain for a predetermined set of make span, $w = (w_1, w_2, \ldots, w_m)$, whether there is a feasible overall schedule $A = (A_1, A_2, \ldots, A_m)$ which satisfies

$$W_p(A_p) \leq w_p \quad (1 \leq p \leq m) \tag{5}$$

based on the optimum-seeking backtracking method. Figure 6 shows a flow chart of the algorithm. The limited backtracking method is the one which limits the number of backtracks within a stage at the cost of losing optimality. In the chart, it is meant to find a feasible overall schedule when it seems to be quite possible to obtain a schedule for the predetermined value of w_m. But, actually, it is used for practical cases where strict optimality is not required, or for the case where the number of jobs, n, is too big to be tackled by backtracking.

The algorithm starts with w_1 which was calculated from w_m through w_{m-1}, $w_{m-2}, \ldots, w_2$, by inequalities (4). It proceeds with the generation of a feasible subschedule a_1 of stage 1. The generation of a_1 is made in such a way that the next job is assigned to the machine which became disengaged earliest.

For the example, suppose that jobs 9, 2, and 1 have been assigned on M_{11}, M_{12}, and M_{13} in that order. Then machine 3 becomes disengaged earliest. Thus, the next job will be assigned to M_{13}.

Suppose that a subschedule a_1 has been generated in which jobs $i_1, i_2, \ldots, i_{u-1}, i_u$ were assigned in that order and that the schedule fixes up to the q_1th position, q_2th position, $\ldots$, q_{m_1}th position in sequence on the 1st, 2nd, $\ldots$, m_1th machine, respectively. If

$$t_{i_u, 1} > w_1 \tag{6}$$

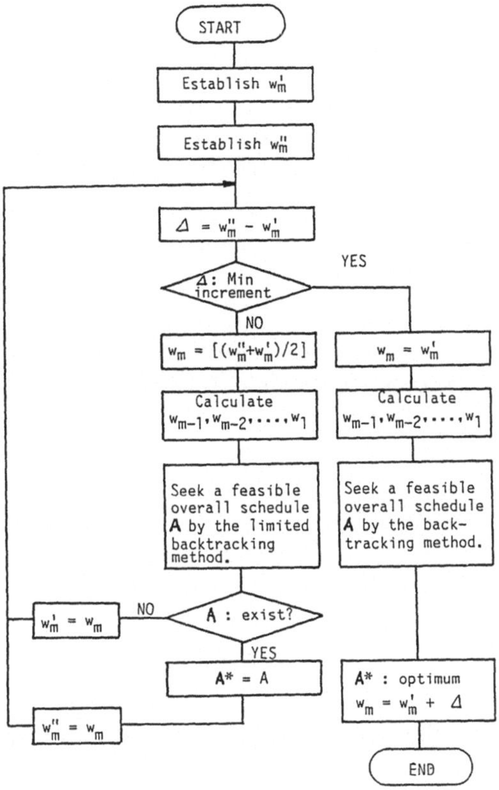

3. Just-in-Time Production

Fig. 6. A flow chart of the algorithm

then subschedule a_1 does not lead to a feasible complete schedule. Job i_u is cancelled and another unassigned job is going to be assigned to make a new subschedule. If

$$t_{i_u, 1} \leq w_1 \tag{7}$$

then the following value is calculated:

$$X = \sum_{i=1}^{m_1} (w_1 - t_{i[q_i], 1}) . \tag{8}$$

The value X gives an available allowance to complete the rest of the jobs, $N - \{i_1, i_2, \ldots, i_{u-1}, i_u\}$. Now, let Y give the minimum time needed to complete the rest of the jobs. The time consists of processing times of these jobs and their setup times. Then, if

$$X \geq Y \tag{9}$$

there is a possibility that a_1 leads to a feasible complete schedule and the subschedule can be extended until all jobs are assigned within make span w_1. If

$$X < Y \tag{10}$$

then it can be concluded at this point that a feasible subschedule a_i does not lead to a feasible complete schedule. The algorithm applies a backtracking procedure, where the last assignment, i_u, is cancelled and a systematic attempt is made to extend the feasible sequence until one of the following situations occurs.

- A solution cannot be found for w_1, implying that every complete sequence that occurs is generated and discarded. There is no solution which has the length of make span of w_m.
- A feasible complete schedule A_1 is found.

For the example, setting $w_2 = 11\,068$[1] results in $w_1 = 10\,958$ from inequality (4). For this value, the following feasible complete schedule A_1 can be obtained:

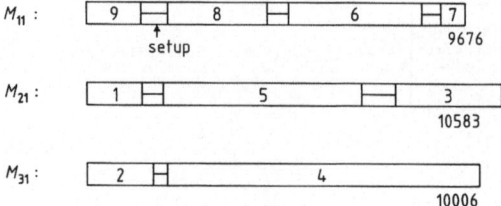

Now it is necessary to look for a feasible complete schedule of the jobs into the front buffers of stage 2, which will automatically give a complete schedule of the second stage.

The generation of a feasible schedule of the front buffers is somewhat different from that of the first stage and is made in such a way that the job whose initial product has the earliest release time among the unassigned jobs is assigned first to a disengaged front buffer of stage 2.

The release time of the initial product of job j after job i is completed at stage 1 under schedule A_1 is denoted by $r[A_1]_{ij,1}$. The notation of schedule A_1 is omitted unless necessary. It is given by the following:

$$r_{ij,1} = t_{i,1} + s_{ij,1} + c_{j1} . \tag{11}$$

Thus, the job which has $\min_j (r_{ij,1})$ among the products kept in the back buffers of stage 1 will be assigned next.

The disengagement time of front buffer $F_{k2} (1 \le k \le m_2)$ which has just been disengaged with job i under schedule A_1 is denoted by

$$d[A_1]_{F_{k2},i} .$$

The notation of schedule A_1 is omitted as usual unless necessary. It is given by the following:

---

[1] Incidentally, this is the optimal value of make span w_2.

3. Just-in-Time Production

$$d_{F_{k2},i} = t_{i,1} . \tag{12}$$

If there are more than one disengaged front buffer, then there occurs a problem of priority for backtracking. In the method, the priority for backtracking is given to the front buffer connected to the machine which has become or will become idle first.

The time that machine M_{k2} ($1 \leq k \leq m_2$) becomes idle after being engaged with jobs i and j in this order under schedule A_1 is denoted by

$$e[A_1]_{M_{k2},ij} .$$

Notation $[A_1]$ is omitted unless necessary. It is given by the following:
If

$$t_{j,1} \leq \text{Max } [t_{j,1} - c_{j1}(l_j - 1), \; t_{i,2} + s_{ij,2}] + c_{j2}(l_j - 1) \tag{13}$$

then

$$e_{M_{k2},ij} = \text{Max } [t_{j,1} - c_{j1}(l_j - 1), \; t_{i,2} + s_{ij,2}] + c_{j2}(l_j) . \tag{14}$$

Otherwise,

$$e_{M_{k2},ij} = t_{j,1} + c_{j,2} . \tag{15}$$

For the example, the release times of initial products of jobs 9, 1, and 2 are 469, 494, and 562, respectively. Thus, job 9 is put into F_{12}; job 1 into F_{22}; job 2 into F_{32}. The release times of initial products of jobs 8, 5, and 4 are 2907, 3662, and 4676, respectively. Thus, job 8 will be assigned to a disengaged front buffer. Front buffers F_{12}, F_{22} and F_{32} become disengaged at 1407, 1482, and 1686, respectively. Thus, by time 2907, all the front buffers have become disengaged. Thus, job 8 is put to any front buffer of stage 2. Priority of backtracking is given by checking the times that machines M_{12}, M_{22}, and M_{32} become idle. They become idle at 882, 3397, and 7682, respectively. Thus, job 8 will be put into F_{21}.

Each time a job has been assigned to a front buffer, similar relations to Eqs. (7) and (9) for stage 2 are checked to see whether the subschedule will lead to a feasible complete schedule. If these relations are satisfied, the subschedule is going to be extended until all jobs are assigned within make span w_2. If these relations are not satisfied, the last assignment is cancelled and a systematic attempt is made to extend the feasible sequence until one of the following situations occurs.

- A solution cannot be found for A_1, implying that another solution for stage 1 must be sought. If there is no other solution for stage 1, then there is no solution which has the length of make span of w_m.
- A feasible complete schedule A_2 is found.

For the example, the following solution can be obtained:

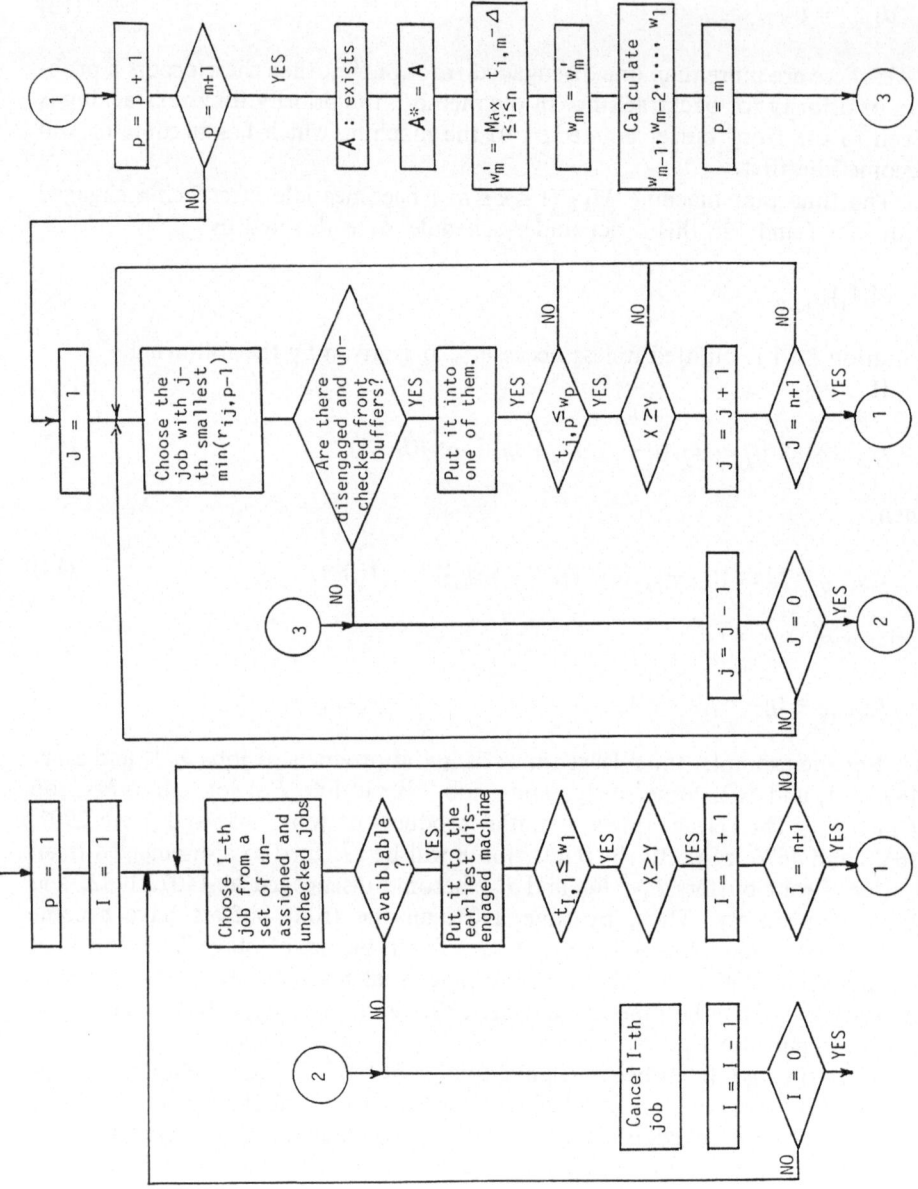

Fig. 7. A flow chart of the backtracking method

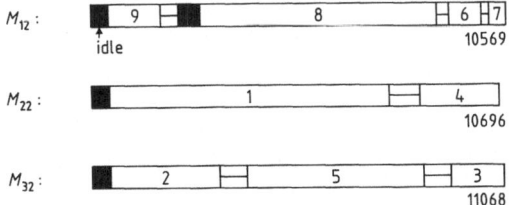

For stage p ($3 \leq p \leq m$), the same procedure applies as the procedure for the second stage. As is noticed easily, the number of backtracks is much fewer for stages from 2 to m, compared with stage 1, because of the restrictions of the buffer area (cf. Section 2 (d)).

Based on this concept, the optimum-seeking backtracking method was established. A flow chart of the backtracking method is shown in Fig. 7. Obviously, the number of jobs, n, limits the use of the backtracking method. In practical cases, strict optimality is not required. Thus, in addition to the backtracking method, a limited backtracking method combining the above backtracking method with a heuristic interchange method was developed, which is to be used when n is too large to be handled by the backtracking method, as shown in Fig. 6.

4 Practical Experiences

Implementation of just-in-time production requires new physical arrangements such as product-oriented layout and short setup times. Once such arrangements have been made, the just-in-time production scheduling problem can, in many cases, be formulated into the flow shop problem this paper defines[2]. To solve it on the basis of the method developed above, a computer is needed. It is advisable and economically justifiable to dedicate one microcomputer to solving it, since nowadays cheap and efficient microcomputers are available on the market. The author consulted various Japanese manufacturing companies on acceptable computation time for its solution by such a microcomputer. The time limit accepted lies between two hours and eight hours. Based on this investigation, a computer program for solution of the problem, to be run on a microcomputer, was developed. The program can handle up to about 50 jobs with high-quality solutions by a microcomputer like the NEC PC98XA. The program has been used by various just-in-time production systems producing such products as radiators [3], print circuit boards, and kitchen units – with successful results.

---

[2] The most important point in operating this flow shop is taking workload balance between the stages [2]. It is assumed in this paper that workloads have been balanced between the stages.

5 References

1. Yamashina, Hajime, Japansk Tillverkningsfilosofi och Kanban-Systemet. Sveriges Mekanforbund, 1982
2. Yamashina, Hajime, Analysis of the effect of buffer storage capacity in transfer line systems. AIIE Transactions, June 1977
3. Yamashina, Hajime et al., Operation of the flow type FMS. FMS-5 International Conference, Stockholm, 1985

Chapter 4
The Drum-Buffer-Rope (DBR) Approach to Logistics

Oded Cohen

Oded Cohen has a Bachelor degree from the Technion, Israel Institute of Technology (IIT), Haifa, 1973, in Industrial and Production Engineering, and an M. Sc. in Operations Research, IIT, 1977. While studying for his master degree, he lectured in Data Processing and MIS for manufacturing. Since 1977 he has worked in MIS, Industrial Engineering, Production Management, Marketing, and Customer Service for several industrial companies. In 1982 he joined Creative Output and worked in implementing the OPT® system in several locations all over the world and was associated with the establishment and management of a Creative Output office in the UK. In 1987 he joined the Avraham Y. Goldratt Institute. The Institute was established by Dr. E. Goldratt who recently developed the Theory Of Constraints (TOC). Oded Cohen has been involved in implementing the Theory Of Constraints in several organizations and he is currently in charge of the Institute's activities in the UK and Southern Europe.

1 Introduction

Since 1970, a host of industries in the USA and Europe have lost their predominant position in manufacturing. Millions of jobs have been lost and many companies have disappeared or have shrunk dramatically. In 1970 the West lost its predominant position in smokestack industries like steel, brass, and textiles. In 1975 the consumer electrical industry recognized that they were losing the race when cheaper and better stereos, TVs, and microwave ovens flooded in from the Far East. In 1980 it was the automotive industry, and in 1985 it was the high-technology electronics industry. Now it seems as if the aerospace-defence industries will lose their leadership before 1990 in spite of governmental protection.

The first reaction of the Western world was to ignore the signs. Later, people started to explain why it happened. Recently, the "Zero inventory" campaign was set up in an attempt to copy the Just-in-Time (JIT) philosophy. The increasing pace of international competition entails that the current methods of running manufacturing business are grossly inadequate for today's competitive environment. A better way is needed!

Dr. Goldratt developed on overall management approach for running a system known as the Theory Of Constraints (TOC), from which he derived the logistical system for the material flow called the *DBR* – *Drum Buffer Rope*. The uniqueness of Dr. Goldratt's approach is not only the technicalities that can allow one to implement the ideas immediately, but the search for the understanding of WHY does it work – the Cause and Effect relationships and the warnings of the psychological obstacles one needs to overcome if he really wants to be successful in implementing those ideas. The shop floor, being an integral part of the manufacturing environment, requires an acceptable methodology expressed as rules to enable its management to be consistent with the overall business objectives.

The Goal [1] was written with the desire to convey to managers that there are some assumptions by which we can explain phenomena in the shop floor and that we can derive some "rules" that can assist us in the decision making process. Some of these statements were formulated as the "OPT® Rules". The development of these rules can be found in References [3 – 8]. The main line of this article is to show how these rules of manufacturing can be derived from efficient modes of production like the Assembly Line (AL) and Just-in-Time (JIT), for the beneficial implementation of the entire manufacturing community.

The AL and JIT approaches are real breakthroughs in manufacturing: not only in the sense that they found a new systematic approach to production, but also in that they succeeded in making a dramatic impact on the market-place. Henry Ford, at the beginning of this century, reduced the production cost of his Model T cars to a level at which the middle class could afford to buy. The market grew substantially. The Kanban, the first JIT technique, was developed by Dr Ohno in Toyota in the mid-1960s. It enabled Japan to beat the American motor industry and to capture, in the late 1970s, a remarkable market share by producing cheaper and better-quality cars.

Today, most of the companies which dominate the market – those which have more than 50% of the world market in their products – are using either assembly lines or JIT to assist product flow. However, those production modes are not applicable to the vast majority of plants, for the following reasons.

1. One must be in very high-volume, repetitive business in order to justify the enormous investment in setting up the assembly line.
2. It is impractical to build assembly lines for batch production.
3. The logistical part of JIT is not generally applicable. (JIT is not very popular, even in Japan.)
4. The very long implementation period (measured in years) makes it impractical, since time is a scarce resource.

One of the major differences betwen JIT/AL and the Just-in-Case companies is a performance indicator – the inventory turnover (IT), which is most commonly expressed as:

$$IT = \frac{\text{Cost of goods sold}}{\text{Average inventory}}$$

While the industry average is less then 10 ITs per year, the JIT/AL companies are turning their inventories between 50 and 100 times per year, and they strive to increase it more and more.

Information about the difference in IT performance was available at the beginning of the current decade [4]. However, the ramifications of the difference were not greatly appreciated. Today we know that the above-mentioned characteristics have the utmost importance in deriving a production management system.

At the same time, most of the Just-in-Case systems rely on computerized systems that release materials to the shop floor in methods related to MRP (materials requirements planning). Those methods assume fixed manufacturing lead times while ignoring capacity constraints and bottlenecks. However, when reality hits them and due dates are not met, they solve it by instructing the software to increase the slack time, causing a further increase in manufacturing lead time and in inventory.

The DBR approach suggests that all efforts should initially be focused on inventory reduction since it has maximum impact on all aspects of running a manufacturing business. This approach can be translated to a full production management (PM) system. But before rushing, let's make sure that we are making the right decision, since the only reason for building a new PM system, as for every management decision, is that this move is going to be good for the company. This raises a fundamental question: What is the impact of a local decision on the global performance?

2 Prerequisites for a Production Management System

In devising a new PM system a measurement mechanism is needed to guide us in the right direction. This mechanism should have three elements:

- A Goal – the company's goal
- Global Measurements – to indicate whether or not we are moving towards our goal
- Operational Measurements – to bridge between local activities and the global measurements.

The DBR approach assumes the Goal to be the following.

The Goal of a manufacturing company is to make money in the present and in the future.

For the Global Measurements we can use the current set of two "Bottom line results" (the financial measurements):

NP – net profit, an absolute measurement
ROI – return on investment, a relative measurement
Besides these two measurements we have a necessary condition:
CF – cash flow, a "red line" of survival.

Improving the whole three simultaneously brings us closer to the goal:

THE GOAL: TO MAKE MONEY

Bottom line measurements

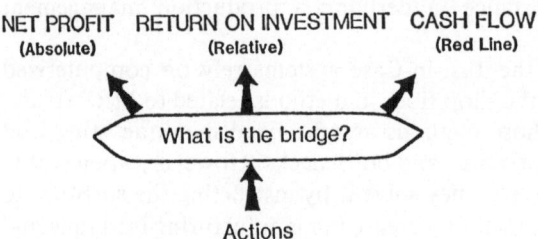

Fig. 1. The relationship between actions and results

$$\left.\begin{array}{l} \text{NP} \nearrow \text{(up)} \\ \text{ROI} \nearrow \text{(up)} \\ \text{CF} \nearrow \text{(up)} \end{array}\right\} \begin{array}{l}\text{together}\\ \text{simultaneously}\end{array} \Rightarrow \text{Good for the company!}$$

The Global Measurements are inadequate to allow us to judge the impact of specific actions on the goal. For questions like "In what batches should we release material and process material into our plant?" or "Should we buy the new robot?" we need a *bridge*.

Today, the bridge used in most companies is a set of procedures based on the cost concept. Behind the cost concept stands the assumption that reduction in unit cost improves the financial measurements. This assumption is not necessarily true. It is definitely not one of the basic assumptions of JIT and the assembly line. The cost as a bridge, although overruled many times by management intuition, is directing the Just-in-Case approach (JIC) in the wrong way [2, 9], and "cost accountingis enemy number one of productivity" [10]. The bridge is the well-established measurements that are already used by nearly every plant as performance indicators: total sales, inventory, and total operating expenses.

The Operational Measurements can be defined as follows.

Throughput (T): the rate at which the system generates money through sales (output which is not sold is not throughput but inventory)
Inventory (I): all the money the system invests in things the system intends to sell
Operating Expenses (OE): all the money the system spends in turning inventory into throughput.

(A broader explanation can be found in Ref. [1].) We now examine how the bridge works.

3 The Role of Inventory

If an activity increases throughput (T) without impacting the other two then

4. DBR

$$T \nearrow \text{(up)} \Rightarrow \begin{array}{c} NP \nearrow \text{(up)} \\ ROI \nearrow \text{(up)} \\ CF \nearrow \text{(up)} \end{array} \Rightarrow \text{Good!}$$

If we can reduce operating expenses (OE) without harming the other two then, by the same logic used for T, all three financial measurements will be improved, which means that this move is good for the company.

The impact of inventory is, however, not so simple. That is probably the reason why the role of inventory was not discovered until recently. At first glance it looks as if inventory impacts only ROI and CF:

$$I \searrow \text{(down)} \Rightarrow \begin{array}{c} NP \text{ (no change)} \\ ROI \nearrow \text{(up)} \\ CF \nearrow \text{(up)} \end{array} \Rightarrow \begin{array}{l} \text{The impact on the} \\ \text{global performance} \\ \text{is not clear!} \end{array}$$

There is another indirect link through the reduction of carrying costs which are operating expenses:

$$I \searrow \Rightarrow OE \searrow \Rightarrow \begin{array}{c} NP \nearrow \\ ROI \nearrow \\ CF \nearrow \end{array} \Rightarrow \text{Good!}$$

However, this link is not strong enough, as it weakens as inventory goes down, and therefore cannot explain why the Japanese put enormous emphasis on further reducing inventory when they have already the lowest inventory in the world.

There are two major characteristics that differentiate JIT/AL from the other approaches:

– Very short Manufacturing lead time (mnf-LT)
– Very low work-in-process (WIP).

These two characteristics are interrelated and directly proportional to each other.

$$\text{mnf-LT} \propto \text{WIP}$$

An example to demonstrate this relationship would be to produce 1000 parts that need five production stages. Let's examine two options: running one batch of 1000 pieces (Fig. 2) versus running five batches of 200 pieces each (Fig. 3) In the first case manufacturing lead time (for the first part) will be about 2000 hours and in the second case it will be around 500 hours – a reduction of 75%. Assuming a purchasing value of $ 1 per piece, the average WIP value will be $ 750 in the first case and $ 300 for the second, resulting in a reduction of 60%. So, in this example 60% reduction in WIP yields 75% reduction in manufacturing lead time.

The competitive-edge race is being fought in areas of product, price, and responsiveness. Each one of these categories can be dissected into two distinct branches. The elements that participate in the competitive-edge race are:

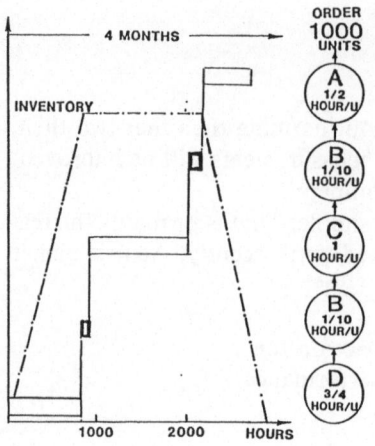

Fig. 2. High-inventory manufacturing

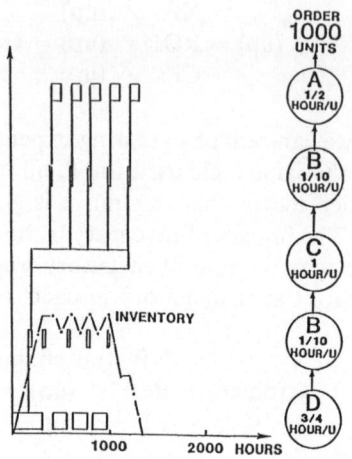

Fig. 3. Low-inventory manufacturing

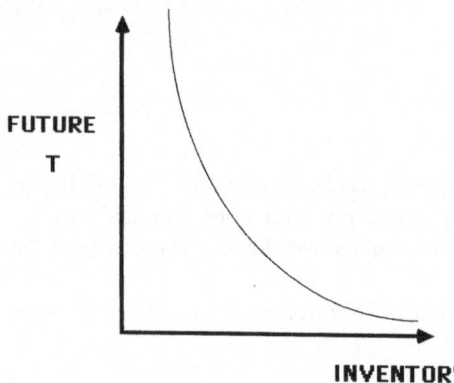

Fig. 4. Future throughput as a function of inventory

Product: Quality
                Engineering features
Price: Higher margins
                Lower investment per unit
Responsiveness: Due date performance
                Shorter quoted lead time.

As shown in Ref. [2], shorter manufacturing lead times, resulting from the low-inventory mode of production, improve dramatically all the above six elements. As one or more of these elements dominate the market of a specific product, these improvements will secure the current market share and can lead to more sales in future.

$$I \searrow \text{(down)} \Rightarrow \text{mnf-LT} \searrow \Rightarrow CF \nearrow \Rightarrow \text{future } T \nearrow \text{(up)}$$

There is, probably, a connection between T and I that can be illustrated qualitatively by Fig. 4.

4. DBR

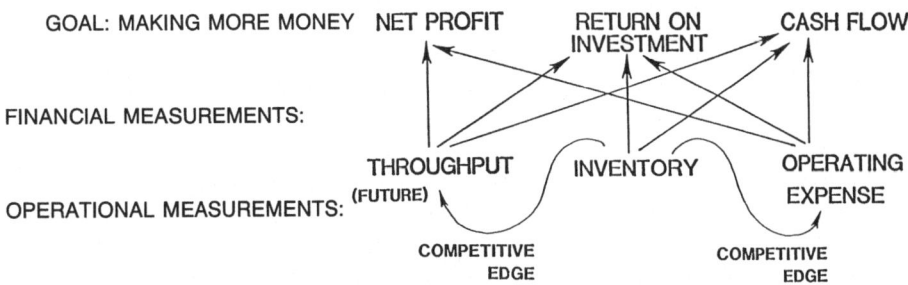

Fig. 5. Operational measurements and bottom line

Conclusion: Inventory reduction has a remarkable impact on the bottom-line results of the company. That means that Throughput (T), Inventory (I), and Operating Expenses (OE) can be our bridge and be used as our Operational Measurements (Fig. 5).

We have stated that we want to improve our performance. We can choose to concentrate on increasing T or reducing I or reducing OE. Direct increase in throughput cannot be achieved by a PM system: someone has to go out, market and sell these products. Cutting operating expenses is recommended as long as we are talking about cutting waste. Reducing head count is extremely dangerous as this move, apart from naturally being very unpopular, can result in lost throughput due to lack of flexibility. A reduction in inventory, on the other hand, if done properly, will yield improvements in the other measurements. So, we should concentrate our efforts in *Inventory Reduction*. But, how do we do it?

4 Basic Principles of Inventory Reduction and Product Flow

The principles governing product flow were known as the "nine rules of OPT®". The general feelings are that they are "just common sense"; those feelings plus the fact that several phenomena, in the shop floor and especially in the JIT/AL environments can be explained in the light of these rules, proved their validity. Both JIT and AL focus attention on the flow; they do not attempt to balance capacity. It can be easily proven that they have some excess capacity in their lines. From that observation we can derive a general rule that will guide us in devising the solution.

Rule 1: Do not balance capacity – [11].

Now our major concern is the flow. That means that anything that may disrupt the flow should be considered as a *constraint!* There are four types of constraints:

- Capacity
- Market
- Material
- Management policy.

The OPT® rules are derived for the capacity constraints and especially for the Bottlenecks (BNs). Both capacity and market constraints should be handled by the logistical system (materials flow). Materials may be temporary constraints that we have to recognize and work within accordingly. Management policy should not be a constraint in the long run. However, changing management policy is a crucial area within the implementation procedure and is the focus for a lot of attention using a variety of education, training, and behavioural tools in order to present an acceptable agent of change.

There are two types of resources in a plant:

- Capacity Constraint Resource (CCR) – a resource that causes disruptions to the material flow. A Bottleneck (BN) is a resource that does not have enough capacity to fulfil the market demand. A BN is a CCR.
- Non Capacity Constraint Resources (NCCR) – a resource that needs very little management attention as it has enough productive capacity for the production programme as well as protective capacity to overcome Murphey's disruptions. Most of the Non-bottleneck resources are NCCRs, however, some of them may be CCRs.

In JIT/AL there is excess capacity, but there is also a mechanism to prevent the NCCR from being overactivated. It is well accepted that if a workstation breaks down in a line the whole line may stop even if there are stations that have the capacity to continue to work. In the Kanban system, a worker will not dare to work if he has not go the "production card" even though he has the raw material, the machine is available, and the parts are needed for the same day.

Rule 2: The level of utilization of a non-bottleneck is not determined by its own potential but by other constraints within the system.

Let's check the validity of Rule 2 by using our bridge: $T - I - OE$. The BN limits T as it has given finite capacity to fulfil partially the market demand. The NBNs should gear their level of activity to supply the materials needed for the BN to work on or for assemblies which are fed by a BN. Any attempt to activate the NBNs above the required level will cause no increase in T, but I will increase!

Conclusion: Overactivation of resources is working against our operational measurements, and therefore it is an undesirable action.

Rule 3: Activation and utilization are not synonymous.

Although Rule 3 is a direct derivative of Rule 2 it has its own importance. In establishing the directions for solutions we should not ignore obstacles that may jeopardize the implementation of the solution. Rule 3 indicates direct conflict with the current performance measurements of "efficiencies" and deviation from

4. DBR

standards. A solution for that conflict should be found prior to the implementation [12].

Now, as the BN is controlling our throughput, any disruption means losing time on the BN. As there is no additional capacity on the BN, we will not be able to recover from that loss, leading to loss of sales (T drops). In the AL everyone is aware about the BNs. In the debugging period the line engineers spend a lot of time smoothing the flow and handling the BNs. In JIT when a work-centre cannot work, the "red button" is hit for everyone to know that another constraint has been discovered and for the process improvement team to know where to focus next. It is crucial that an infrastructure is in place that will allow a fast response to overcome the problem. If this infrastructure is not fully operational the disruption ensuing will jeopardize the use of JIT.

Rule 4: An hour lost on BN is an hour lost on the system.

We can use Rule 4 to highlight actions that we take today which do not make sense and actions that we had better take if we want to get closer to our goal. For example, first of all make sure that the BN does not run out of work; then make sure that it gets the top priority from the maintenance people when it breaks down; make sure that it does not work on scrapped parts; allow offload to "inefficient" machines, etc. In short, any activity that will protect throughput or will release the capacity of the BN is going to be good news. On the BNs every available minute is utilized either for production or for setting-up the machine for the next part. The situation with the NBNs is different because there we have excess capacity, which means that the resource must have some idle time when it is not used for production or setup. More production will create more inventory and not more throughput. Therefore an attempt to save time on an NBN when we still have idle time will be a waste!

Rule 5: An hour gained on a non-bottleneck is a mirage.

There are two major ramifications of Rule 5, as follows.

- We had better re-evaluate "cost reduction" investments which aim to reduce operation time on NBN resources. Under the current practices they may yield additional inventory.
- We should utilize the idle time of the NBNs to speed up the flow and further reduce the inventory. The existing excess capacity should be considered as an asset. Thus we can rephrase Rule 5 to be: An hour saved on a non-bottleneck resource is an opportunity hour.

The assembly line and the Kanban approaches have recognizes that you need to protect the flow by using inventory buffers. In the AL it is the amount of space on the conveyor belt that was planned between two successive operations. In the Kanban it is the number of cards you allow in the raw-materials and finished-goods areas of a work-centre [13].

DBR recognized the need to protect the flow from disruptions by putting in some buffers. The DBR approach will suggest handling the buffers in a more advanced way and to focus most of the protection at the most critical points, namely in front of the BN and in front of an assembly or subassembly operation which is fed by a BN. These buffers should be considered as an insurance policy to protect throughput and due date performance. However, different locations of the BN will require different patterns of inventory. This leads us to the next rule.

Rule 6: BNs govern both inventory and throughput.

Rule 6 will guide us not tu rush to remove a bottleneck once we have found it. Yes, if we eliminate the constraint then T may go up. But at the same time this action may be dangerous since the new structure, dictated by the new BN, may need a huge investment in inventory which can drain our cash because it was not planned. On the other hand, if that move will reduce inventory as well then it is desirable one. The safest way to handle such decisions is to run several simulations and to check their impact in terms of $T - I - OE$.

Up to now we have dealt with the resources. But a manufacturing environment has products to produce. What rules can we derive for the product flow? These are the rules associated with the *Batch Size*.

5 The Batch Size and the Theory of Constraints (TOC) Approach to Solving a Problem

Current practice of production planning is still based on the fixed batch approach. In order to understand the inherent problems that can arise we can look at the example of the well-known EOQ (Economic Order Quantity) formula.

The EOQ approach assumes two tpyes of costs: the first is associated with setup considerations which indicate that the larger the batch is the less setup cost is imposed on the unit; the other cost is associated with the carrying cost — this implies that the smaller the batch the less carrying it will have.

There is a perceived conflict where the EOQ method is used to compromise by taking the total cost and finding the minimum total unit cost — this point is the EOQ (Fig. 6). The major reservations about this method are described in Refs. [3], [5], and [9]. Here we want to concentrate on one aspect only, and that

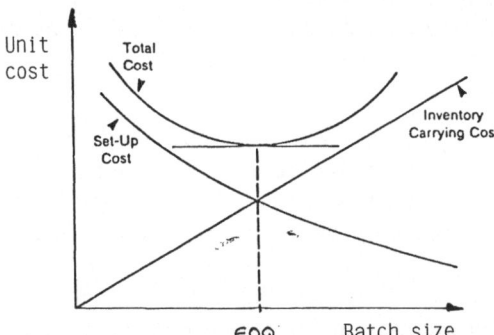

Fig. 6. Economic Order Quantity

4. DBR

is: What does this graph represent and do we want this type of solution? Let's go back and examine the conflict using the basic way of thinking of the Theory of Constraints (TOC). The above conflict can be presented as

A (implies) ⇒ that C should exist
Set up cost → Batch should be as big as possible

but

B (implies) ⇒ that C should not exist
Carrying cost → Batch should be as small as possible

This bring us to a conflict: small batch or big? To handle the conflict we have to check our basic assumptions under which we have made those statements. In this case we have to check: What do we mean by batch? A search for the batch size in the assembly line will yield two correct numbers: 1 and approaching infinity. As 1 is not equal to infinity it means that there are two different batches which are defined as

Transfer batch (Trb): a batch from the part point of view, indicating how many parts we produce before transferring them to the next production stage

Process batch (Prb): a batch from a resource point of view, indicating how many identical parts we produce before changing over to another part.

In the assembly line, transfer batch is 1 and process batch is as long as the life of the product. In JIT the transfer batch is a container which contains in most cases a fraction of a day's consumption of the assembly. The process batch of a specific part in a specific work-centre is a series of containers (transfer batches) that are processed together. The process batch is totally arbitrary and is determined by the sequence of the arrival of the "production cards" [13]. Now we can state the following rules.

Rule 7: The transfer batch may not (and many times should not) be equal to the process batch.

Rule 8: The process batch should be variable, not fixed.

By defining precisely what a batch is, we have practically solved the so-called dilemma that we had. As we have two types of batches we no longer have to compromise. We have now just to determine the size of the transfer and build a procedure to establish the process batches.

As was shown in Figs. 3 and 4, smaller transfer batch (Trb) yields lower inventory. Thus:

Trb ↘ ⇒ I ↘

The major considerations in setting up the Trb size are practical ones regarding handling and control. The process batch (Prb) should be defined, as in the Kanban, as a sequence of consecutive transfer batches. Prbs should not be predetermined but constructed dynamically in the process of scheduling.

Scheduling is a part of the control. Neither the assembly line nor the Kanban need a scheduling system. They have mechanisms to take care of who is doing what, when, and how many. In order to be able to build a logistical system we need to establish a principle that will convert the JIT and AL mechanism to a workable, programmable concept. This is the ninth rule.

Rule 9: Schedules should be estimated by looking at all the constraints. Lead times are results of a schedule and cannot be predetermined.

To run the plant along the lines of the above rules we need a logistical system. This is the Drum – Buffer – Rope (DBR) logistical system. This system may be implemented manually or with the assistance of computerized software.

6 Using the TOC to Build the Logistical System

There are several steps which have to be followed when we come to build a systematic approach to handle an environment with constraints.

Constraint: anything that limits a system from achieving higher performance versus its goal.

The steps of the process are:

1. Identify the system constraint(s).
2. Decide how to exploit the system constraint(s).
3. Subordinate everything else to the above decision.
4. Elevate the system constraint(s).
5. If, in the previous steps, the constraint(s) have been broken, go back to Step 1, but don't let inertia become the system constraint.

Step 1: Identify the system constraint(s). The two major constraints that the logistical system has to accommodate are: the market demands (the amount of each product that we can sell), and the capacity of the CCRs (capacity constraint resources). The procedures for identifying the CCRs are associated with the general characteristics of the plant, such as due date performance, finished-goods inventory profile, WIP profile, overtime expenses, etc. These characteristics are dominated more by the type of product flow rather than the types of industry. The implementation procedures are explored in Ref. [1].

Step 2: Decide how to exploit the system constraint(s). There are very few CCRs in any plant. The DBR way recognizes that such a constraint should dictate the rate of production of the entire plant – the *Drum* beat. Utilizing the CCR is done by determining the schedule of the CCR, taking into account only its limited capacity and the market demand that it is trying to satisfy. The CCR limits the throughput of the plant and controls the due date performance. The schedule should project forward in time from the present. In determining the jobs on the CCR the market due dates can give a rough sequence. There are four cases that might alter these sequencing decisions:

- Different lead times from CCR to due dates
- One CCR feeding another
- Setup on CCR
- A CCR feeding more than one part to the same product.

In each one of the above situations a trade-off decision must be made. Let's take one for example: setup on a CCR. The trade-off concerns the size of the process batch on the CCR. Large process batches will yield:

- more thoughput in the long run (T ↗)
- more inventory (I ↗) and
- worse due date performance in the short term (DDP).

Now we can see that the trade-off is between short term and long term, and between throughput and inventory. As one is at the expense of the other, management should be involved. Management should express clearly its trends and policies and the software system should translate them to workable schedules.

Determining the schedule (the Drum beat) is not enough. As the CCR is crucial we have to protect if from disruptions and to make sure that it does not run out of work. Any plant is subject to fluctuations, e.g. variation in process times, machine breakdown, scrap, absenteeism, inaccurate forecasts. To protect each CCR we will establish an inventory buffer in front of it. This buffer will contain only the inventory needed to keep the CCR busy during the next predetermined time interval. This is the *"Time Buffer"*. It protects the throughput of the plant against any disruption that may occur in operations preceding the CCR and that can be overcome within the above time interval.

Beating the *Drum* and building the time *Buffer* will ensure high utilization of the CCR, secure throughput (T) and due date performance (DDP); however we still have to control the inventory (I).

Step 3: Subordinate everything else to the above decision. Once we have built the Buffers in line with our policy, any additional inventory is not going to give us more protection. It will be a waste of money and may jeopardize T and DDP. The JIT/AL approaches have succeeded in preventing inventory's growing beyond the level that was predetermined. Each one of the production stages can work until the buffer behind it is full. And when the buffer is full the instruction is simply "stopworking!". This is a *"Rope"* that connects the buffer behind the operation with material being released from the buffer in front of the operation (Fig. 7).

The DBR approach demonstrates that putting a "Rope" between every two successive operations is excessive protection that might even reduce throughput. Controlling the gating (first) operations in every route is enough. The Rope

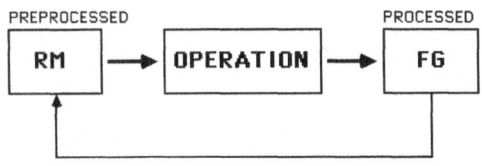

Fig. 7. The Rope concept for a single operation

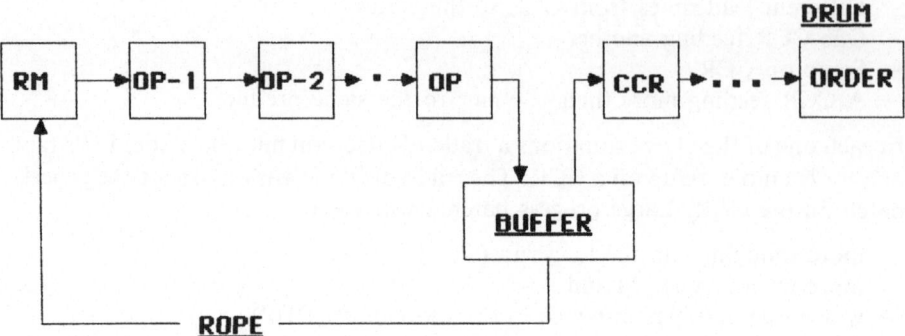

Fig. 8. The Rope concept for a single route

should be between the Buffer and the "raw-material" (release) area in front of the gating operation. In other words the production rate of the gating operation will be dictated by the rate at which material in the buffer is consumed by the CCR (Fig. 8).

The Drum, Buffer, and Rope (DBR) are the basic elements of synchronized manufacturing, since they provide all we need to maintain the production flow with a given predetermined inventory level [1, 2, 14]. Now we need to translate those concepts to a computerized system.

7 The Role of the Software

The DBR software is based on the principles of synchronized manufacturing.

- The capacity constraint resource (CCR) dictates the schedule based on market demand and its own potential.
- The schedules of the succeeding operations (including assembly) are derived accordingly.

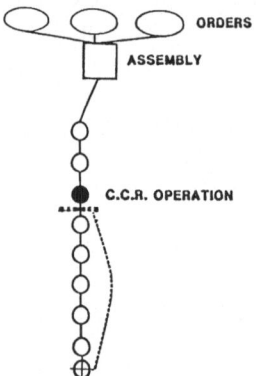

Fig. 9. Synchronized manufacturing for a single-route product

4. DBR

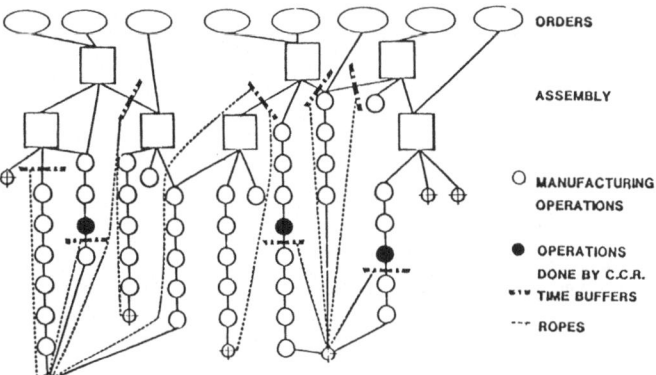

Fig. 10. Synchronized manufacturing in a plant

— The schedules of preceding operations support the time buffer and are derived backwards in time from the CCR schedule.

Every plant produces many parts. In some plants every product has just one single route (Fig. 9); in others, parts go into subassemblies and then into assemblies, parts may be sold as spare parts, and parts or assemblies may be common to several products. This is reality for production people, and can be represented as a "Spaghetti diagram" (Fig. 10).

Each plant has few CCRs. Every one of them should be protected by a time buffer and so should the assemblies fed by them. Constructing a schedule along the lines of DBR involves a huge amount of calculations that need to be performed in a short time and should be flexible to changes. This calls for the support of the computer.

The ability to produce schedules in a reasonable time depends on the data structure, its availability and fast access for the scheduling modules. The diagrams shown in Figs. 9 and 10 represent the product flow from raw materials to finished goods. There are several home made DBR systems based on the concepts described in The Race [2]. A new DBR software is under development, but the only software which is close to the DBR approach and is available in the market is the OPT®.

The Scheduling Modules of OPT®

Three modules are used to produce realistic, workable schedules (see Fig. 11).

1. SPLIT   This locates the time buffers in the appropriate places and separates the data needed for scheduling the CCRs.
2. BOPT®   This is a finite forward scheduler that beats the drum of the CCRs and the market constraints. This module creates the realistic master schedule (as it is based on the true constraints). The major output of this module is:

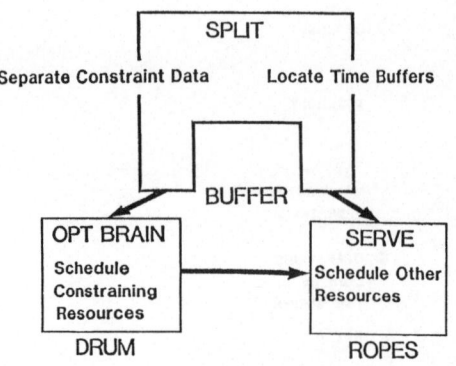

Fig. 11. Schematic structure of the scheduling modules of the OPT® software

- Forecast of order completion
- Work schedules report for the CCRs
- Load profile for CCRs (to ensure maximum utilization).

3. SERVE This is an infinite backward scheduler that backs off from the master production demands, ensuring that the operations feeding the buffers are completed in time. SERVE also establishes the Ropes for the gating operations dictating the earliest allowed material release. (This permission to work is similar in principle to the production Kanban cards, authorizing withdrawal of raw material.) The resource load profiles produced by the SERVE module identify potential CCRs, whether BNs or just temporary BNs which, if we ignore schedules, may be in danger. Analysing these resources and the way to handle them is a part of the implementation procedures.

Once the system is up and running and everyone works to the schedules, the whole approach starts to gain credibility as a result of smoother running of the entire plant, better relationships betwen different departments, and the achievement of operational improvements. The time is right now to go to the next step of improvement. The software plays an important role by offering its simulation capabilities.

Step 4 calls for elevating the constraints. What shall we do? Should we buy more machines, hire workers for another shift, reduce setup on CCR? All of these alternatives can be easily simulated by the software, producing results in terms of T and I to be presented to management for decision making. When a decision is made the software is used to adjust the schedules, the buffers, and the ropes to the new situation.

However, installing a software – is that enough?

8 Implementation Considerations

The experience gathered from hundreds of implementations all over the world tells us that installing a software is in many cases easy and that the real issue is the successful management of change. In order for positive results in bottom-line performance to become a permanent feature we have to fairly and squarely address the issue of change, not just attitudes but also habits.

Building the schedule is just one facet of the change needed in the plant. This is the *logistical change*. This is quite straightforward change involving the production control and materials management people that will produce the new schedules, release material, and expedite according to new rules.

One important element is an expected clash (at least in the first period) between the DBR schedules and the measurements imposed by management on the shop floor, like efficiencies and variation from standards. This clash calls for management to evaluate the way they measure their work-force, just to make sure that the shop floor can follow the schedules. On top of that, the new understanding policies, e.g. finished goods levels, market product mix, pricing, etc. This is the *managerial change* required.

Last but not least are our workers and shop floor management. Dr. Ohno from Toyota said that he spent most of his life chasing the cost concept out of his people's heads. Everyone on the shop floor is a financial wizard who can tell you how to do things "more economically". Some times they are right, but sometimes they are totally wrong as in cases when they want to produce big batches, save transportation, or to activate machines because they feel more secure when there is work to do. We have to change people's habits, and for that we need to manage a *behavioural change*. This is done through a two-day Functional Education Workshop in which the students find for themselves, with the aid of a PC simulator and a tutor, how to perform better within their span of control and improve the overall performance.

Those are the three elements of the implementation procedures taught in ten day course for the internal tutors – the Jonah Training (TM). This takes place after a plant or a division has already decided to embrace the TOC as the governing policy. As mentioned above, TOC is associated with change, and that means potential resistance. So, in order to assist companies in exploring the applicability of TOC to their business and the expected bottom-line benefits, a set of *introductory procedures* were developed and are conveyed in the Executive Decision Workshop (EDM) [15].

There is one more fundamental issue: With TOC we are not talking about one-off improvement!

9 There is no Finish Line

Our competitors do not sit still. They are improving all the time. They have the mechanism to invoke more improvements. JIT can locate the areas needing to be improved by quality circles or process improvement teams before they can further reduce inventory or increase throughput.

DBR provides a better mechanism. Buffer analysis, which is a whole new system, unfortunately beyond the scope of this article [2], provides the tool to

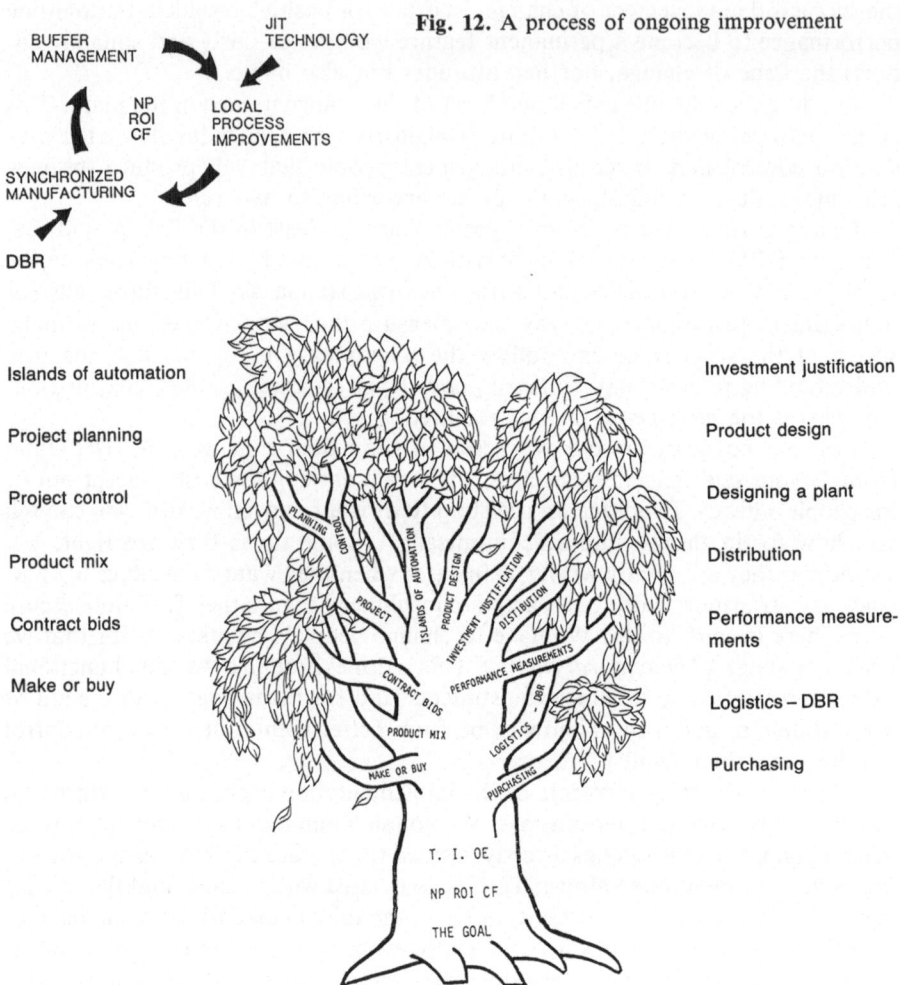

Fig. 12. A process of ongoing improvement

Fig. 13. Future system needing to be developed

locate the areas that we need to focus our attention. These are the areas where the payback or our efforts will be greatest in terms of the operational measurements (T – I – OE). In these areas we will use process improvement techniques that in return will yield further reduction in inventory without hurting throughput and overall performance. Once a local improvement is accomplished and a constraint removed, this is the time for us to look for another constraint. This is *Step 5* in running a constrained environment and it completes one round in a *process of ongoing improvement* (Fig. 12).

Using the TOC approach we are able to locate and quantify the sensitive areas of our business. By focusing on these areas we can use the compan's limited resources in terms of time, effort, and money to give us the best overall impact by increasing our competitive edge. This Pareto approach provides us with a natural pecking order, the result of which means global benefits come quickly. The prerequisite to achieving these benefits is constantly driving the inventory down in our plants. An inventory reduction of 80 – 90%, apart from being highly desirable, is also feasible within many environments. There is no finish line!

The Drum – Buffer – Rope logistical system is just the first step. In order for us to compete on equal terms in the global market-place of tomorrow we need to cover other aspects of running a manufacturing business. The TOC concepts are applicable for these facets but systems with detailed implementation procedures still need to be developed (Fig. 13).

10 References

1. Goldratt, E. M. and Cox, J., The Goal. North River Press Inc.
2. Goldratt, E. M. and Fox, R. E., The Race. North River Press Inc.
3. The OPT Way of Thinking. Executive video course module 1
4. Fox, R. E., MRP Kanban, or OPT. *Inventories and Production*, July/August 1982
5. Fox, R. E., OPT – an answer for America – Part II. *Inventories and Production*, November/December 1982
6. Fox, R. E., OPT – an answer for America – Part III. *Inventories and Production*, January/February 1983
7. Fox, R. E., OPT – an answer for America – Part IV. *Inventories and Production*, March/April 1983
8. Fox, R. E., OPT vs. MRP – thoughtware vs. software. *Inventories and Production*, November/December 1983
9. The Fallacy of Cost Accounting. Executive video course medule 3
10. Goldratt, E. M., Cost accounting – the number one enemy of productivity. APICS Conference 1983
11. Goldratt, E. M., The unbalanced plant. APICS Conference 1981
12. Goldratt, E. M. and Klarman, *Embracing A Process of Ongoing Improvement*
13. The Just-in-Time System and the OPT Rules. Executive video course module 2
14. The logical Ropes of OPT. Executive video course module 4
15. Theory of constraints Journal. Avraham Y. Goldratt Institute
16. Goldratt, G.M., Computerized shop floor scheduling, International Journal of Production Research, Vol. 26 No. 3. March 1988

Chapter 5
Period Batch Control
John L. Burbidge

Professor John L. Burbidge was educated in Larchmont, N.Y., USA, Wellington School, Somerset, UK, and Cambridge University, UK. He is now a Visiting Professor at Cranfield Institute, UK. He started as an apprentice at the Bristol Aeroplane Co. and worked for 26 years in the UK Engineering Industry – from apprentice to Managing Director. Later he was for 14 years with the ILO – Poland, Cyprus, Egypt, and Turin. He was Professor of Production Management at Turin International Centre. His research covers Group Technology and Production Control. He is the author of 12 books and has published many papers. Prof. Burbidge is a member of IFIP WG 5.7.

1 Introduction

Period Batch Control (PBC) is the oldest of the "Just-in-time" (JIT) methods of Production Control. With minor variations, it only assembles this period products which can be sent on completion to customers; it only makes this period parts needed for assembly in the next period; and it only accepts this period delivery from suppliers of materials needed for processing, and of bought parts needed for assembly, in the next period.

PBC was developed in England in the 1930s by the late R. J. Gigli. During his lifetime he installed the method in thirty different factories, but his greatest success was during the 1939 – 45 war, when he worked in the Ministry of Aircraft Production and used PBC to control the manufacture of Spitfire fighter aircraft. At a time of limited factory capacity and of materials shortages, he increased Spitfire output by only making each month the parts needed for assembly in the following month.

R. J. Gigli was obliged in his day to introduce PBC in factories with traditional Process organization. The introduction of Group Technology has completely changed the situation. Where Gigli hoped to achieve about 10 stock turns per year, today with GT and PBC one can expect at least 25 stock turns, even with low-volume multiproduct batch production. Again, where Gigli needed a large number of clerical workers to operate PBC, today the computer makes PBC a relatively cheap system to operate.

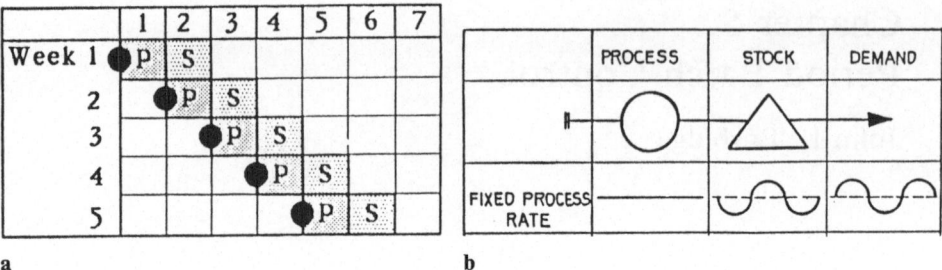

Fig. 1. Flexible progamming: **a** programming; **b** smoothing. Key: ●, programme meeting; S, sales programme; P, production programme

2 Programming with PBC

Programming is the first level of progressive Production Control. At this level, Production programmes are produced which are used directly to regulate final assembly and indirectly as the basis for regulating the manufacture of made parts and the delivery of bought materials and parts by suppliers.

With PBC the first need is for a series of short-term programmes based normally on one-week periods for simple products or two-week periods for more complex products, such as machine tools for example. As illustrated in Fig. 1, new sales and production programmes are prepared each period, making it possible to follow changes in market demand with a minimum investment in finished product stock. If, however, the periodic sales variations are large, production output can be "smoothed" by making a few of the more popular products for stock when the period sales are low. This method of programming has been used for many years to regulate assembly in mass production and is called Flexible Programming.

There is also a need with PBC for an annual programme which is used for planning purchase contracts and for financial planning. There must always be some method for reconciling the actual cumulative output up to a given date, which is shown by the series of short-term programmes, with the planned cumulative output at the same date, which is shown by the annual programmes. When they differ, the annual programme must be amended to bring it into line with the series of short-term programmes.

3 Ordering Made Parts with PBC

Ordering is the second level of progressive Production Control. At this level, orders are produced which regulate the manufacture of components and the delivery of bought materials and components by suppliers.

5. Period Batch Control

	1	2	3	4	5	6	7	8	9
Week 1	O	M	P	S					
2		O	M	P	S				
3			O	M	P	S			
4				O	M	P	S		
5					O	M	P	S	

Fig. 2. Ordering with PBC. Key: O, data processing; M, make parts.
Note: workshop orders and supplier's call-off notes are produced by explosion from the period production programmes (P)

As is shown in Fig. 2, ordering with PBC is a simple extension of Flexible Programming. Additional periods are added for data processing and component manufacture. Immediately after the programme meeting each period, the requirement schedule for made parts is calculated for each group from the chosen production programme. This schedule is issued in the form of list orders for the period to each group.

It is not difficult, if one wishes to do so, to add additional quantities to the production requirement schedule, to provide a scrap allowance, a spare parts allowance, and/or to maintain a buffer stock. The only difficulty is that it is difficult and costly to control such stocks.

Ideally, in a perfect situation where there were no shortages, one would make the exact requirement quantity for production, plus any additional parts needed for sale as spares. Such perfection is always impossible. If, however, one can achieve a state where shortages are the exception (say less than 1% of the parts ordered) rather than the rule, then it will generally be more economical to take special emergency action to replace shortages than it will be to maintain buffer stocks for all items.

4 Loading with PBC

A loading system is needed with PBC in order to compare the load imposed by each period production programme on each group, with the capacity available in the groups.

The Nett Load method is to be preferred. This compares the Nett Load, or sum of the operation times for all parts to be made each period, with the standard running time, or planned capacity less ancillary, idle, and down time, followed by adjustment for above- or below-standard performance rating. The standard running time is found using either random observation studies or machine-monitoring equipment.

A "Period load summary" is produced each period in the form illustrated in Table 1. This shows first if the production programme is feasible, and second it shows to the group foreman which are the critical — most heavily loaded — machines in his group, so that he can arrange to keep them manned at all times and can if necessary plan some overtime.

Table 1. Period load summary

Period Load Summary			Period 5		Starts 26.2.86		Group 6
No.	Code	Machine	Machine hours				
			Cap	Load	−	+	Critical
1.	01657	Lathe	62	72	−	10	X
2.	01658	Lathe	60	61	−	1	X
3.	01702	Lathe	66	52	19	−	
4.	04330	Mill	62	52	10	−	
5.	04331	Mill	70	51	19	−	
6.	04333	Mill	60	45	15	−	
7.	06066	Sur. Grind	55	40	15	−	
8.	07168	Drill	60	39	21	−	
9.	07213	Drill	75	30	45	−	
10.	07222	Drill	75	25	50	−	
11.	04873	Slotter	70	20	50	−	
12.	−	Bench	−	80	−	−	
		Totals	715	567	244	11	

Man hours: $567 \div 74 = 7.65 = 8$ men.

5 Dispatching with PBC

Dispatching is the third level of progressive Production Control. It is concerned with planning, directing, and controlling the completion of shop orders in manufacturing workshops. It deals with such tasks as materials handling, tool storage, setting-up, inspection, and operation scheduling. With Group Technology these types of task are normally delegated to the groups.

The most difficult of these tasks with traditional Process organization is Operation Scheduling. This is very much simpler with GT and PBC. In this case the group foreman receives at the beginning of each period a "List order" showing the parts he must make, and a "Period load summary" showing the load they impose on the machines in his group. He knows which are his critical, or most heavily loaded, machines and attempts to keep them operating all the time and to increase their capacity by scheduling the work to minimize setting-up time.

He knows also which are the critical parts he must make − parts with long throughput times − and he will attempt to shorten these time by close-scheduling, or starting following operations before the previous operations are finished. The loading of the remaining parts can be based on a simple decision rule such as: load first the part with most outstanding operations.

One other problem which must be considered may arise if a machine type is only used for first operations and is seldom used for a single operation on its own. There is normally no problem unless the machine is "critical" or heavily loaded. In this case it may be necessary to allow some overlapping of the period orders. With one-week periods for example, if the list orders and period load summaries can be prepared by Wednesday evening, materials can be issued on the Thursday, and the critical machines can be kept working at the end of each period on parts for the next period.

6 Purchase Deliveries with PBC

Ordering – the second level of Production Control – not only regulates the issue of shop orders, but also regulates purchase deliveries from suppliers. With PBC, purchasing is based on the call-off method. The supplier is given a contract, together with a forecast of probable future requirements, but he may only deliver each period in the quantities authorized by periodic call-off instructions.

The forecast of future requirements is found by explosion and implosion from the annual programm. The call-off quantities of each part are found by explosion from the period production programmes. In order to prevent an inflation of transportation, it is generally desirable to reduce the number of suppliers, so that each provides a larger number of different parts, and/or to arrange for the periodic collection of deliveries by one carrier from a number of suppliers in the same district.

The first objective is to get the Class A or top-value purchased items on call-off. As 10% of these items will often cover over 50% of the cost of purchases, this change can have a major effect in reducing the investment in stocks and work in progress. There will also be many purchased items which are purchased from stockists, or are made for stock by their manufacturers, which can also be obtained without difficulty using the call-off method. The general experience is that it is not difficult to get the first 50% of purchases on call-off, but the second 50% is much more difficult.

7 Conclusion

PBC is the oldest of the Just-in-time (JIT) Production Control systems. It is particularly suited for use with Group Technology, because it tends to provide an even load on the groups, to increase their capacity, and to simplify operation scheduling.

With PBC and GT, it is possible to achieve rates of stock turnover of over 25 times per year, even with multi-product engineering batch production.

PBC is comparatively simple to introduce. Each of the stages here described can be introduced progressively, and be tested in practice, before abandoning the system previously in use.

PBC is a very simple system to use. Once it is established, it will operate reliably, with a low investment in stocks and work-in-progress, at a low cost of operation.

PBC has been used successfully in a wide range of industries. They include not only engineering assembly industries but also implosive industries such as foundries, potteries and glass works, jobbing engineering, and potentially most other types of industry as well.

Chapter 6
All-Embracing Production Control

Gideon Halevi

Dr. Gideon Halevi received his B.Sc. in Mechanical Engineering from the Technion – Israel Institute of Technology, Haifa, Israel (1952), his M.Sc. in ME from the University of Pennsylvania (1959), and his Doctor of Science in Technology from the Technion, Haifa, Israel (1973). Dr. Halevi is an adjunct Professor at the Technion (1981) teaching mostly graduate students on the CAD/CAM field. For more than 30 years he was employed by the Israel Military Industries in production development, combined technical operations, Data Processing Manager and Director of CAD/CAM Research and Development. In 1983 he formed a company, Hal Robotics, for the purpose of developing products required to implement All-Embracing Technology. He has published a book on the role of computers in manufacturing processes. He is the Chairman Elect of CASA-RI/SME, Israel Chapter, a member of the CIRP, the Israeli representative to IFIP TC 5 and many other public activities.

1 Introduction

Technology is affected by the capability of the available tools, and the imagination of those who build and drive it. For many years production management was built around a chain-of-activities philosophy. In this chain-of-activities approach there are hierarchical levels, where each level has a specific task to perform. Each level receives input that acts as constraints, performs its algorithms by some method, and generates output. The output of a certain level (stage) acts as an input to the next level. The main production planning stages and their tasks are as follows.

Master Production Schedule (MPS)

The master production schedule transforms the manufacturing objectives of quantity and due date for the final product, which are assigned by the non-engineering functions of the organization, into an engineering production plan. The decisions in this phase depend either on the forecast or on confirmed customer orders, and the optimization criteria are meeting due dates, minimum level of work-in-process, and plant load balance. These criteria are subject to the constraint of plant capacity and to the constraints set in the routing phase. The master production schedule is a long-range plan. Decisions concerning lot size, make or buy, addition of facilities, overtime work and shifts, and confirm or change due dates are made until the objectives can be met.

Materials Requirements (Resource) Planning (MRP)

The purpose of this phase is to plan the manufacturing and purchasing activities necessary to meet the targets of the master production schedule. A quantity and a due date are set for each part of the final production. The decisions in this phase are confined to the demands of the master production schedule, and the optimization criteria are meeting due dates, minimum level of inventory and work-in-process, and department load balance. The parameters are on-hand inventory, in-process orders, and on-order quantities.

Capacity Planning System (CPS)

The goal here is to transform the manufacturing requirement, as set forth in the requirement planning phase, into a detailed machine-loading plan for each machine or group of machines in the plant. It is a scheduling and sequencing task. The decisions in this phase are confined to the demands of the MRP, and the optimization criteria are capacity balancing, meeting due dates, minimum level of work-in-process, and manufacturing lead time. The parameters are plant capacity, tooling, on-hand materials and employees.

The division between the stages is straightforward and logical. However, it suffers from a practical problem. In a way all three stages are doing the same thing but with a different degree of accuracy. The MPS assumes infinite inventory and capacity. The MRP realizes that there is a finite inventory, but assumes an infinite capacity. The CPS assumes finite capacity. But introducing the actual, or finite, available resources, the infinite planning might become unrealistic. This blindness to capacity is one of the main MRP problems. Several methods were used to get around this problem. The most universal one is rough-cut capacity planning in the MRP. Rough cut is typically used to measure the fit of a master schedule. This technique translates a master schedule (end-item requirements) into a common resource unit of measure such as work hours. Available capacity at *key* resources is compared witch the plan to check for potential overload. Resolving the overload is a *manual* process. There are many philosophies, methods, and ideas of how to resolve the practical problems. Some of them are of a mathematical nature, but the trend is toward "supplying information to the person in charge and letting him make the manual decisions based on the information supplied to him by the computer system".

One dominating feature in any of the suggested systems is the separation of technical data from the production planning phases, i.e. the bill of materials and the routing are externally given axiomatic data. One may have alternatives, but even they are externally fixed — given beforehand without knowing if they are required or not. Thus, production planning loses some of the flexibility inherent

6. All-Embracing Production Control 79

in the manufacturing process and ignores many possible solutions. All-embracing Technology opens up those options and increases the flexibility.

Developments in the computer field brought forward a new generation of computer systems that are more powerful and faster, with a great increase in the size of memory and secondary storage. Today, computers introduce capabilities that were not possible before; thus we may set more advantageous objectives, with a higher probability of being able to achieve them. Our exploration of such possibilities has led us to the conclusion that a total reassessment of the manufacturing process is needed. This has resulted in a new, all-embracing technology that views the manufacturing process as a single, unique system.

2 What is All-Embracing Technology?

All-embracing Technology is a computer-oriented manufacturing philosophy whose general objectives are to increase productivity and reduce manufacturing costs.

The fundamental method to achieve this goal is by utilizing the flexibility inherent in the manufacturing process, and in striving to achieve overall plant optimum performance rather than optimum solutions to isolated stages or products. Achieving an optimum in a single activity does not necessarily lead to overall optimum performance.

All-embracing is referred to for shortness as "Hal" – the computer robot in *2001 – a Space Odyssey*.

Hal considers the manufacturing process as a nucleus and satellites rather than as a chain-of-activities. The engineering activities are the nucleus and the other activities are the satellites. Thereby, Hal introduces engineering, value engineering and technology to all phases of the manufacturing process. It broadens the scope of alternate solutions and eliminates the artificial constraints used as an interface between the engineering and production phases in the chain-of-activities. Hal is an all-embracing technology which is superior to an integrated system, since it does not contemplate the relationship between individual areas of activities, but rather dissolves them into one single system.

The Hal approach makes use of the following notions.

– There are infinitely many ways of producing a workpiece.

– Any component can be produced by any available facility. It can be produced with a 5D DNC machine, a universal machine, or manually with the aid of a chisel and file.

– The cost and lead time required to produce a component are functions of the process used.

– There are infinitely many ways of meeting design objectives.

– In any component about 75% of the dimensions (geometric shape) are non-functional (fillers). These dimensions can vary considerably without affecting the component performance.

- The cost and lead time required at produce a component are functions of its design. A minute change in fillets or dimensions to suit a standard tooling or an existing setup on a machine can result in significant cost variations.

The main concepts of Hal are as follows.

- Engineering phases are incorporated into the production phases.
- All phases of the manufacturing cycle work toward a single objective. Each phase considers the problems and difficulties of the other phases.
- The objective is to increase productivity and decrease the manufacturing cost of the product mix required at any period rather than to optimize any single product, component, or operation.
- No artificial constraints are created or considered.
- The manufacturing cycle is kept dynamic and flexible until the moment production starts.
- The objectives and approach of GT are adapted.
- Creating part families is not an objective, but only a means to achieve savings in manufacturing and shortening of lead times.
- Efforts are made to use standard tooling and existing machine setups in product design and process planning.
- A computer is a working machine. Automatic decisions are reached by employing a computer in every possible case. Interactive human-machine use is reserved for cases in which data or an algorithm are not available.
- Work-in-process is regarded as shaped raw material, and its original destination can be changed.

Hal is based on two powerful tools:
1. Part Description System (PD)
2. Generative Process Planning Program (GPPP).

Part Description System is a computer-aided design (CAD) system that was designed to serve all manufacturing requirements, i.e. design, process planning, structure analysis, industrial robot vision. Its prime objectives are to enable the computer (memory and not graphic terminal) to:

- "see" shapes
- manipulate shapes
- change shapes
- compare shapes.

It uses a specially developed file organization that is economic in storage space and has fast retrieval of any drawing by key or by attributes.

The Part Description System has the following features.

- Retrieves existing drawings by key or attributes (random unpredictable inquiries).

6. All-Embracing Production Control

- Automatic design of subsystems and tooling.
- Checks and enforces company standards.
- Checks and recommends design for ease of production and ease of assembly.
- Checks and recommends tolerances.
- Performs strength and stress analysis.
- Provides for technological transfer.
- Displays the shapes of a part under design on a graphics terminal.

Generative Process Planning Program (GPPP) is a computer program that reads part description records and generates automatically an optimum process plan (routing), including machine selection, gripping location and type, and tooling. Moreover, it indicates the production lead time, including handling time and item production cost.

The extended features of GPPP include:

- Forced process planning. This forces the system to select a given machine with its existing setup. Program will indicate cost deviation with respect to the practical optimum process plan.
- Manufacturing quantity increase. Grouping items for similar premachining operations.
- Forced components flow. This forces the program to select machines in such a manner as to result in a flow line, or work cell on the shop floor.

3 The Manufacturing Cycle

Under All-embracing the manufacturing cycle is divided into three main modules as shown in Fig. 1.

3.1 Master Production Planning

The master production planning combines, among others, the MPS, MRP, and CPS into one module and does it as accurately and realistically as possible, thus overcoming the drawbacks as detailed in the Introduction section. One may rightfully argue that management and MRP can do without such great accuracy as we propose to furnish. However, such accuracy will be a by-product of the production system and, consequently, it will cost less to supply accurate data than rought estimates by the use of separate applications.

The objectives of the Hal production planning module are to construct a realistic master production plan. Specifically, it will:

a) consider inventory and machine loading, which will be useful for sales people in promising delivery dates

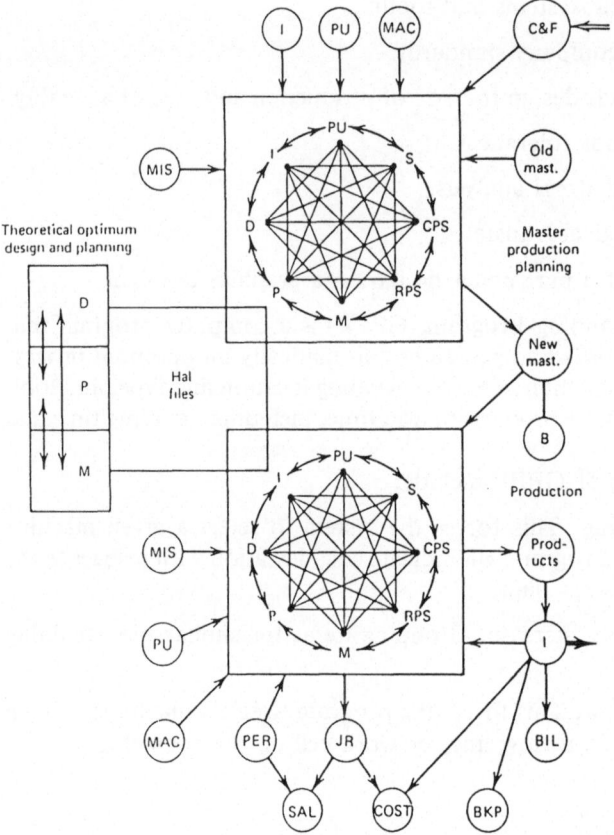

Fig. 1. Hal manufacturing cycle. Notation: D, product design; P. process planning; M, methods, time, and motion study; RPS, requirement planning; CPS, capacity planning; S, shop; I, inventory; PU, purchasing; MAC, machine file; C & F, customer orders and forecasting; MIS, miscellaneous; B, budget; PER, personnel; JR, job recording; SAL, salaries; COST, costing; BKP, bookkeeping; BIL, billing

b) compute requirement planning which considers machine load status

c) attempt to utilize dead stock, slow moving items, and rejected components

d) schedule the requirements to meet due date while balancing load without pulling jobs forward or backward

e) work out the production plan as closely as possible with product mix optimization.

This module is carried out as shown in Fig. 2. The concepts and methods used are as follows.

1. The planning is carried out in a time-period – product-network matrix.
2. The product network lead time is regarded both as elastic (i.e. it can be compressed or stretched within allowable boundaries) and rigid (i.e. it can be pulled forward or backward on a time scale as a rigid unit).

6. All-Embracing Production Control

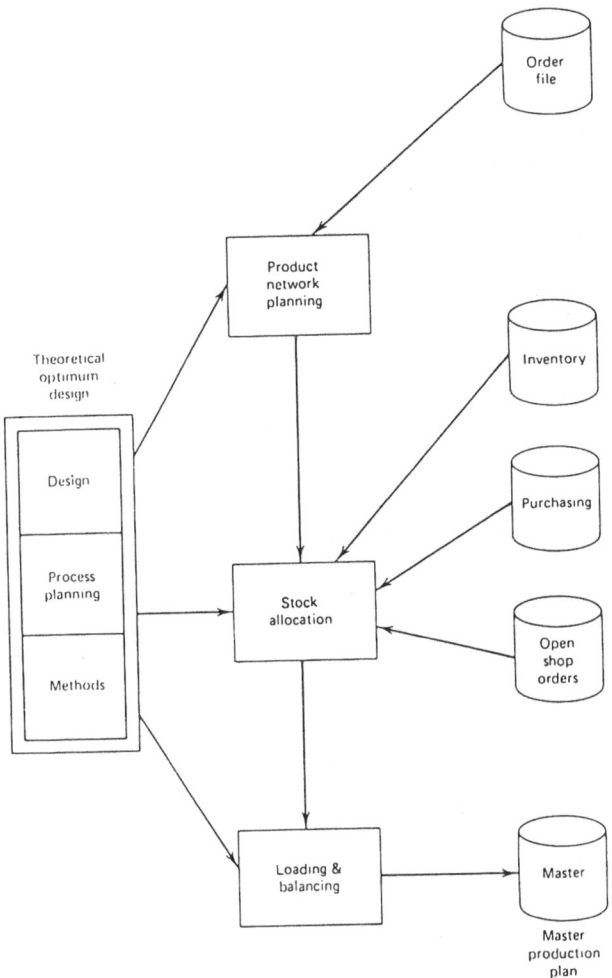

Fig. 2. Master production planning module

3. On-hand and on-order items are regarded as shaped raw material. They are allocated according to optimum requirements.

4. The allocation is treated as a feedback completion entry.

5. Stock allocation is performed period by period rather than by the allocation of an item through all periods.

6. The term "routing file" does not exist in the system.

7. The initial time periods are considered as a frozen zone. The allocation is fixed and the network planning is forward.

8. In other time periods the allocation is temporary and the network planning is backward.

9. Past periods are treated as regular periods but with zero available capacity.

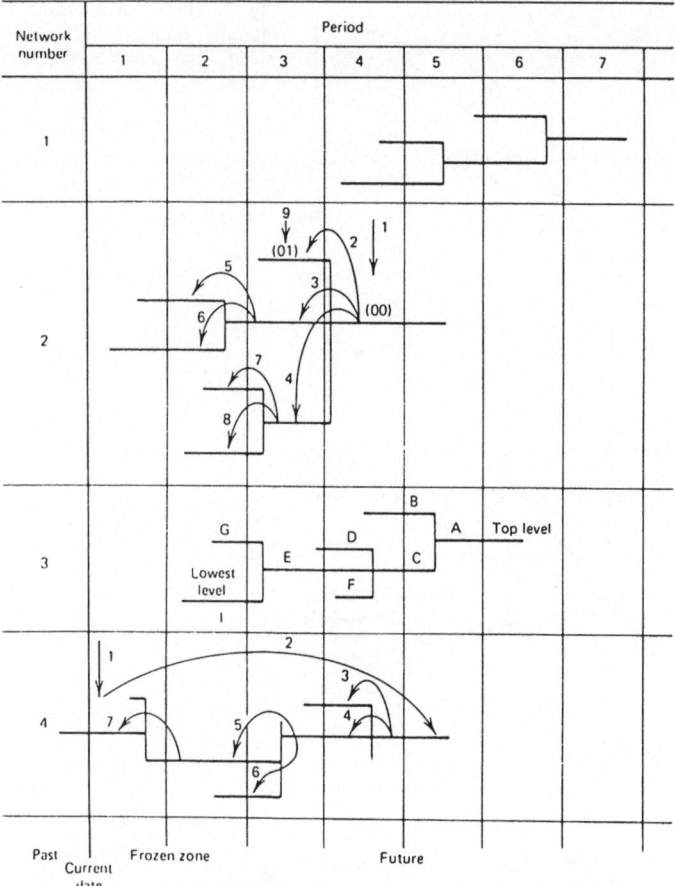

Fig. 3. Allocation by periods

Product Network Planning

The Hal production planning module is carried out using a time-period – product-network matrix. To construct the matrix the product network is spread over the time period with a due date at a fixed point (backward scheduling). If the network falls into the past the elastic feature is used to compress it (reduce lead time) in an attempt to have the early start date coincide with the current date. The elastic limit is the slack, i.e. between early start (ES) and latest start (LS). If, when using full compression, the network still falls into the past, it will be accepted, and this product has no elasticity. Other measures will be taken in order to meet due dates. Figure 3 shows the time-period – product-network matrix.

Stock Allocation

Stock allocation is carried out in three steps as follows.

First Step. Stock in inventory and on order is considered as free stock and can be allocated as needed to any open order. The allocation takes the form of a "completion" feedback entry which marks the job of producing the component "completed". The allocation model uses the minimum lead time product network, i.e. the latest start dates are used for each component.

The allocation model examines the requirement of components for all orders per period, starting with the first and advancing consecutively to the last period.

Three types of periods are recognized by the system. Each one is treated differently.

1. Periods that fall in the past. Components required in such periods present critical orders and get priority in stock allocation. The program scans the period column till it reaches a row with requirements (Fig. 3, row 4, line 1). This row represents an order. The program walks through the row (the product network in the ascending period direction), till it reaches the top level of the order, i.e. the product ordered (row 4, line 2). If the above product is available in stock, the required quantity is allocated to that particular order. The product order is marked as completed. The program then walks through the row, in the descending periods direction, marking all the branches of the network as completed (row 4, lines 3, 4, 5, 6, 7). If only partial quantity is available in stock, the order is split into two lots: one with the available quantity, marked as completed, and the other with the remaining quantity. The lead time and product network (now having a shorter duration) are adjusted to fit the new lot size.

   In such cases and when no stock is available for the product, the program walks through the product network in the descending periods direction till it reaches a product assembly level lower than the one handled (row 4, line 3). An attempt to allocate stock to this level is made as before. The entire procedure above is repeated until the whole network is marked "completed" or the initial period has been reached.

2. Periods that fall in the frozen zone. These are allocated on a single-item basis.

3. Periods that fall in the future. The allocation is carried out level by level, starting with low level code 00 components (the low level code indicates the lowest level at which a particular item is found in any bill of material) and period by period in ascending order.

Second Step. As a result of the first step two lists ar available: a list of the required items and components, and a list of dead stock (where dead stock is defined as an available stock not required by any of the open customer orders). By using the part description system an attempt is made to utilize the dead stock, rejected items, by transforming these items into the required ones, if such transformation is economically justified. Furthermore, the system checks whether it is possible and profitable to tear down dead stock assemblies and subassemblies and use their components as-is or by transformation.

The algorithm is as follows. In the first stage a test is made of the geometrical equality and inclusion of the required item within the surplus item. Required items that do not pass the test are released to shop and are excluded from further

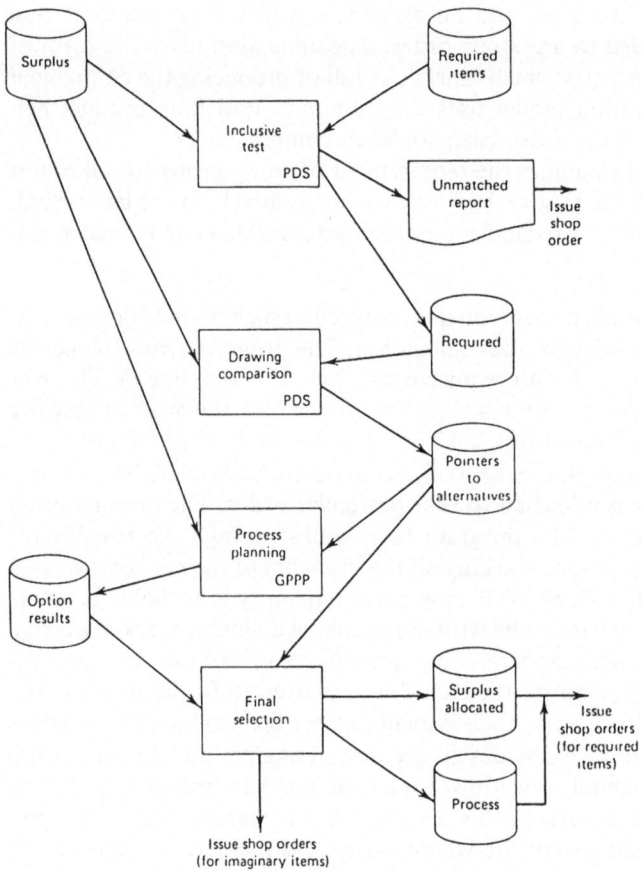

Fig. 4. Utilization of surplus items module. Notation: PDS, part description system; GPPP, generative process planning program

consideration. The remaining required items are compared not only for inclusion but also according to the required special features and their location, thereby reducing the number of alternatives. All alternatives are examined as to the conversion cost from the surplus item to the required item. This is done by defining the surplus items as shaped raw material to the GPPP. The output of the GPPP is an optimum process, including time and cost. An economic algorithm analysis decides which of the alternatives to choose, i.e. which surplus item is to be remachined in order to obtain a required item. More than one alternative may be chosen if the quantity of the surplus items is insufficient. Special shop orders are issued for such machining operations. If the economic module decides that it is not economical to transform items, a normal shop open order is issued for such items. Figure 4 shows the block diagram of this feature.

Third Step. A critical order is defined as an order whose portion of the network falls in a past period. This means that if normal manufacturing practice is

6. All-Embracing Production Control

used, the delivery date cannot be met. To solve this situation the first attempt is made by utilizing slow-moving items, where a slow-moving item is defined as an available item that will not be required for a long time (six months plus its lead time). Such an item is transformed into the required item, and thus its lead time is reduced. The method used is similar to the one described in the second step. If this attempt fails, the use of a substitute design is examined. By using the part description system the program checks whether any of the dead stock or slow-moving items can be substituted for the required item. This is done by replacing the required item with the available item and checking for possible assembly, clearance in movement, and functions.

All the above allocations are temporary and subject to change.

Loading and Balancing

Load balancing. The workload required to complete all customer orders on time is arrived at by reviewing the present state of the time-period – product-network matrix. The total load and work-centre load are considered separately for each period. Components marked as completed or as purchased items are ignored. Figure 5 shows the matrix (Part b) and the load profile (Part a). The required load includes critical orders: that is, periods that are in the past, and disregards overloaded or underloaded periods. In order to arrive at a master production plan, the load should be balanced and the past period abolished. For laod-balancing purposes, past periods are considered to have zero available load. Any required load in these past periods transforms them into overloaded periods. A single technique can be used to resolve overloaded and past periods.

Overloaded cases can be resolved by network shifting if the average load is equal to or less then the available load. Load profiles reveal such information. If the required load over all periods is above the available load, as shown in Fig. 6a, the system will supply information and recommendations to management as to which course of action to take, but the decision rests with the management. The available capacity can be increased by working overtime, extra shifts, or by the purchase of new equipment. The required load can be reduced by subcontracting work, turning down orders, or reducing order quantity. In such cases, the work-centre load profiles are used to pinpoint the bottleneck in production. A similar situation may occur if the initial periods are overloaded and not preceded by underloaded periods. Load profiles are prepared of the orders whose due dates fall within the overloaded zone. This is shown in Fig. 6(b). If such a profile shows overload, or even 70% load, the overloaded zone should be treated by management as described before.

Technological means. Initially, an attempt to resolve the overload and underload situation is made by technological means, i.e. generating a substitute process thereby transferring jobs from an overloaded work-centre to an unloaded one. By definition the substitute process would not be as efficient as the original one, but on the overall product-mix manufacturing it might give better results.

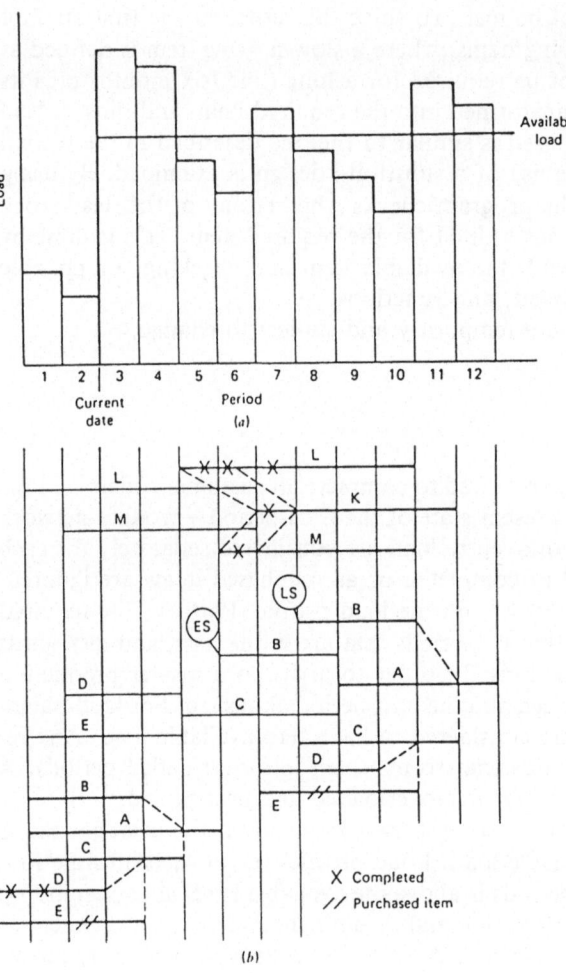

Fig. 5. Load profile and matrix: (a) load profile; (b) matrix

The substitution is done by a procedure where an unloaded machine is searching a job for itself. The initial search field is in overloaded work-centres of the same time period. GPPP is used to generate processes and those with minimum deviation of cost or time of the original work-centre have the priority. If after this search there are still underload situations, then the search field is broadened but utilizes the same decision rule.

If the technological attempt to solve the overload/underload did not supply a complete solution, the remaining overloaded situation is resolved by shifting the planned dates of the components forward or backward to fill underloaded periods. The networks are shifted in a telescopic manner, where the order due date is fixed.

Load balancing is done in two steps: (1) resolving overload throughout all the periods, starting with the initial periods, and (2) resolving underload throughout all the periods, starting with the initial periods. When an overloaded period is not preceded by an underloaded period, forward shifting is used, while when an

6. All-Embracing Production Control

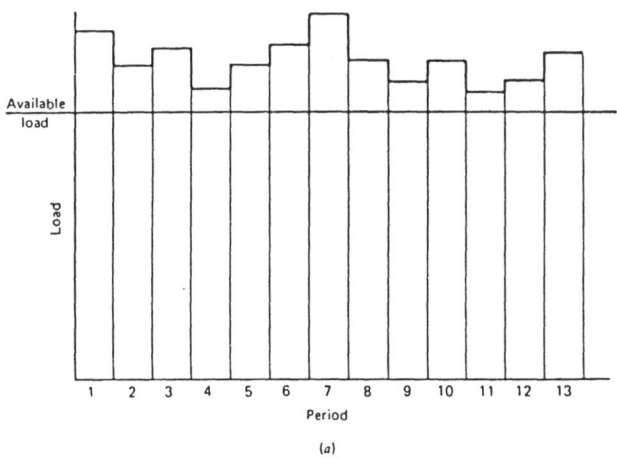

Fig. 6. Load profiles

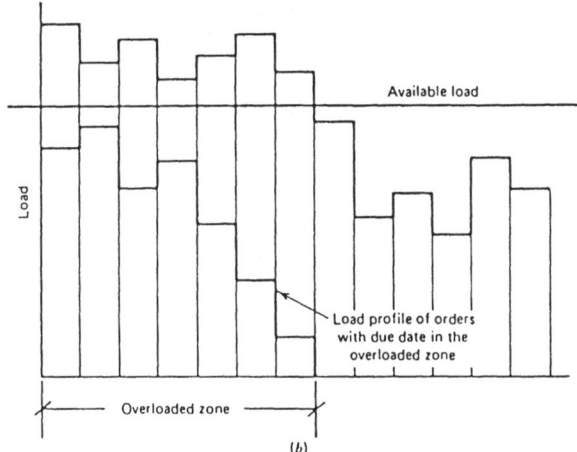

overloaded period is preceded by an underloaded period, backward shifting is used, as described below.

(1) Forward shifting. Orders having the latest due date are handled first. Although the order network was planned on the basis of the latest start date (LS), as shown in Fig. 7(a), several methods of shifting can be employed, such as the following.

- *Splitting.* An initial attempt is made to shift forward items that do not affect any part of the network. If the lead time of an item is included in an overloaded and underloaded period, a split is used in the underloaded period, thus pulling the job forward. Such a method used on an order for product A is shown in Fig. 7(b). It could be applied only to components E, G, and C. By using this method, the overall lead time of Product A was reduced from 20 to 28 time units.

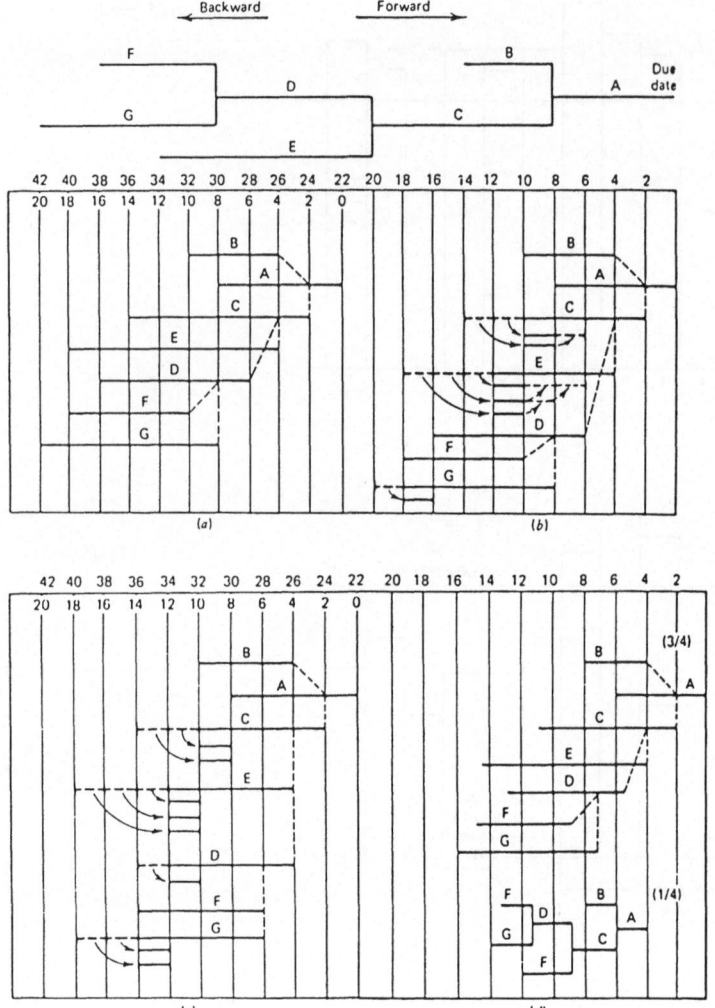

Fig. 7. Forward shifting methods: (a) forward shifting; (b) splitting; (c) splitting and shifting; (d) order splitting

- *Splitting and shifting.* When a split is made on an assembly, all its components can be shifted forward. This situation is shown in Fig. 7(c). Since assembly C was split, subassembly D can be shifted forward on the condition that a sufficient quantity is available for assembly. By using this method, the overall lead time of product A can be reduced to 14 time units.
- *Order splitting.* When an overloaded zone is followed by an underloaded zone that includes the order due date, the order quantity is split into two or more orders. This is shown in Fig. 7(d).
- *Reducing lead time.* The system can absorb up to 30% of the lead time by improving the basic data coefficient, operation overlapping, and work effi-

6. All-Embracing Production Control

ciency. A network that ends in an overloaded period and starts (due date) in an underloaded period can be pulled forward by being recomputed with a three-step reduction in lead times of 10% each. Such orders will be marked for special treatment in the capacity planning phase.

(2) Backward shifting. Orders having the earliest due date are handled first. Backward shifting is allowed within the range of early start (ES) and latest start (LS) dates. Several methods are used, as follows.

- *Network spreading.* When an overloaded period is preceded by substantial underloaded periods, the whole network is spread in a proportional way. This is done by increasing the overlap and overhang limiting factors. The increase is carried out in steps until no overlap is used at all: that is, the loading is by ES. Figure 8(a) shows this method (compare Fig. 7(a)).

- *Assembly shifting.* When an overloaded period is encountered in only one or more work-centres, it is sufficient to shift only those jobs that are in the

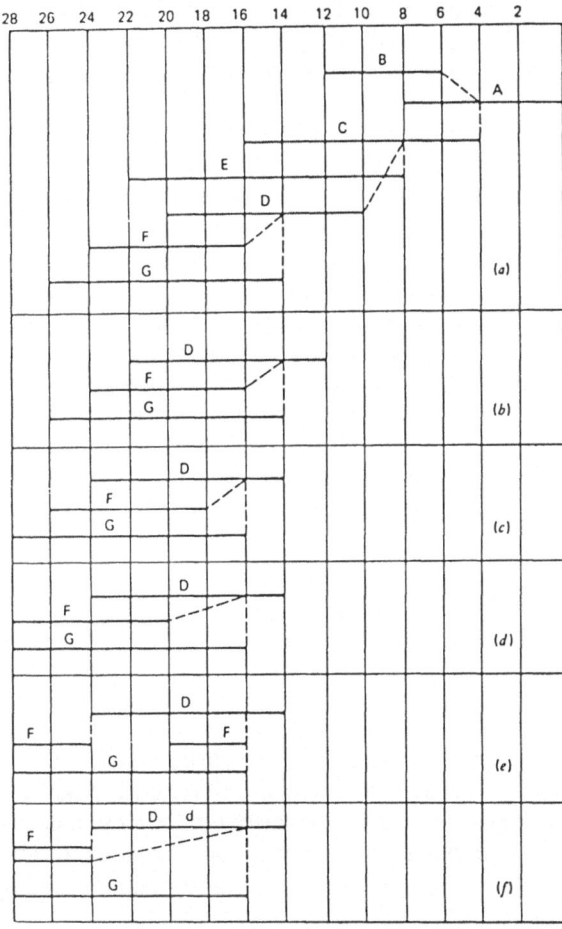

Fig. 8. Backward shifting method

overloaded work-centres. An assembly is free to move as an individual item (i.e. not affecting any other part of the network) within the boundaries formed by the present overlapping factor and the minimum overlapping factor. Subassembly D in Fig. 8(b) was shifted backward without affecting its components G and F. From this point the components are locked to their assembly and are moved as a unit. This is shown in Fig. 8(c).

- *Item shifting.* An item can be shifted backward to any location between its ES and LS without affecting the network (see item F in Fig. 8(d)). Items B and E in Fig. 8(a) can also be shifted.

- *Item shift and split.* When a narrow underloaded zone is encountered in a single work-centre followed by an overloaded zone, the method of shifting and splitting the quantity shown in Fig. 8(e) and (f) can be used.

3.2 Production Module

This module covers the actual manufacturing. The input is the product mix that must be produced in a given period, as planned in the master production planning module. The output is the list of products ready for delivery.

The purpose of this module is to make sure that the plan will be worked out, and in the most economical way possible. To achieve the purpose of this module, total flexibility concerning routings, product design, and available inventory is assumed. This means that deviation from the theoretical optimum input data is allowed.

The flexibility lies in the sequence and method of production and in the allocation of available raw material and semi-finished items. The first scheduling attempt is according to the theoretical optimum data. If this fails to produce satisfactory results, the measures taken are: alter process, contrive process to suit an existing setup of an unloaded machine, use and modify in-process items, change filler dimensions, and so on. The limiting factors in the use of the above measures is the economic consideration, i.e. the increase of the product mix cost due to the deviation from the theoretical optimum must remain below the expenses resulting from idle machines, work-in-process, and penalties for not meeting due dates.

4 Conclusion

Developments in the use of computers as an aid to manufacturing have proceeded in a modular, disjointed fashion. Systems designed and developed to solve a specific problem as expediently as possible were necessarily limited in scope. Integration of these systems is often attempted, but only as an afterthought. The resulting proliferation of disjointed computer systems tends to magnify manufacturing problems.

The rapid growth of the capabilities of digital computers, together with the increased computational power that exists today, have made it possible to set more advantageous objectives to the role of computers in manufacturing processes than ever before. A total reassessment of the manufacturing process, which has evolved from manual systems, is required, and computer-oriented systems have to be built.

All-embracing Technology is a computer-oriented manufacturing philosophy. If has an approach that can practically be implemented by using Hal Technology tools, i.e. part description system, generative process planning programs, and other feasibility study programs and algorithms that have been written and tried out. However, one need not necessarily use the above programs and techniques. It is not the actual programs but the philosophy behind them that is the message of Hal Technology.

5 Bibliography

1. Halevi, G., New approach to CAM production planning. In Proceedings 1975 AIIE Spring Annual Conference, pp. 287–296
2. Halevi, G., Adaptive production process planning. In CAM-I Proceedings, 76-MM-02, 1976, pp. 261–283
3. Halevi, G., Production planning module of All Embracing Technology. Proceedings of Autofact West, Vol. 1, CAD/CAM VIII, 1980, pp. 451–465
4. Kimremia J. and Gershwin, S. B., An Algorithm for the Computer Control of Production in a Flexible Manufacturing System, MIT Cambridge Report, January 1982
5. McCartney, J. and Hinds, B. K., Tooling economics in integrated manufacturing systems. Int. J. Prod. Red., Vol. 20, No. 4, 1982, pp. 483–505
6. Chen, Ming-Fong and Ito, Yoshimi., Investigation on the engineering's thinking flow in the process planning of machine tool manufacturing. 13th NAMRC, 19–22 May 1985, Berkely, CA, pp. 418–422
7. Savage, E. and Mikovak, M., Finite scheduling staging a comeback? CIM Technology, Spring 1986, pp. 26–31

Part III
Fundamental Techniques

Chapter 7
Graph Theoretical Approaches

Peter Falster

Peter Falster is Associate Professor in the Electric Power Engineering Department, Technical University of Denmark. His Ph.D. is in Systems Science and Graph-oriented Production Models in APL. Dr. Falster is Chairman of IFIP WG 5.7: Computer Aided Production Management, and the Danish Representative to IFIP TC 5. He is also a member of the Danish Data Processing Association's TC on CIM.

1 Introduction

For the production engineer one great virtue of graphs is their primitive simplicity and therefore their ability to represent a variety of structures of production planning and control problems.

To trace the origins of the applications of graphs for this problem area is not easy. Limiting ourselves to the past 30 years a number of authors, often apparently unaware of one another's work, have discovered similar graph concepts and ideas. The two main areas of graph applications are related to the classical conception of production planning and control as being divided into a materials planning problem based on bill-of-materials and a machine planning problem based on routings. These two fundamental structures – the bill-of-materials and the routings – are therefore the basis for this chapter's treatment. However, many questions we wish to ask about the graphs can be analysed with the aid of conventional matrix algebra and a generalized minimax algebra, which therefore also is the topic of this chapter.

The first papers which seem to have had a major impact on the formulation of the materials requirements and scheduling problems are the papers by A. Vazsonyi in the Management Science Journal 1954–55 [1, 2]. In the first paper he introduced the graph representation of a product's bill-of-materials under the name "The Gozinto Graph", the next-assembly matrix representation, its power series, and a recursive equation for the calculation of the total requirements matrix. In the second paper he discussed the problem of time-phasing materials requirements and introduced, corresponding to the total requirements matrix, a set-back matrix representing production lead times. He then presented the combined materials requirements and scheduling procedure using an indexed matrix notation. Unfortunately, it was not possible to give the problem a short and concise formulation by means of matrix algebra as is the case for the static materials requirements problem. The present author has shown in his Ph.D. thesis how the problem can be given a simple formulation using the programming language APL [3, 4].

Giffler further developed the work of Vazsonyi, in particular considering more effective algorithms for gross and net requirements calculation and a new schedule algebra [5]. At the same time (1959) the Pert and CPM scheduling techniques were developed for project planning or single-piece production [6].

Cuninghame-Green developed in 1962 a so-called minimax algebra very similar to Giffler's scheduled algebra and applied it to various job-shop scheduling problems, shortest routes problems, and general minimax problems [7]. The paper by Yoeli from 1961 [8] on a generalization of Boolean matrix theory should also be mentioned in relation to the work by Cuninghame-Green. Both Giffler and Cuninghame-Green have given a more updated state-of-the-art report of their work Refs. [9] and [10].

Gleiberman made in 1964 an important contribution concerning an effective method for the updating of the total requirements matrix, e.g. when engineering changes are introduced [11]. In Refs. [3] and [12] the present author generalized the method to Boolean and set-back (lead time) matrices and also related the method for the famous Warshall algorithm for generating the reachability matrix [13].

Thompson and Giffler gave in Ref. [14] characteristic properties of the n-job, m-machine scheduling problem. The original Pert and CPM techniques for project planning assumed unlimited resources, and Balas considered in 1966 [15] the problem of resource-constrained project networks and their optimization with respect to minimum project duration. To solve the problem he introduced so-called disjunctive arcs to represent the resource constraints and found a solution by a branch and bound technique.

The work by Conway et al. from 1967 on a theory of scheduling is also important to mention when we discuss scheduling problems in industry [16].

Franksen (1968) has shown in an unpublished paper [17] the relation between the Walrasian micro-economic model of production, which is the basis for linear programming, and the materials requirements calculation based on bill-of-materials and its dual costs calculation problem. This work was the basis for the present author's own Ph.D. thesis [3].

A special scheduling problem is the shortest-route problem, and the work by Bellman [18] from 1956, Ford and Fulkerson [19] from 1962, Dantzig [20] from 1966, and Carré [21] from 1971 must also be mentioned. Basically, the problem is based on the minimax algebra.

A recent paper (1985) by Cohen et al. [22] also uses minimax algebra for discrete-event processes and performance evaluation in manufacturing. They showed that the periodical behaviour of a set of repeatedly performed activities

7. Graph Theoretical Approaches

can be totally characterized by solving an eigenvalue and eigenvector problem, and they developed an effective algorithm for the problem which basically consisted of finding the shortest paths from one node to all other nodes in a graph.

From a computer-oriented point of view some contributions should also be mentioned besides those given above. The early bill-of-materials algorithms were all oriented to magnetic tape solutions and based on a topological sort of structure. Here the papers by Loewner [23] from 1965, Langefors [24] from 1962, Dzubak and Warburton [25] from 1965, and Francis [26] from 1979 all considered effective sparse matrix techniques for determination of part levels, the generation of the total requirements matrix, and the computation of parts requirements. Some of these techniques have been given an explicit formulation in APL by the present author in Ref. [12].

The purpose of this chapter is on the one hand to give a graph theoretical representation of the materials requirements and costs problem, and the activity sequencing and scheduling problem, respectively. On the other hand the aim is also to give an explicit and concise matrix formulation using the array-oriented language APL [27].

2 Some Basic Digraph Concepts

A directed graph, or digraph for short, is a geometrical diagram consisting of a finite non-empty set of nodes N and a finite set of directed arcs E each joining or bounded by a node pair (n_i, n_j) not necessarily distinct (Fig. 1) [28, 29].

A digraph is a very useful tool for the representation of measurements of the entities of production and the associated mathematical relations of equivalence and of partial order among the entities. Basically, since a digraph consists of two kinds of objects (nodes and arcs), entities and relations can be represented in either of two ways:

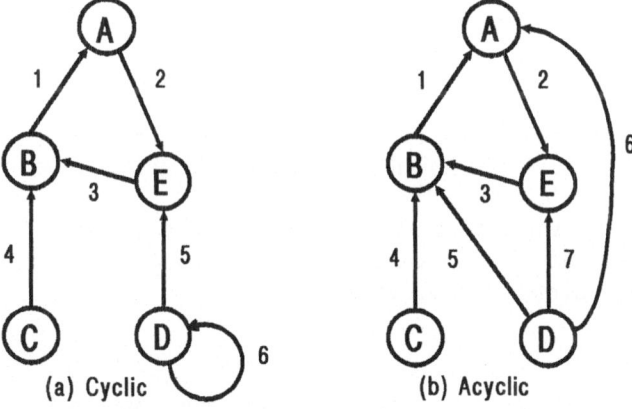

Fig. 1. A cyclic and an acyclic digraph: (a) cyclic; (b) acyclic

a) The nodes represent the entities and the arcs the relations. This is the classical and most popular representation, utilized for instance in requirements and cost planning.
b) The arcs represent the entities and the nodes the relations. This representation is used for example in some project planning methods like CPM and Pert.

In our representation of activities and events we shall use exclusively the first representation: "the entity-on-node representation". A few words are necessary about measurement, which constitutes the only link between the observations of reality and the corresponding graph theoretical model. More precisely, measurement is defined as the assignment of numerals to observable quantities of objects according to rule [30]. Measurements, and in particular production measurements, can be classified according to four scale forms:

- The **nominal** scale measurements are those from which we assign unique identifications to products, activities, resources, etc.
- The **ordinal** scale measurements are those from which we assign priorities to e.g. the products and the activities.
- The **interval** scale measurements are those from which we assign e.g. ready dates, due dates, prices, etc. to the products and the activities.
- The **ratio** scale measurements are those from which we assign demands, requirements, costs to the products and activities.

The diagram itself describes the qualitative properties of the measurements on a nominal and ordinal scale. Now the quantitative measurements, i.e. measurements on a ratio or interval scale, are assigned to the nodes or the arcs of the digraph.

Measurements on a ratio scale are characterized by a fixed zero point, but such that we can change the unit, e.g. two measurements x (kg) and y (gram) are related by:

$$x = ay$$

We can depict the measurements x and y by the nodes and the ratio by an arc:

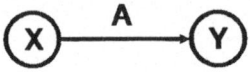

This is a fundamental representation for the quantitative properties of the product structure, where we assign the technical coefficients to the arcs and the requirements and demands to the nodes.

Measurements on an interval scale are characterized by the relaxation of the fixed zero point assumption for the ratio scale, e.g. a measurement v (time event) is related to another measurement w (time event) by:

$$w = ay + v = x + v$$

where x is e.g. the lead time (hours, days) between the two events. We can depict the measurements v and w by the nodes and the difference x by an arc:

7. Graph Theoretical Approaches

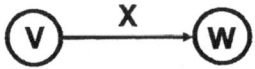

This is the fundamental representation for the quantitative properties of the activity structure, where we assign the lead times to the arcs and the time events to the nodes.

The outdegree of a node v is the number of arcs directed away from it, and the indegree is the number of arcs directed to the nodes. For example, in Fig. 1(a) the indegree of node B is two and the outdegree is one. A path is an unbroken route alternating through nodes and arcs beginning and ending at a node and with no arc repeated: e.g. D5E3B is a path. A directed path is a path following the arc orientation. The length or topological distance of a path is the number of arcs in it: e.g. along the path D5E3B1A the distance is three. We say that node y is reachable from a node x if there is a directed path from x to y: e.g. B is reachable from node D along the directed path D5E3B. A cycle is a directed path from a node to itself: e.g. the directed path A2E3B1A.

A digraph without cycles is called an acyclic digraph (Fig. 1(b)). These graphs have properties to be distinguished from graphs with cycles. The latter are called cyclic digraphs and will be discussed in relation to the requirements and scheduling calculation.

From Fig. 1(b) it is seen that an acyclic graph has at least one node with indegree zero, called source, i.e. C, D, and at least one node with outdegree zero, called sink, i.e. A.

To each node in an acyclic graph may be assigned a unique level number which is the largest distance from any of the sources. Dually, to each node may be assigned a unique inverted level number which is the largest distance from the sinks. The level numbers (from the sources) of nodes A, B, C, D, and E in Fig. 1(b) are 3, 2, 0, 0, and 1. These numbers are utilized in most product requirements routines.

We now introduce two basic unary operations establishing the following digraph:

- Converse
- Transitive.

The converse digraph D' to a digraph D is obtained by reversing the direction of every arc of D. In spite of its simple definition this concept has many useful applications, in particular related to the rather powerful principle of directional duality. Examples have already been given of dual concepts: indegree vs. outdegree, source vs. sink, and level vs. inverted level.

The transitive closure digraph D^t to a digraph D is obtained by adding arcs to the digraph D according to the use of the transitive law on the digraph.

The graphical representation of relations is useful in order to visualize and provide information about the structure of a problem. However, for computing purposes, matrices or, in general, arrays have for a long time served as an organized machine-readable representation of sets of data.

There are two basic Boolean topological matrices: the outnode matrix, ON, and the innode matrix, IN, which together represent a digraph. Both describe the continuity properties between the nodes and the arcs in a graph, so that the out-node matrix defines the arcs which are "directed out from the nodes" and the innode matrix the arcs which are "directed into the nodes". For the example in Fig. 1(b) the two matrices are:

```
  ϱ□←ON                    ϱ□←ON
  0 1 0 0 0                1 0 0 0 0
  0 0 0 0 1                1 0 0 0 0
  0 0 0 0 1                0 1 0 0 0
  0 0 1 0 0                0 1 0 0 0
  0 0 0 1 0                0 1 0 0 0
  0 0 0 1 0                1 0 0 0 0
  0 0 0 1 0                0 0 0 0 1
7 5                      7 5
```

The row indices and the column indices of the two matrices represent respectively the arcs and the nodes of the digraph. Strictly speaking, besides being topological representations of the graph they also represent two relations. It is well known in mathematics that the composition of two relations with a common domain is a new relation defined by their relative product. The two relations ON and IN can be combined in either of two ways:

a) One which eliminates the arcs and defines a node-to-node relationship PN
b) One which eliminates the nodes and defines an arc-to-arc relationship PA.

In matrix terms the two relative products are expressed as a Boolean matrix product in APL:

```
    ϱ□←PN←(⌾ON) v.∧ IN
    0 0 0 0 0
    1 0 0 0 0
    0 1 0 0 0
    1 1 0 0 1
    1 1 0 0 0
  5 5
```

and

    PA←IN v.∧ ⌾ON

Read row-wise (alternatively, column-wise) the node-to-node relationship PN expresses a predecessor (alternatively successor) relation between nodes in distance 1 in the graph and is also denoted PN1. For small examples the predecessor matrix PN can easily be established directly by inspection of the digraph.

Similarly, the arc-to-arc matrix PA represents a predecessor/successor relation between arcs in the digraph. The node-predecessor matrix PN is used when the entities are represented by the nodes and the relation by the arcs in the digraph. The arc-predecessor matrix PA is used in the dual case.

7. Graph Theoretical Approaches

Below, we assume that the node-predecessor matrix is a representation of a digraph and, for shorthand notation, P is written only.

The predecessor matrix P defines precedence between nodes in distance 1. Procedence between nodes in distance 2, 3, etc. is simply computed by taking the power of P. In distance J the predecessor relationships denoted PJ are determined as follows:

$$PJ \leftarrow P \vee \wedge \ldots \vee \wedge P$$

For an acyclic graph the maximum distance in a graph is limited therefore

$$P^{\delta+1} = 0$$

if the maximum distance is δ. The matrix P is then said to be nilpotent. However, for a cyclic graph the power will not home in on any definite value, but will repeat itself by the least common multiple of the largest distance of the strong components (cyclic parts of digraphs).

We define P^0 as being the unit matrix I, assuming implicitly that all nodes have a loop to represent that all nodes can be reached in distance zero. By taking into power and using the transitive law of the partial order relation we establish the transitive closure of the graph and thereby draw all conclusions from the first-order precedence relation.

The total reachability properties of a graph are determined as the union of the individual power matrices:

$$R = P^0 \vee P^1 \vee P^2 \vee \ldots \vee P^\delta$$

or, because it is a Boolean sum, then also:

$$R = (I \vee P)^{2\delta}$$

and stopping when:

$$(I \vee P)^{2\delta} = (I \vee P)^{2\delta+1}$$

It should be noticed that for both cyclic and acyclic graphs R is finite and determined by the expression.

The reachability matrix is:

```
☐ ← R ← P0 v P1 v P2 v P3
1 0 0 0 0
1 1 0 0 0
0 1 1 0 0
1 1 0 1 1
1 1 0 0 1
```

because it is an acyclic digraph and thereby $P^4 = 0$.

There exist various methods more effective than the above-given for generating the reachability matrix [12]. A very effective method which seems related to the well-known so-called Warshall algorithm [13] is given below in matrix terms. Originally, it was formulated for the engineering change of the bill-of-materials, [11]. Let us illustrate its structural aspects in connection with an updating of the reachability matrix.

If a new set of arcs represented by ΔP is added to a digraph, the predecessor matrix P is updated by:

$$P \leftarrow \Delta P \vee P$$

and, conversely, if a set of arcs in a digraph has to be removed, the predecessor matrix P is updated by expressing the fact at ΔP is included in P:

$$P \leftarrow P > \Delta P$$

The updating of the reachability R by ΔP can easily be done if we assume that ΔP contains non-zero elements in only one column or row.

It can be shown that the changes to R are determined by:

$$\Delta R \leftarrow R \vee . \wedge \Delta P \vee . \wedge R$$

and

$$R \leftarrow R \vee \Delta R$$

3 Production Requirements and Costs

The main discussion in this section originates in the firm's bill-of-materials. An assembly product, e.g. a diesel engine, is made up of several hundred parts and details. The set of parts may be described in terms of a digraph where the nodes represent the parts and the arcs the causal predecessor or successor relationships in terms of "where used" or "bill-of-materials". A product structure may be represented, for example, by the digraphs of (Fig. 2(a) and (b)).

We shall asume that the primal digraph in Fig. 2(a) represents a "where used" relation, while the dual or converse graph in Fig. 2(b) represents a "bill-of-materials" relation. It can also be said that a product is made up by an implosion (synthesis) process and the details are determined by an explosion (analysis) process. From the primal digraph one can easily determine the final products as the sinks, the details as the sources, and the subassemblies as the nodes with an indegree and outdegree different from zero.

In general, most manufacturing processes are acyclic, i.e. the corresponding digraph representation will be an acyclic digraph. On the other hand, if the process is cyclic, e.g. represented by the dotted arc in Fig. 2(a), the graph can be

7. Graph Theoretical Approaches

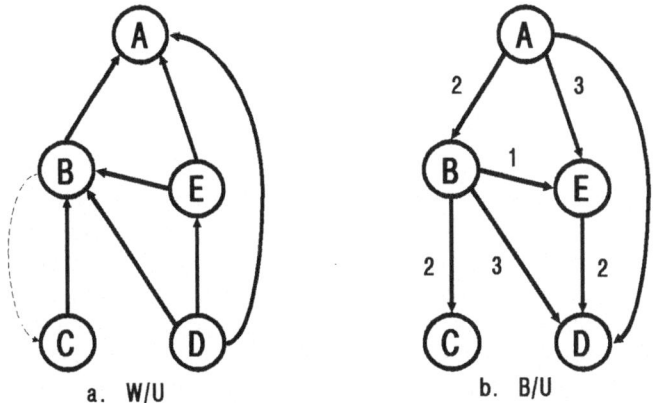

Fig. 2. Product digraph and technical coefficients: (a) primal digraph; (b) dual or converse digraph

changed back into an acyclic graph by breaking the cycle and defining a new node in the digraph. However, nothing prevents us from representing cyclic processes. One example is the cyclic digraph from Leontief's input–output model, where the nodes depict sectors and the arcs the interaction between the sectors. Another example is an electric generator, where the generated electricity is consumed in creating the magnetic field necessary to generate the electricity.

Confining ourselves to acyclic digraphs and assuming "where used" orientation the next step is a quantification by assigning measurements from the ratio scale to the digraph. The technical coefficient (QTY) is a non-negative number defining the number of items of parts that are used to produce a single item of a part in the topological distance 1. Thus, the technical coefficients chosen arbitrarily are assigned to each arc in the graph (Fig. 2(b)).

The technical coefficients are represented in general by a vector:

$QTY = (qty_1, qty_2, \ldots, qty_i, \ldots, qty_A)$

where A denotes the number of arcs in the digraph.

A graph and its converse are open to two interpretations and, thereby, two applications. The first interpretation is based on the primal graph and concerns the product requirements planning. To each node in the digraph are assigned two algebraic properties representing the demand of a product and the calculated requirements of a product, respectively. The vector of product demands is:

$DMD = (dmd_1, dmd_2, \ldots, dmd_i, \ldots, dmd_N)$

and the vector of product requirements is:

$REQ = (req_1, req_2, \ldots, req_i, \ldots req_N)$

where N denotes the number of nodes in the digraph.

Fig. 3. A node loop

The product requirements are calculated by an explosion of the product demand. That is, the planning process is an explosion backwards in the graph contrary to the arc orientation. The arc orientation is in accordance with the physical flow of products out of the firm. The calculation works as follows. We multiply the technical coefficients on the same directed serial path between any two nodes and add the resulting technical coefficients in case of parallel paths between the two nodes in question: e.g. the total technical coefficient between product D and B in Fig. 2(b) is calculated as the product of 2 (D, E) and 1 (E, B) and adding to 3 (D, B), which gives 5. By means of the total technical coefficients between any two products it is then a simple matter to calculate the requirements from a given demand.

Now, relaxing the assumption of an acyclic graph for a moment it can be shown that the computation for a cyclic graph is in principle the same. However, because of the cycles, a path between two nodes can be infinite. For example, having a loop at a node (Fig. 3) with technical coefficient 0.1 and following the path around the loop gives an endless power series for the total technical coefficient:

$$0.1^0 + 0.1^1 + 0.1^2 + \ldots + 0.1^P + \ldots$$

which, however, converges to 10/9.

That is, we say that a production system is feasible if it satisfies the following three assumptions [5]:

- The technical coefficients are non-negative.
- The technical coefficients for a direct consumption represented by a loop at a node should be non-negative and less than or equal to one.
- The total technical coefficients, i.e. the sum of indirect and direct consumption, should for each product be greater than or equal to one; cf. the examples above.

It should be made clear that all nodes in an acyclic product graph have a technical coefficient of value 1. It means that to produce one unit of a product it is necessary to order the manufacture or purchase of one unit of that product.

The second interpretation is based on the dual digraph and concerns the costs calculation. Concerning the costs calculation, we assign to each node in the digraph two algebraic properties representing the direct assembly unit costs for a product and the direct and indirect accumulated assembly costs of a product, respectively. The assembly unit costs are: for the details the costs of raw materials, and for an assembly the costs of assembling one unit from its immediate subassemblies. The vector of assembly unit costs is:

$$ASC = (asc_1, asc_2, \ldots, asc_i, \ldots asc_N)$$

7. Graph Theoretical Approaches

and the vector of accumulated assembly unit costs is:

$$ACC = (acc_1, acc_2, \ldots, acc_i, \ldots, acc_N)$$

The accumulated unit costs are calculated by an implosion of the assembly unit costs. That is, the planning process works as an implosion forward in the digraph contrary to the arc orientation. The arc orientation is chosen in accordance with the physical cash flow of payments into the firm.

To begin with, let us change the representation of QTY by redefining it to a diagonal matrix with QTY in the main diagonal and zero-elements outside the diagonal.

By means of the two fundamental topological matrices ON and IN we establish the "next-product" matrix NP by:

$$NP \leftarrow (\oslash ON) + . \times QTY + . \times IN$$

Each entry $np^1_{i,j} \neq 0$ in NP is a technical coefficient qty which represents the number of units or product "i" that are directly used to produce one unit of product "j" in distance 1. It is well known that the entry $np^k_{i,j} \neq 0$ in the kth power of NP, i.e. NP^k, represents the requirements in distance k of the digraph. Thus we can define the "total product" matrix TP or total requirements matrix with each entry $tp_{i,j} \neq 0$ as the total requirements of product "i" in order to produce one unit of product "j" and calculated by the sum of requirements over all distances k in the digraph.

The basic equations relating NP, NP^k and TP are as follows:

$$NP = NP^{k-1} \times NP$$

$$TP = I + NP^1 + NP^2 + \ldots = \sum_{k=0}^{\infty} NP^k$$

It the power series converge, i.e. $NP^k \to 0$ as $k \to \infty$ then the production system is called feasible and is equivalent to:

$$TP = (I - NP)^{-1}$$

The NP-matrix is also said to be nilpotent. The reader should notice that there exists no Boolean analogy to the last equation.

Below, we shall assume acyclic production processes, that is:

$$TP = \sum_{k=0}^{\delta} NP^k$$

where δ is finite and equal to the maximum distance in the graph.

The generation of TP by direct inversion of $(I - NP)$ or by the sum of the power series of NP is out of the question for large-scale problems. However, there are different methods which are more economical in storage and which should be mentioned.

Rewriting

$$TP = (I - NP)^{-1}$$

gives:

$$TP = I + TP \times NP$$

The last equation can be programmed as a simple iteration as follows:

$$TP_i = I + TP_{i-1} \times NP$$

$$TP_0 = I$$

There exists another method which is basically the same as the updating procedure on the simplex tableau for linear programming. The method was originally developed by Giffler [5] as a technique for updating TP after a change has been made to NP. Let us represent the changes to the NP matrix by ΔNP, then it can be shown after a few matrix manipulations that the new TP_1 and NP_1 matrices can be derived from the old TP_0 and NP_0 matrices by:

$$TP_1 = TP_0 + (I - \Delta NP_0 \times TP_0)^{-1}$$

$$NP_1 = NP_0 + \Delta NP_0$$

Furthermore, if the entries in the ΔNP_0 matrix are confined to a single row the inversion of the matrix $(I - \Delta NP_0 \times TP_0)$ is a simple pivoting.

An alternative formulation developed by Gleiberman [11] lead, however, to a more general formulation which already has been given its Boolean counterparts. Substituting the Boolean matrices R and ΔP with TP and NP gives the general equations:

$$\Delta TP_i = TP_i \times \Delta NP_i \times TP_i$$

$$TP_{i+1} = TP_i + \Delta TP_i$$

assuming only one engineering change to a single product of NP.

The essence of this approach is also a more efficient method for generating TP from NP by an iterative updating of TP from a set of single bills-of-materials represented by a sequence of $\Delta NP_1, \Delta NP_2, \ldots, \Delta NP_p$ matrices. We refer to Ref. [11] for a further discussion of the method in particular in relation to cyclic digraphs.

A gross requirements calculation for a certain period of production can be based on either the total requirements matrix TP or the next-product matrix NP. From a specified demand, DMD (output) of any product "i", the gross require-

7. Graph Theoretical Approaches

ments REQ (input) are calculated by an explosion of the different products required to produce product "i".

$$REQ = TP \times DMD$$

Instead of first inverting $(I - NP)$ or developing the power series in order to establish TP, there exist a variety of methods based on the next-product matrix NP which can speed up calculation. Most of the explosion functions implemented today are in fact based on the NP matrix.

$$REQ_i = DMD + NP \times REO_{i-1}$$

$$REQ_0 = DMD$$

So much for the gross requirements calculation by the explosion of product demands. Turning instead to the cost calculation it is performed as a dual calculation to the requirements calculation by an implosion of the assembly unit costs ASC. The accumulated assembly unit costs ACC is based on the where-used digraph. That means we transpose the TP matrix and perform all the calculations isomorphic to the requirements calculations.

The total manufacturing costs caused by a certain production are determined by means of the so-called **adjoint** condition which can be defined between the requirements model based on the primal digraph and the unit costs model based on the dual digraph.

The direct costs DC are determined from the direct demand DMD and the accumulated assembly unit costs ACC, or the requirements REQ and the direct assembly unit costs ASC, as follows:

$$ASC + . \times REQ$$

or

$$ACC + . \times DMD$$

or

$$DC \leftarrow ASC + . \times TP + . \times DMD$$

The total product matrix TP appears in the last equation of Ref. [28] as an adjointness operator between ASC and DMD.

The value of the indirect capital IC tied up in products which are consumed by the production system itself is determined from the indirect requirements $(REQ - DMD)$ and the accumulated assembly unit costs, or the indirect accumulated assembly unit costs $(ACC - ASC)$ and the requirements REQ, as follows:

$$ACC + . \times (REQ - DMD)$$

or

$$(ACC - ASC) + . \times REQ$$

or

$$IC \leftarrow ACC + . \times REQ$$

The inventory carrying costs, i.e. the indirect costs, are calculated on the basis of the above capital tied up in the inventory. In order to calculate the indirect costs, we must take the time-phasing of the products into consideration.

The total value of the manufacturing TC is then determined from an equation which expresses Walras's law as "the total value of the input is equal to the total value of the output", i.e.:

$$TC \leftarrow ACC + . \times REQ$$

$$TC \leftarrow REQ + . \times ACC$$

$$TC \leftarrow IC + DC$$

In the above examples we have only considered aggregated values for the total manufacturing system. It is very easy to obtain information about the actual contribution of each product to this total value if we replace the inner product by an element-by-element vector multiplication.

4 Activity Sequencing and Scheduling

The time planning or activity sequencing and scheduling problem is probably the most complicated subproblem within the production planning area partly because of its many interpretations of the time concepts and partly because of its many feasible solutions. The problem originates in the routings or process charts of a firm.

A discussion of the time planning problem must clearly distinguish between problems with:

1. Non-repeated activities
2. Repeated activities.

We shall first discuss problems of a non-repeated nature. A manufacturing process made up of sequences of activities can be conceived from two viewpoints:

- The discrete events of starting/stopping one or several activities
- The activity of performing an operation.

Thus, an event is described by a certain instant of time measured on an interval scale, while an activity is described by a time duration measured on a ratio scale. In the discussion of the interval scale we represented the events by nodes and the activity duration by arcs.

7. Graph Theoretical Approaches

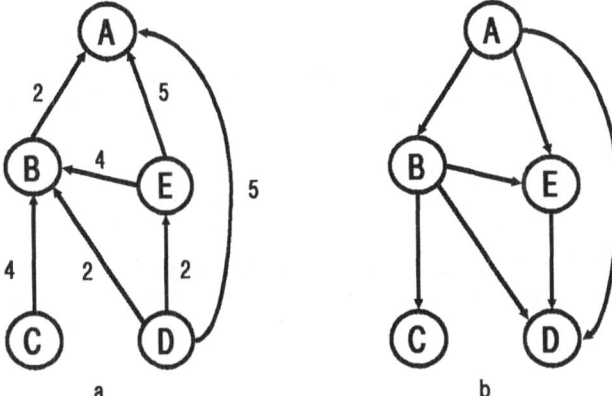

Fig. 4. Event-on-node digraphs: (a) forward-oriented; (b) backward-oriented

The event-on-node representation can be described either in terms of a predecessor relation "goes before" or a successor relation "follows after" between the nodes. In the former case the digraph is said to be forward-oriented and in the latter case backward-oriented as illustrated by our standard digraph (Fig. 4(a) and (b)).

Each of the technical time coefficients, conceived as lead times, specifies a lower bound on the time that elapses between two next-following events. Accordingly, the lead times have been depicted by the arcs in the graph (Fig. 4(a)). A vector of arc lead times is:

$$LDT1 = (ldt\,1_1, ldt\,1_2, \ldots, ldt\,1_i, \ldots, ldt\,1_A)$$

We make the following assumptions concerning the lead times:

- Lead times should be integer numbers, and we allow for negative values.
- Lead times should be deterministic.

The lead times can be caused by processing, setup, queue or waiting, teardown, change-over or transport. A useful interpretation of the lead time could be that it is the time that elapses from the event of taking a product from a stock for manufacturing to the events of having the product at the next stock.

Corresponding to the explosion and implosion for the requirements and costs calculation there exists a forward and a backward time calculation.

The backward planning is based on the forward-oriented graph (Fig. 4(a)) where each node is assigned two algebraic properties: a given due date and a calculated late date, respectively. The vector of due dates is:

$$DUE = (due_1, due_2, \ldots, due_i, \ldots, due_N)$$

and the vector of late dates is:

$$LD = (ld_1, ld_2, \ldots, ld_i, \ldots ld_N)$$

The late dates are calculated from the given due dates by backward planning in the graph by subtraction of lead times from the time events. The calculation works as follows. We add the lead times on the same directed serial path between any two nodes and take the maximum value in case of parallel paths between the two nodes in question. For example, the total lead time between nodes D and B in Fig. 4(a) is calculated as the sum of 2 (D, E) and 4 (E, B) and taking the maximum of this results in 6 and 2 (D, B); this gives 6 as the lead time.

The forward planning is based on the backward-oriented digraph (Fig. 4(b)) where each node is assigned two new algebraic properties measured on an interval scale: a given ready data and a calculated early date, respectively. The vector of ready dates is:

$$RDY = (rdy_1, rdy_2, \ldots, rdy_i, \ldots, rdy_N)$$

and the vector or early date is:

$$ED = (ed_1, ed_2, \ldots, ed_i, \ldots, ed_N)$$

The early dates are calculated from the given ready dates by forward planning in the digraph by addition of lead times to the time events. It is customary to add a dummy start event and a dummy finish event to the sources and the sinks, respectively. This gives the planner the opportuniy to assign an overall ready date and due date to the whole digraph by means of these dummy events.

Recalling that the lead times express lower bounds on the events means that a next-following event is allowed to start after a preceding event at any time, only requiring that the time that elapses between the two events is greater or equal to the specified lead time. Figure 5 illustrates an acyclic digraph with three events and with a negative lead time between nodes B and C. Hence, the digraph represents qualitatively that event C is dependent on the event B, which has nothing to do with the actual point in time that separates the two events. This in fact is represented by the negative lead time (-3) which expresses that even C should take place not more than 3 time periods before the event B takes place. That is, if A starts at relative time 10, then B can start at the time 20 and C the maximum of 16 (and $20-3$) which is 17. Clearly, if B is delayed from the time 20 to the time 22, then C will be delayed two periods to the time 19.

In the preceding we have assumed acyclic digraphs and positive/negative lead times. However, cyclic scheduling digraphs can be useful in order to cope with the following situation:

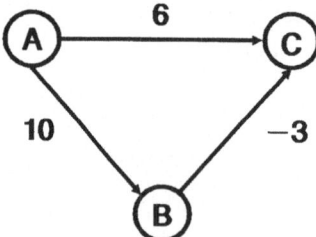

Fig. 5. Acyclic digraph with negative lead time

7. Graph Theoretical Approaches

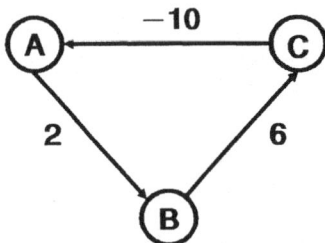

Fig. 6. Cyclic digraph with negative lead time

- A set of events has to be performed after each other or in parallel within certain time limits.

Now introducing a cyclic digraph with positive/negative lead times reveals the following interesting observations. Taking again a simple example with three nodes and a negative time between nodes C and A shows that we can walk around this cycle for ever (Fig. 6). However, a walk around the cycle and adding the lead time gives -2, which says that event A shall start after an event which is two periods earlier. Clearly, this is in fact possible.

On the other hand, if it were positive, it would mean that an event has to start before it possibly can. We can therefore state the following rule:

"No cycles can be in such a way that the sum of the lead times around is positive."

This rule states the condition for a non-repeated scheduling system to be feasible. Obviously, only one walk around a cycle is necessary to reveal whether the sum is positive or negative.

In summary, the combined effect of negative times and cycles is a constraint on the events so that events within the cycles (strong components) have to be performed within certain time limits. In other words, the nodes within a cycle (strong component) are given limits on its slack (defined below).

It is hereby shown that the approach with cyclic event digraphs is in fact isomorphic to the approach with cyclic product digraphs [3]. In the **activity-on-node** representation we pay on the first hand attention to the activities instead of the events by assigning to each node in the digraph (Fig. 7) an activity and a lead time defined as the activity duration.

The arcs represent both the fabrication constraints originating in the **technological** precedence constraints on the activities and the resource constraints originating in the **sequences** in which the activities are served by the resources (facilities, machines, tools, men, etc.). These sequences are not fixed in general, and finding the optimal sequences is a combinatorial problem which is outside the scope of this chapter.

The vector of node lead time is:

$$LDT0 = (ldt0_1, ldt0_2, \ldots, ldt0_i, \ldots, ldt0_N)$$

The fact, however, is that from a planning point of view the event of starting and finishing an activity is more important than the activity itself. We can therefore

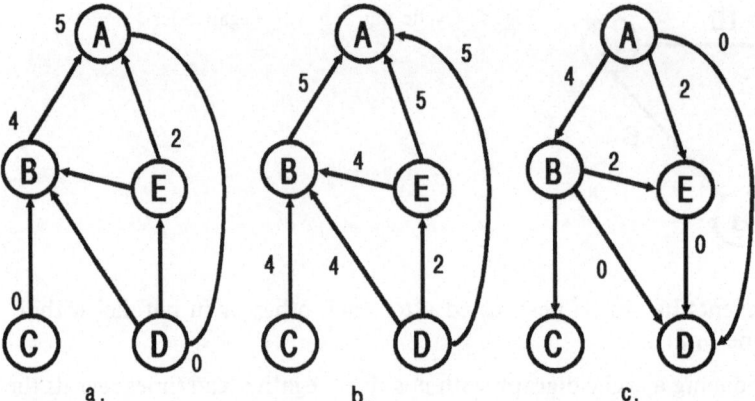

Fig. 7. Activity-on-node digraphs: (a) primal; (b) dual, forward planning; (c) dual, backward planning

establish two dual digraphs (Fig. 7(b) and (c)) isomorphic to the primal digraph (Fig. 7(a)) by assigning the node times to the arcs. Accordingly:

- In case of forward planning we can assign to each outgoing arc the node lead time corresponding to a "from start to start" interpreation (Fig. 7(b)).
- In case of backward planning we assign to each incoming arc the node lead time corresponding to a "from finish to finish" interpretation (Fig. 7(c)).

Clearly there exist two interpretations of the lead times:

- Either the lead time is the time that elapses from the start event of an activity to the start event of the next-following activity, briefly from start to start.
- Or the lead time is the time that elapses from the finish event of an activity to the finish event of the following activity, briefly from finish to finish.

The essence of this very brief presentation of the scheduling problems is that in spite of the different interpretations of the time concept, there exists basically one computational model which structurally is the same as the requirements and costs model, as will be illustrated below.

To begin with, let us redefine the arc lead times $LDT1$ to a diagonal matrix with the elements outside the diagonal being negative infinity. (Clearly, from a computer viewpoint we represent infinity by the largest number in the computer.) Corresponding to the next-product matrix, we define a next-activity (or next-event) matrix NA:

$$NA \leftarrow (\Delta \otimes ON) \lceil . + LDT1 \lceil . + (\Delta\, IN)$$

where the function Δ performs a mapping of the Boolean elements $(0,1)$ into $(-\infty, 0)$. The mapping function Δ is an example of a general mapping function from the nominal scale to the interval scale. The mapping is an injection, i.e. the converse mapping (surjection) of numbers from the interval scale to the nominal scale can be performed by any of the primitive relational functions " = ", " < ",

7. Graph Theoretical Approaches

etc. Another example of a mapping function is $\varDelta$ which performs a mapping from $(0,1)$ into $(\infty, 0)$. The basic idea of these mappings is that the numbers $-\infty$ (∞) and 0 are the identity numbers of the functions "$\lceil$", "$\lfloor$", and "$+$", respectively. A more extensive discussion of this so-called minimax algebra is given by Cuninghame-Green in Ref. [10].

Each entry $na_{i,j}^1 \neq -\infty$ in NA represents the lead time between two next-following events (activity start/finish) in distance 1. The entries $na_{i,j}^1 = -\infty$ represent the fact that we cannot come from event "i" to event "j" in topological distance 1. Obviously, this would have caused trouble in the generation of the power series of NA. The entries $na_{i,j}^k \neq -\infty$ in the power series of NA, i.e. NA^k, represent the maximum lead times in distance k of the digraph. Thus we can define a "total activity" matrix TA with each entry $ta_{i,j} \neq -\infty$ equal to the maximum lead time between any two events and calculated by the maximum of lead times over all distances k in the digraph.

The basic equations relating NA, NA^k, and TA are as follows:

$$NA^k = NA^{k-1} \lceil . + NA$$

$$TA = Z \lceil NA^1 \lceil NA^2 \lceil NA^3 \ldots = \bigsqcup_{k=0}^{\infty} NA^k$$

Comparing these equations with the equations for NP^k and TP reveals that we simply have substituted the operations "$+$" and "$\times$" with "$\lceil$" and "$+$" in the inner product while the form of the equations is the same. The Z-matrix is a diagonal matrix with zero-values in the diagonal, representing that an event precedes itself with a zero magnitude and with $-\infty$ outside the diagonal.

If the power series converge to minus infinity, i.e. $N^k \to -\infty$ as $k \to \infty$, then the production system is said to be feasible (i.e. non-repeated) and it can be written:

$$TA = Z \lceil NA^1 \lceil NA^2 \lceil NA^3 \lceil \ldots \lceil NA^\delta$$

where δ is taken to be the maximum unique topological path.

Let us turn to the actual scheduling of the events in the digraph from a given set of external influences specified as a set of due dates and ready dates on the nodes. The calculation can either be based on the total activity matrix TA or the next-activity matrix NA and is isomorphic to the requirements and cost calculations. Therefore, we will restrict ourselves to illustrating the approach by using the TA matrix.

The earliest event dates ED (early start of activities) are determined from the ready dates RDY by a forward calculation on the digraph:

$$ED \leftarrow (\oslash TA) \lceil . + RDY$$

or

$$ED \leftarrow RDY \lceil . + TA$$

The latest event dates LT (late start of activities) are determined from the $DUE\text{-}LDT0$ dates by a backward calculation on the digraph:

$$LD \leftarrow (DUE\text{-}LDT0) \lceil . + (- \lozenge TA)$$

The term $- \lozenge TA$ is called the conjugate of TA according to Ref. [10]. The above equations may also be written as:

$$LD \leftarrow -TA \lceil . + -(DUE\text{-}LDT0)$$

where we reverse the time axis as proposed in Ref. [22].

An important property of a scheduling system is its lack or float. The float is defined as the difference between event late date and event early date:

$$FLOAT \leftarrow LD - ED$$

If we can identify a path of zero float activities it is called a critical path. Let us generalize the above formulation such that it is similar to the state-space representation of linear systems:

Primal: $\quad X = X \times A + U \times B$
$\qquad\qquad Y = X \times C$
Dual: $\quad\; Q = A \times Q + C \times V$
$\qquad\qquad Z = B \times Q$

X is a vector of early start times for the activities, and A is the next-activity matrix. Assuming we have r resources then U is a vector of times specifying when resources are available for the first activity of their sequence. B is a matrix of dimension number of resources times number of activities, such that:

$$b_{ij} = 0$$

if the first activity in the sequence of resource j is activity i and the resource is immediately available,

$$b_{ji} = -\infty \quad \text{otherwise.}$$

Y is a vector of earliest times when resources are released from their activities. C is a matrix of dimension number of activities times number of resources, such that

$$c_{ij} = ldt0_i$$

if activity i with duration $ldt0_i$ is the last activity of resource j,

$$c_{ij} = -\infty \quad \text{otherwise.}$$

7. Graph Theoretical Approaches

Q denotes a vector of latest starting times of activities; V denotes a vector of latest time when resources must be released once activities have been performed; and Z the latest time when resources must be available so that activities can be performed in due time. The equation can now be reformulated using minimax algebra in the APL notation:

Primal: $X = (X \lceil . + A) \lceil (U \lceil . B)$
$Y = X \lceil . + C$
Dual: $Q = (A \lceil . + Q) + C \lceil . + V$
$Z = B \lceil . + Q$

assuming that the dual formula are based on a reverse time scale (all time vectors are negated).

In the discussion until now we have assumed non-repetitive activities. However, most manufacturing systems perform repetitive tasks. Such systems are represented by a cyclic digraph for which we relax the assumption that the lead times in a cycle must never be non-positive. In relation to repetitive tasks, we often want to simplify planning by scheduling activities so that they process in regular steps at maximum speed (maximum throughput) with the same constant time λ elapsing between the initation of consecutive cycles on every machine, with λ as small as possible. That is, we are concerned with a problem of the type.

Find X, λ such that $(NA \lceil . + X) = \lambda \lceil . + X$

We have an eigenvector-eigenvalue problem: e.g. an industrial process of four machines which interact according to the following NA matrix and the digraph of Fig. 8.

```
ϱ□←NA
    1      2    -∞      6
    1      3    -∞    -∞
    1    -∞      1    -∞
  -∞    -∞      2      2
    4      4
```

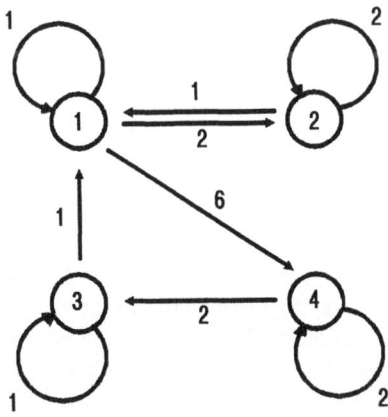

Fig. 8. Cyclic digraph of repetitive production system

Reference [10] shows how to calculate the eigenvalue λ as the greatest of the "cycle-average", which for the example is seen to be $\lambda = (1+6+2) \div 3 = 3$. Now, calculating the total activity matrix:

$$\varrho\square \leftarrow TA = (NA - \lambda)^\delta$$

```
    0  -1   2   3
   -2   0   0   1
   -2  -3   0   1
   -3  -4  -1   0
4
4
```

All columns of TA are fundamental eigenvectors. If we choose the eigenvector of the first column as the basis for defining the earliest start dates, then X could be, for example:

$$X = 4\ 2\ 2\ 1$$

and we see:

$$(NA \lceil . + X) = 3 + X$$

i.e. the systems move forward in regular steps of 3 time units.

Above we studied the longest path in a digraph (max-algebra). The shortest routes in a digraph are calculated using min-algebra using the same equations as above for the TA matrix:

$$TA = Z \lceil NA^1 \lceil NA^2 \lceil \ldots \lceil NA^N$$

$$TA = (Z \lfloor NA^1)^N, \quad NA^i = NA^{i-1} \lfloor . + NA$$

$$\lambda \lfloor . + TA = (Z \lfloor NA^1) \lfloor . + TA$$

where NA is the distance matrix. The columns of the TA matrix are the eigenvectors for the matrix $Z \lfloor NA$ and have eigenvalue 0.

6 Conclusion

This chapter has been centred on an identification of the production model structures with a common small set of primitive operations.

The set of primitive operations can be described from two viewpoints: geometrical by means of digraphs, and algebraic by means or matrices of arrays. A digraph on one hand provides a visual formulation of the basic measurements of a production system, while a matrix or array formulation on the other hand is the operational counterpart if implemented in a generalized matrix-oriented

7. Graph Theoretical Approaches

Table 1. Comparison of the basic models

	Boolean calculation	Time calculation	Requirements and costs calculation
Basic topological and algebraic matrices	ON, IN, I	$LDT0, LDT1, Z$	QTY, I
External influences	S	RDY, DUE	DMD, ASC
Next-predecessor matrices	$P \leftarrow (\emptyset ON) \vee . \wedge I \vee . \wedge IN$	$NA \leftarrow (\Delta \emptyset ON) \lfloor . + LDT1 \lfloor . + (\Delta IN)$	$NP \leftarrow (\emptyset ON) + . \times QTY + . \times IN$
Power series	$R \leftarrow I \vee P^1 \vee P^2 \vee \ldots \vee P^\delta$	$TA \leftarrow Z \lceil NA^1 \lceil NA^2 \lceil \ldots \lceil NA^\delta$	$TP \leftarrow I + NP^1 + NP^2 + \ldots + NP^\delta$
Total matrices	$R \leftarrow (I \vee P)^{2\delta}$	$TA \leftarrow (Z \lceil NA)^{2\delta}$	$TP \leftarrow (I - NP)^{-1}$
Recursive	$R \leftarrow I \vee P \vee . \wedge R$	$TA \leftarrow Z \lceil NA \lfloor . + TA$	$TP \leftarrow I + NP + . \times TP$
Feasibility and consistency test	$R.$ Convergent	$NA^k \rightarrow \infty$ $TA \rightarrow$ convergent $\Big\} k \rightarrow \infty$	$NP^k \rightarrow 0$ $TP \rightarrow$ Convergent $\Big\} k \rightarrow \infty$
Interaction with environment	$T \leftarrow R \vee . \wedge S$	$ED \leftarrow RDY \lfloor . + TA$ $LD \leftarrow (DUE - LDT0) \lfloor . + (- \emptyset TA)$	$REQ \leftarrow TP + . \times DMD$ $ACC \leftarrow (\emptyset TP) + . \times ASC$
Adjointness			$DC \leftarrow ASC + . \times TP + . \times DMD$

language like APL. This language has been chosen because it provides an explicit implementation of many of the primitive operations which has been found useful in the formulation of production problems.

In the chapter it has been shown how the production requirements calculation can be given a complete dual formulation to the production costs calculation. Furthermore, the scheduling problem can be given a formulation with the identical structure to these calculations. This implies that the same primitive operation, in fact a matrix inner product, can be used for different purposes just by exchanging the primitive scalar functions. The structure of the basic measurements for these calculations is represented by Boolean matrices from which all the reachability properties, i.e. all conclusions, of a production model can be derived; see Table 1.

7 References

1. Vazsonyi, A., The use of mathematics in production and inventory control. Management Science, Vol. 1, 1954 and Vol. 1, Nos. 3–4, 1955
2. Vazsonyi, A., Scientific Programming in Business and Industry. John Wiley, 1958
3. Falster P., Graph-Oriented Production Models in APL. Electric Power Engineering Department, Technical University of Denmark, Publ. 7710, October 1977
4. Falster, P., Dataoperations in production systems. In Ø. Bjørke and O. I. Franksen (eds): System Structures in Engineering – Economic Design and Production. Tapir Publishers, 1978
5. Giffler, B., Mathematical Solution of Production Planning and Scheduling Problems. IBM Technical Report 09.01028.026 (ASDD), White Plains, NY, October 1960
6. Kelley, J. and Walker, M., Critical-path planning and scheduling. In Proceedings of the Eastern Joint Computer Conference, 1959
7. Cuninghame-Green, R. A., Describing industrial processes with interference and approximating their steady-state behaviour. Operations Research Quarterly, Vol. 13, 1962 pp 95–100
8. Yoeli, M., A note on a generalization of Boolean matrix theory. American Mathematical Monthly, Vol. 68, 1961 pp. 552–557
9. Giffler, B., Schedule algebra: a progress report. Naval Research Logistic Quarterly, Vol. 15, No. 2, June 1968
10. Cuninghame-Green, R. A., Minimax Algebra, Lecture Notes in Economics and Mathematical Systems, Vol. 166. Springer-Verlag, New York, 1979
11. Gleiberman, L., The engineering change of the total requirements matrix for a bill of materials. Management Science, Vol. 10, No. 3, April 1964, pp. 488–493
12. Falster, P., On datastructures and algorithms in factory management systems. In Ø. Bjørke & O. I. Franksen (eds): Structures and Operations in Engineering and Management Systems. Tapir Publishers, 1981
13. Warshall, S., A theorem on Boolean matrices. Journal of the *ACM,* Vol. 9, 1962
14. Giffler, B. and Thompson, G. L., Algorithms for solving production-scheduling problems. Operations Research, Vol. 8, No. 4, July 1960
15. Balas, E., Finding a minimaximal path in a disjunctive PERT network. Theory of Graphs, Int. Symp., Rome, Dunrod, 1966
16. Conway, R. W. et al., Theory of Scheduling. Addison-Wesley, Reading, MA, 1967
17. Franksen, O. I., Numerical Examples and Comments to Vazsonyi's Two Articles. Lecture Notes, Spring Semester, 1968, 27 pp. and A Graph Theoretic Approach to Production Problems. Special Seminar, Dept. of Systems Design, Univ. of Waterloo, Canada, 2 June, 1975, 14 pp

7. Graph Theoretical Approaches

18. Bellman, R., Some mathematical aspects of scheduling theory. J. Soc. Ind. and Appl. Math., Vol. 4, No. 3, September 1956
19. Ford, L. R. and Fulkerson, D. R., Flows in Networks. Princeton University Press, Princeton, NJ, 1962
20. Dantzig, G. B., All shortest routes in a graph theory of graphs. Int. Symp., Rome, 1966
21. Carré, B. A., An algebra for network routing problems. J. Inst. Math. Appl., Vol. 7, 1971 pp. 273–294
22. Cohen, G. et al., A linear-system theoretic view of discrete-event processes and its use for performance evaluation in manufacturing. IEEE Trans. on Automatic Control, Vol. AC-30, No. 3, March 1985
23. Loewner, P. G., Fabrication and assembly operations, Part III: matrix methods for processing configuration data. IBM Systems Journal, Vol. 4, No. 2, 1965
24. Langefors, B., Computation of Parts Requirements for Production Scheduling. *BIT,* Vol. 2, 1962 pp. 91–111
25. Dzubak, B. J. and Warburton, C. R., The organization of structural files. Comm. of the ACM, Vol. 8, No. 7, July 1985, pp. 446–452
26. Francis, N. D., Computation of material requirements for production planning. *BIT,* Vol. 19, 1979 pp. 34–38
27. IBM Corporation APL Language, GC 26-3847, 1983
28. Berge, C., The Theory of Graphs and its Applications. Methuen & Co., 1962
29. Harary F., Norman, R. Z., and Cartwright, D., Structural Models: An Introduction to the Theory of Directed Graphs. John Wiley, New York, 1965
30. Franksen, O. I., Mr. Babbage's Secret: The Tale of a Cypher – and APL. Strandberg Publishers, Birkerød, Denmark, 1984. International Edition by Prentice-Hall Inc., Englewood Cliffs, NJ, 1985

Chapter 8
Simulation and Simulation Models

Jim Browne

Jim Browne is as Lecturer in Production Engineering and Director of the Computer Integrated Manufacturing Research Unit (CIMRU) at University College Galway (UCG) in the Republic of Ireland. He was educated in engineering at UCG, and took his Ph.D. from the University of Manchester in the UK. He has worked in the electronics industry in Ireland, UK and Canada, and is a member of IFIP TC 5 and WG 5.7 and of CIM EUROPE. His research interest is in the area of CIM, in particular in systems simulation and AI techniques for production systems within a CIM environment.

1 Introduction

Modern manufacturing systems are complex. They tend to involve a mix of technologies and subsystems which are to some extent interrelated if not integrated into a coherent whole. The emphasis on CIM (Computer Integrated Manufacture) is enforcing integration. Clearly, the design and operation of such systems present a challenge to manufacturing systems analysts. Our concern in this paper is with Production Management Systems (PMS) in modern manufacturing systems and with the role that digital computer simulation can play within PMS. Our view is that simulation can play two roles within PMS:

1. Simulation as a design tool for PMS.
2. Simulation as a decision support tool to aid the operation of PMS.

We explore each of these roles in this chapter using examples to illustrate our ideas.

The structure of the chapter is as follows. Firstly we look at a definition of digital computer simulation and outline the simulation methodology in a general way. We then review examples of simulation models which address each of the areas identified above. We consider the problems associated with simulation, including model validation and the long lead time necessary to develop realistic models. We review some of the simulation languages and packages which are available to support the development of models. Finally, we look at the future of the simulation technique and outline some possible alternative approaches.

2 What is Simulation Modelling?

A model is a representation of a system of objects or ideas in a form other than the system itself. Simulation is an operations-research-based technique used in the definition, development, and analysis of a model. Digital computer simulation has been defined by Pritsker [1] as "the establishment of a mathematical logical model of a system and the experimental manipulation of it on a digital computer". A simulation model is constructed which represents the essential feature of the system under review. By loading the model with suitable input data the reaction of the system and its output may be observed. As with all models, the value of the model is determined by how well it "mirrors" the behaviour of the system it represents and how accurately it predicts the behaviour of the system. We can build very simple models which depend on major simplifying assumptions about a system. For example, elementary mathematical queueing models can be used to represent elements of systems − say, for example, an individual CNC machine with an FMS (Flexible Manufacturing System) − while at a higher level a queueing network can be used to provide a simplistic model of the complete FMS.

Such mathematical models are simple, very generalized in form, and therefore easy to set up and quick and cheap to use. However, to incorporate such everday subtleties as direct labour, split batches, work scheduling and shift working, a mathematical approach rapidly becomes cumbersome and inadequate. More detailed logic-based models can, however, be constructed of complex systems for which no mathematical description exists. Demanding a detailed understanding of the system to be studied, such models are more expensive in their consumption of time, expertise, and computer resources. In the most detailed cases such models can reflect operational reality to a very high degree, with extensive use being made of production control and engineering data files, and user-defined distributions reflecting the historical occurrence of phenomena such as machine breakdown. These most sophisticated and expensive representations have sometimes been termed "emulation" models to differentiate them from the more general interpretation of the term simulation.

Simulation has generally been applied in problem areas where the variables and their relationships are clearly understood but where no efficient analytical solution method exists. The major advantage of the simulation methodology, and this applies in particular to the simulation of production systems, is that it facilitates controlled experimentation on a system without disturbance of the actual system. The disadvantages of simulation are related primarily to the lead time to design and implement a model and to the fact that simulation is not an optimizing technique.

3 Simulation and Production System Design

Over the past ten years or so a number of authors have reported the use of simulation as a design tool. In particular, simulation has been widely used in the design of Flexible Manufacturing Systems (FMS). A Flexible Manufacturing System is an integrated computer-controlled complex of automated materials handling devices and computer numerically controlled (CNC) machine tools that can simultaneously process medium-sized volumes of a variety of part types [2]. Browne and Chan [3] have indicated the role of simulation within a integrated procedure for FMS design − simulation is extremely valuable at the detailed design stage of FMS, when the overall architecture of the FMS has been determined. The type of design questions which simulation can help to answer include:

− Questions of resource capacity.
− Questions on the relative locations of machines.
− Questions on the WIP (work-in-progress) philosophy − whether to use a centralized stock buffer or localized buffers, the size of buffers, etc.
− Questions on the scheduling/sequencing strategy to adopt.

P. O'Gorman et al. [4] describe a simulation model of a hypothetical flexible transfer line. The model was used to compare the performance of various scheduling systems in an FMS environment. The scheduling systems involved include Johnson's algorithm [5] for the three-machine case and a number of well-known sequencing heuristics, including FIFO and Shortest Processing Time (SPT). The flexible transfer line under study involves three CNC machines, each with local buffers and arranged in the form of a line. Each job visits each machine in a fixed order and the job times are such as to conform to the special conditions for Johnson's rule in the "3-machine" case. Johnson's rule is optimal on the criterion of minimum total throughput time, and this criterion was the basis for comparing the various scheduling systems. Results from the model indicate that simple heuristic priority rules such as FIFO and SPT perform almost as well as Johnson's algorithm in the flexible transfer line situation. As these rules are much simpler to implement in practice it is unlikely that Johnson's algorithm is of any practical benefit in the context of FMS. This study of scheduling systems is typical of the type of analysis which can be carried out at the design stage of an FMS, using simulation. Rathmill et al. [6] describe a more detailed FMS simulator written in GPSS and used as a "detailed systems design aid", while Carrie [7] concluded that "Simulation was effective. The results lead to amendments to the control systems software".

Simulation modelling is also widely used in the design of conventional manufacturing systems. Browne and Davies [8] developed a detailed model to aid in the design of a production control system. The model was of a large machine shop which machined components for assembly into a range of machine tools. A finished product typically contained up to seven hundred machined components, and given that the company involved manufactured a range of five basic models of machine tool, there were typically three thousand different items to be machined. The model, which was implemented in GASP IV, was based on actual company data and simulated the work flow through the machine shop. The machine shop contained eighty machines and involved one hundred machine operators working two shifts. The model included consideration of raw material input, machine breakdown, operation absenteeism, batch sizing, and finished parts stores. The model was intended to be used by machine shop management to assess the effects of various management strategies and production control policies on the operation and performance of the machine shop. Specifically, the model was used to analyse the effects of various batch sizes [9], mechanisms for releasing batches into the machine shop [10], and the effects of various sequencing heuristics on the performance of the shop.

An indicating of the role of this model can be gained from consideration of the results derived from it. For example, based on experiments reported in Browne and Davies [10] the following advice was issued to management: "Phase release individual batches into the shop in order of decreasing expected throughput time, where throughput time is defined as the time taken from issue of a batch to completion and deposition into finished parts stores... it is not really important how priority is assigned to individual batches although there is some marginal advantage associated with the Shortest Processing Time heuristic". This might not seem like a startling result, particularly to those readers familiar with MRP and the lead time offsetting associated with it. However, to a management team which up to then had simply accepted the lead times used in their MRP system and were putting major effort into pushing work through the shop it was an important result. It indicated to the management team that it was more important to have realistic throughput or lead times and to release batches based on these than to be over-concerned as to which short-term bach priority system to use.

4 Simulation as a Decision Support Tool to Aid in the Operation of PMS

When we refer to the operation of Production Management Systems we are thinking of online production scheduling. To date, applications in this area have been few. Classen and Malstrom [11] report a CAM-i (Computer Aided Manufacturing – International) survey of commercially available production control software packages which found that the ability to simulate the impact of management decisions on production was not provided by any of 54 packages analysed. Grant [12] summarizes the situation well when he states:

"Historically simulation techniques have been highly successful and used extensively for the planning and analysis of current operations or proposed designs. On-line simulation analyses are feasible and can be cost effective; however, applications of simulation in this mode have been few. On-line applications of simulation for real time shop floor control purposes can only be applied effectively if the data supplied to the simulation models are accurate, organized, and timely. These are constraints that in the past have proven difficult to overcome."

Coll et al. [13] identified two types of online decision support simulators:

" – one for gross capacity planning using an 'aggregated' model and a second for short-term decision making including scheduling performed by a detailed simulation model. The detailed simulation model may be used to compare the projected workload, generated possibly from a materials requirements planning (MRP) package against the machine and labour resources available. The model may assist the process of capacity management by producing profiles of load versus capacity over time. This in turn assists in management decision making regarding overtime allocation and other short-term capacity adjustments."

Thinking in terms of short-term scheduling, a simulation model operating in a "what if" mode, may help to decision maker in areas such as:

- Release of jobs to the production area.
- Batch sizing and batch splitting.
- Sequencing of jobs onto particular machines.
- Use of alternative process routes.
- Possible increase of capacity by overtime working or subcontracting.

Brennan et al. [14] and Grant [12] have identified the difficulties of developing a model to meet these requirements. The issues are:

1. Access to real-time data on the state of the production system.
2. Availability of computing power to ensure that the simulation model can present results to the manager in a reasonable time.

At this point in time, the availability of real-time data is probably the true bottleneck. Further, traditional computer simulation languages and packages have severe limitations in the area of data management.

If a simulation model is to support real-time scheduling, then clearly it must have access to the plant-wide production control database. Typically, this requires integration with the MRP system, given that the simulation model will require access to bills of materials, process routing data, work-centre data, inventory data, and tooling data. Nof and Talavage [15] have identified the type of data structure arrangements for such a simulator. One possibility is for a data structure which is embedded within the simulator. A second possibility is to interface the production database with the simulator data structure via a database management system. The simulator can then access whatever portion of the production database it requires. As Coll et al. [13] point out:

"If the simulation model is to be used as an online decision support system for production scheduling or capacity analysis problems, then a very full use of the production database will be required. If, on the other hand, the model is to be used for strategic decision making at the factory level then only a 'portion' of the production database will need to be connected to the simulation model."

5 Problems Associated with the Simulation of Production Management Systems

Coll et al. [13] have identified the significant problems associated with simulation:

1. The long lead time from initial design to final implementation of a model

2. The problems associated with validation of simulation models
3. The difficulty in defining an objective function against which to measure the performance of a production system and to evaluate the results of various simulation experiments. This applies in particular to the use of simulation models as design tools for conventional production systems
4. The difficulty in defining the interface between simulation models and other production system models — for example the linkage between Manufacturing Resource Planning (MRP II) systems and simulation models
5. The need for a good "interface" between the user and the simulation model.

6 The Lead Time to Develop Models

Wichmann [16] quotes some comments made by Shannon in the June 1985 issue of *Simulation* which are worth repeating here:

"... To use simulation correctly and intelligently, the practitioner is required to have expertise in ... probability, statistics, design of experiments, modelling, computer programming and a simulation language. This translates to about 720 hours of formal classroom instruction plus another 1440 hours of outside study ... just to get basic tools. In order to really become proficient, the practitioner must then got out and gain real world, practical experience (hopefully under the tutelage of an expert)".

Further, even for an expert practitioner, building a useful simulation model is time consuming. The use of simulation languages and packages considerably simplifies the task, and in fact designers have developed simulation packages which address particular simulation needs. Languages such as SLAM 11 and SIMAN for example are oriented towards manufacturing systems simulation. MAST (*M*anufacturing *S*ystems *T*ool) has been classified by Wichmann [16] as a fourth-generation special-purpose simulation tool in that it provides a complete simulation environment: "it is relatively easy to use in the sense that within its capability no programming is needed. The modelling process is data driven and the system prompts the user for much of the input data". Such *data-driven* packages have been produced in order to reduce the time taken to produce a simulation model. They reduce the lead time in two ways:

1. Because they are data-driven, they can be used by people who are not simulation experts and who have minimal training in the simulation methodology.
2. The time to build an individual model is greatly reduced since all that is required is that the model builder input the appropriate data and the appropriate model is generated.

However, as might be expected, there is a price to be paid for this ease of use. Normally such modelling systems are relatively inflexible and are unable to model the small features which make an individual manufacturing system unique. In effect the model user has traded flexibility for ease of modelling. Examples of this type of simulation modelling capability include SAME/AGVS [17] developed by Renault Automation for Automatic Guided Vehicle System design, and VUSIM and PROPHET [18].

It is worth noting that all of the data-driven graphics-based simulators are oriented towards the design of manufacturing systems. The author is not aware of any such tool for building models to be used in an online decision support mode. Further, given the earlier comments in this paper concerning the difficulty of integrating simulation models with production planning and control systems, this is not surprising.

7 Validation of Simulation Models

Validation of simulation models has been and remains a difficult problem. As far back as 1966, Naylor, Burdick, Balintfy, and Chu [19] pointed out that "... the problem of verifying (validating) simulation techniques remains today perhaps the most elusive of all the unresolved problems associated with simulation techniques", and outlined the major methodological positions concerning the problem of validation. These are as follows:

Synthetic a priorism. This position holds that all theory is a system of logical deductions from a series of axioms of unquestionable truth. These axioms are not verifiable by controlled experiment, and their truth must be accepted because it is obvious. Validation then consists of listing these truths and ensuring that the model and its attendant logic are compatible with them.

Ultra empiricism. Ultra empiricism refuses to accept that any postulates or axioms are valid, unless they can be independently verified. Empiricists hold that sense observation is the primary source and ultimate judge of knowledge. No model is considered valid, therefore, unless its assumptions can be verified by independent scientific experiment.

A third position is that put forward by Friedman [20] and discussed mainly in relation to economic models and theories. He criticizes the two positions outlined above on the basis of their preoccupation with the validity of the assumptions of the model and goes on to suggest that the validity of a model rests with its ability to predict the behaviour of the system it purports to represent. Thus, validation of a model consists of "feeding" the model with historical data and comparing the output of the model with observations from real life.

It is notable that the majority of "practical" validation problems encountered in the literature utilize Friedman's methodology. Van Horn [21] points out that "The objective is to validate a specific set of insights not necessarily the mechanism that generated the insights". Kheir and Holmes [22], discussing the validation of simulation models of missile systems, identify a similar validation methodology – "Confidence in missile simulation is generally built by experimental techniques". The idea is to show from the results of many model-checking experiments that when the parts (subsystems) of the model are subjected to the same inputs as the corresponding parts of the missile, the outputs from the model subsystems agree acceptably with the corresponding experiments outputs. Talavage [23] discusses the construction and application of a model of the manufacturing system at Caterpilliar Tractor: "The GASP IV simulation model, called CATLINE, was developed in close association with its potential users at Caterpillar. One result of this close association was a great deal of confidence in the model on the part of Caterpillar personnel. Thus, in contrast to the academic community where a strict and thorough statistical validation of the model would be required, we found that industrial users were satisfied with the model's operation if its behaviour resembled the behaviour that they observed in the real-world system. This is not the say that industrial users are naive".

Despite its widespread adoption as a methodology of validation, there are serious deficiencies associated with this approach. Probably the most serious is the danger of using the same data to validate the model as was used to design and build it in the first place. This arises because of the need to analyse as much data and information as possible about the system prior to constructing the model, and often there is subsequently a lack of suitable and different data.

It is noticeable also that whereas "theorists" talk of "ensuring" that the model is valid, simulation practitioners are more guarded in their comments and look for a "high degree of confidence" in their model's validity. In fact, Van Horn [21] goes so far as to define validation as "the process of building *an acceptable level of confidence* that an inference about a simulated process is a correct or valid inference for the actual process. Seldom, if ever, will validation result in a proof that the simulation is a correct or true model of the real process".

Schlesinger [24] distinguished between model verification and model validation: "Verification refers to a process of confirming that the conceptual model has been correctly translated into an operational programme, and the calculations made with the programme utilize the correct input data. Validation refers to the process of confirming that the conceptual model is applicable or useful by demonstrating an acceptable correspondence between the computation results of the model and actual data". Simulation packages tend to provide error checking routines and trace facilities which greatly simplify the verification procedure. Further, the recent development of animation software to "front end" existing simulation packages facilitates verification and validation. Animation facilitates verification by allowing the model builder to "see" the logic of the model. It facilitates validation by allowing a manager to see the model in operation, thus greatly helping the process of confidence building referred to earlier. Examples of such "animation" products include TESS which supports the SLAM 11 software system and CINEMA which support SIMAN. Tess [25], for example, provides procedures for dynamically presenting the operation of a model, i.e. animation of the simulation results.

8 An Objective Function for a PMS Model

The nature of simulation – in particular, discrete event simulation – is such that it facilitates the easy collection of statistical data on the variables within the model. Thus, in a typical model of the work flow through a production system, the model may be programmed to collect data on resource utilization, work-in-progress levels, ability to meet due dates, overtime costs, etc. This proliferation of data makes for masses of statistical data which must be analysed to understand the performance of the system under study. For many simulation practitioners, particularly those using the early simulation languages and packages, this was a real problem. The approach today seems to be to either use a well-defined objective function (see Browne and Davies [9]) or to allow the user to select variables to be analysed, store the data on these variables, and then present it in the form of information using graphs, histograms, etc.

9 Interface of Simulation to MRP II Systems

In the context of Production Management Systems this is perhaps the single most important difficulty associated with the simulation technique. Simulation clearly is useful in the context of capacity planning and scheduling. Simulation allows a "what if " approach to understanding the effects of various scheduling strategies on the flow of work through a production system. However for this "what if" approach to be effective, the simulation model must have access to up-to-date and even real-time data on shop floor activity, as well as Bill of Materials (BOM) and Process Routing data. In effect the "simulation databases must be a subset of the database for the total production system" (Coll et al. [13]). The author is not aware of any true simulation capability within existing commercially available MRP II systems.

The lack of a true simulation capability within MRP II systems can perhaps be traced to the general weakness of the Shop Floor Control or Production Activity Control (PAC) modules within available MRP II packages. PAC involves data collection and decision making – in particular, scheduling and dispatching – based on this data. Perhaps a more general difficulty with simulation support for the PAC decision-making process is the fact that PAC itself is not completely articulated and there is as yet no generally agreed statement of the elements within PAC and the relationship between them. A major project (funded by the EEC (European Economic Community) is presently investigating this area; see Harley et al. [26].

10 User Interface to Simulation Models

Traditionally, simulation modelling has been the domain of the simulation expert. Packages and languages developed to support the simulation process tended to assume that the user was expert in simulation, computer programming, etc. We have talked earlier about the development of data driven simulation tools such as MAST and SAME/AGVS. They have helped to reduce the expertize and the time needed to create simulation models. Similar developments are taking place in the area of simulation outputs. Traditionally, users analysed the results of various simulation "runs" by reviewing pages of statistical tables. More recent simulation tools (CINEMA, TESS, for example) provide animation facilities as already discussed and also facilities to present results in more manageable form to the user through the use of graphics.

11 Simulation – New Approaches

In many ways the simulation technique has changed very little over the years since it was first applied to analyse production management systems problems. Developments in computer technology and greater understanding of the simulation practitioners' needs has led to the development of more sophisticated simulation packages, including data driven simulators.

In recent years researchers have begun to analyse new concepts in the context of simulation and to develop new forms of simulators. Of particular interest is the application of Artificial Intelligence technology to simulation. Harhen et al. [27] talked about a simulation program which will "reason about the design process, the goals of the user, constraints on computational resources, and will be involved in a search process that seeks adequate answers to the users' needs for information, with the least effort".

A number of researchers are looking at the use of AI tools to support the simulation process, particularly in the areas of model building, experiment planning, and analysis of results. However, to the author's knowledge, no commerically available systems have been developed. Researchers are creating simulation applications using AI tools such as OPS5 and KNOWLEDGE CRAFT. It seems that such tools offer great promise in the area of "rapid prototyping" of simulation models and allow the expert user to create "quick and dirty" models quickly. The authors is engaged in a research project to develop a "fast simulator" based on Petri-Nets and OPS5: the idea is that the production rule paradigm of OPS5 mirrors the firing of transitions within Petri Nets, and thus a simulation model written in OPS5 can be created early from a Petri Net representation of the system under study.

12 Conclusion

This paper has reviewed simulation modelling in the context of production management systems (PMS). It is clear that simulation has an important role to play in PMS, particularly in the context of their design. Simulation is also important as a decision support mechanism for PMS, particularly in the area of production activity control.

Recent developments in simulation include the use of AI techniques to build models in rapid-prototyping mode and also the development of data-driven simulators. These promise to make simulation more accessible to the general user and have facilitated the use of simulation to solve PMS problems.

13 References

1. Pritsker, A. A. B., The GASP IV Simulation Language. Wiley Interscience, 1974
2. Stecke, K. E., Formulation and solution of non linear integer production planning problems for flexible manufacturing systems. Management Science, Vo. 29, No. 3, 1983
3. Browne, J. and Chan, W. W., An integrated FMS design procedure. Annals of O.R., Vol. 1, No. 3, 1985, pp. 207–237
4. O'Gorman, P., Gibbons, J., and Browne, J., Evaluation of scheduling systems for a flexible transfer line using a simulation model. In: Flexible Manufacturing Systems: Methods and Studies. A. Kusiak (ed.). North-Holland, 1986, pp. 209–222
5. Johnson, S. M., Optimal two and three stage production schedules with set-up times included. Naval Research Logistics Quarterly, Q1, 1954, pp. 61–68
6. Rathmill, K., Greenwood, N., and Houshmand, M., Computer simulation of FMS. In: Proceedings of the 2nd International Conference in FMS, London, October 1983. IFS (Publications) Ltd.
7. Carrie, A. S., The role of simulation in FMS. In: Flexible Manufacturing Systems: Methods and Studies, A. Kusiak (ed.). North-Holland, 1986, pp. 191–208
8. Browne, J. and Davies, B. J., The design and validation of a digital simulation model for job shop control decision making. International Journal of Production Research, vol. 22, No. 2, 1984, pp. 335–337
9. Browne, J. and Davies B. J., A preliminary review of batch sizes in job shop using a digital simulation model. Simulation, Vol. 41, No. 4, October 1983, pp. 149–153
10. Browne, J. and Davies, B. J., The phased release of batches into a job shop. International Journal of Operations and Production Management, Vol. 4, No. 4, 1984, pp. 16–27
11. Classen, R. J. and Halstrom, E. M., Effective capacity planning from automated factories requires workable simulation tools and responsive shop floor controls. Industrial Engineer, April 1982
12. Grant, H., Production scheduling using simulation technology. In: Proceedings of the SIM 2 Conference, June 1986, J. Lenz (ed.). IFS (Publications) Ltd.
13. Coll, A., Brennan, L., and Browne, J., Digital simulation modelling of production systems. In: Modelling Production Management Systems, P. Falster and R. B. Mazumber (eds.). North-Holland, 1985, pp. 175–194
14. Brennan, L., Browne, J., and Davies, B. J., The development of an interactive simulation model for management decision making. In: Proceedings of the UKSC Conference on Computer Simulation. Butterworth Scientific Ltd., 1984, pp. 101–107
15. Nof, S. J. and Talavage, J. J., Design approaches to data bases for simulation. In: Proceedings of the 12th International Conference on Systems Science, Hawaii, January 1979

16. Wichmann, K. E., An intelligent simulation environment for the design and operation of FMS. In: Proceedings of the 2nd International Conference on Simulation in Manufacturing, Chicago, June 1986, J. Lenz, (ed.). IFS (Publications) Ltd.
17. Duffan, B. and Bloche, E., AGVS design in automotive industry. In: Proceedings of the 4th International Conference on AGVS, Chicago, June 1986, G. C. Hammond (ed.). IFS (Publications) Ltd.
18. Hill, S. R. and Rogers, M. A. M., Practical experience contrasting conventional modelling and data driven visual interaction simulation techniques. In: Proceedings of the 2nd International Conference on Simulation in Manufacturing, Chicago, June 1986, J. Lenz (ed.). IFS (Publications) Ltd.
19. Naylor, T. M., Balinfy, L. J., Burdick, D. S., and Chu, K., Computer Simulation Techniques. Wiley, New York, 1966
20. Friedman, M., Essay in Positive Economics. University of Chicago Press, 1963
21. Van Horn, R. L., Validation of simulation results. Management Science, Vol. 17, No. 5, January 1971
22. Kheir, N. A. and Holmes, W. M., On validating simulation models of missile systems. Simulation, April 1978
23. Talavage, J. J., Models for the automatic factory. Simulation, March 1978, pp. 80–84
24. Schlesinger, P., Developing standard procedures for simulation validation and verification. In: Proceedings 1974 Summer Computer Simulation Conference, Vol. 1. Society for Computer Simulation, USA
25. Standridge, C. R., The presentation graphics of TESS. In: Proceedings of the 1984 Winter Simulation Conference, December 1984. Society for Computer Simulation, USA
26. Harley, M. et al., A specification for a PAC system in a CIM environment. In: CAPE 86 Conference Proceedings, K. Bo et al. (eds.). North-Holland, 1986, pp. 807–826
27. Harhen, John et al., Artificial intelligence and simulation of manufacturing systems. In: Proceedings of the 10th IFIP World Computer Congress, Dublin, 1986

Chapter 9
Operations Research Models and Techniques

Wing S. Chow, Sunderesh Heragu, and Andrew Kusiak

Wing S. Chow received a B.Sc. degree in Applied Mathematics from the University of Winnipeg, Manitoba, Canada, a B.Sc. degree in Statistics and an M.Sc. degree in Operations Research from the University of Manitoba. He is a Ph.D. student in the Department of Mechanical and Industrial Engineering at the University of Manitoba. His interests are in group technology and scheduling flexible manufacturing systems.

Sunderesh Heragu is a Ph.D. student in the Department of Mechanical and Industrial Engineering at the University of Manitoba, Canada. He received M.B.A. from University of Saskatchewan, Canada in 1985 and B. Engg in Mechanical Engineering from University of Mysore, India in 1982. His interests are in the design and layout planning of flexible manufacturing systems.

Dr. Andrew Kusiak is an Associate Professor of industrial engineering in the Department of Mechanical and Industrial Engineering at the University of Manitoba, Winnipeg, Manitoba, Canada. He received a B.Sc. degree in Precision Engineering, an M.Sc. degree in Mechanical Engineering from Warsaw Technical University, and a Ph.D. degree in Operations Research from the Polish Academy of Sciences in 1978. His main interests are in applications of operations research and artificial intelligence techniques in design and operation of manufacturing and storage systems. Dr. Kusiak is a member of IFIP WG 5.7.

1 Introduction

The development of production management had its beginning in the years of the industrial revolution [1]. The first significant application of operations research

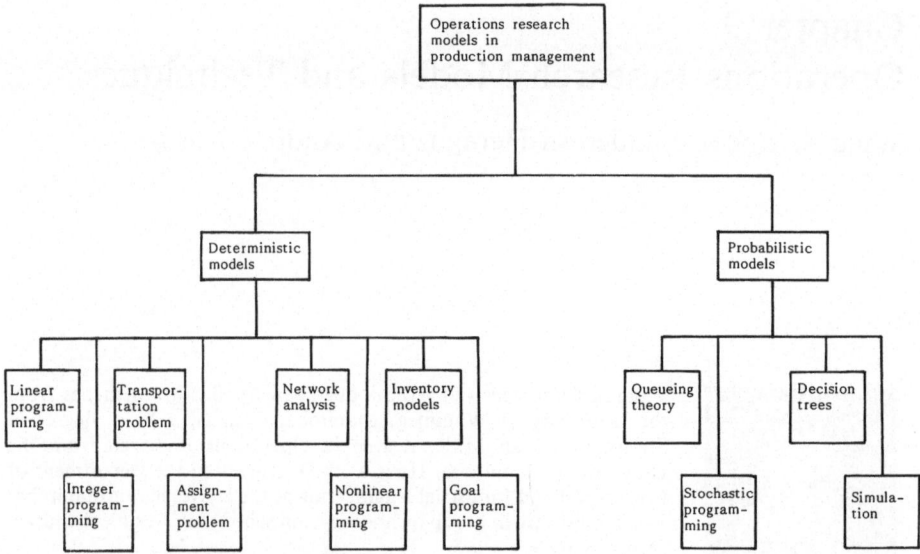

Fig. 1. Operations research models in production management

was the use of the simplex method of linear programming to war operation problems [2]. Since then there have been a number of operations research models which have found their use in production management. These models can be broadly classified into:

1. deterministic models
2. probabilistic models.

In deterministic models, the input data are treated as being invariable, whereas in probabilistic models the input data are variable.

Figure 1 shows some of the operations research models that have been applied to production management. Although a number of such models are available, only a few of them are frequently used. In a survey paper by Shannon et al. [3] it was found that linear programming and simulation were the most widely used models. Over 80% of the respondents to the survey had used these models. Some of the other models that were used to a lesser extent were network analysis, queueing theory, decision trees, and integer programming.

Hence, in this chapter, only the above models, with the exception of simulation, are discussed. Simulation is discussed in a separate chapter of this book. Since linear programming is a widely popular technique, its discussion is extended to the transportation problem, which is a special class of linear programming. A brief review of non-linear programming is also included.

2 Overview of Operations Research Models

2.1 Linear Programming

The linear programming model is concerned with optimizing a linear objective function subject to two types of constraints: resource (technological) constraints and non-negativity constraints. Let

Z = total value of the objective function
x_j = decision variable corresponding to activity j
c_j = coefficient of x_j in the objective function
a_{ij} = coefficient of resource i for variable x_j
b_i = availability of resource i
n = number of decision variables
m = number of resource constraints.

Based on the above definitions, the general linear programming model can be stated as follows:

$$\min Z = \sum_{j=1}^{n} c_j x_j \tag{1}$$

$$\text{s.t.} \sum_{j=1}^{n} a_{ij} x_j \quad (\leq, =, \geq) \; b_i; \quad i = 1, 2, \ldots, n \tag{2}$$

$$x_j \geq 0; \quad j = 1, 2, \ldots, n \tag{3}$$

The variables c_j, a_{ij}, b_i are the parameters of the problem and are assumed to be known constants. To solve the linear programming problem two basic methods can be used:

– graphical method
– simplex method.

These methods are described in introductory books on linear programming, for example Gass [4] and Budnick et al. [5].

2.2 Integer Programming

In the linear programming model it was implicitly assumed that the decision variables could take non-integer values. However, in some applications only integer values are allowed for a decision variable. Consider for example, the problem of determining the number of airplanes to be purchased to carry a given number of passengers from New York to Chicago every week, given that the budget for the purchase is limited. A non-integer solution for the above problem would be absurd. In such cases an additional constraint that all the decision variables be integers is to be added. Then the model becomes an integer programming model. In a case where only some of the decision variables are to be integers, then the model is called a mixed-integer programming model.

A general integer programming formulation is given as follows:

$$\min Z = \sum_{j=1}^{n} c_j x_j \tag{1}$$

$$\text{s.t.} \sum_{j=1}^{n} a_{ij} x_j \ (\leq, =, \geq) \ b_i; \quad i = 1, 2, \ldots, n \tag{2}$$

$$x_j = \text{integer}; \quad j = 1, 2, \ldots, n \tag{4}$$

Zero-one programming is a special case of integer programming. In this model, the decision variables can take the values 0 or 1. To ensure that decision variables take only a value 0 or 1, a new constraint (5), to replace constraint (4), is required:

$$x_j = 0, 1; \quad j = 1, 2, \ldots, n \tag{5}$$

An integer programming problem can be solved by one of the following two basic methods:

- cutting plane method
- branch and bound method.

Readers interested in these methods may refer to Lee et al. [6] and Garfinkel and Nemhauser [7]. An example of zero-one programming known as the set covering problem is presented below.

Example 1. An automated guided vehicle (AGV) is to deliver parts to five different machines. Ten possible routes for the AGV have been specified by a production analyst. The route matrix (6) and the travelling cost associated with each route are given below. An entry 1 in the ith row and jth column of the matrix (6) indicates that machine i can be reached by route j. For example, machines 1, 2, and 4 can be reached by route 6. It is required to determine the travel routes so that all the machines are visited at least once and the travelling cost is minimum.

$$[a_{ij}] = \begin{array}{c} \text{Routes} \\ \begin{array}{cccccccccc} 1 & 2 & 3 & 4 & 5 & 6 & 7 & 8 & 9 & 10 \end{array} \\ \begin{bmatrix} 1 & 1 & & & & 1 & 1 & & & 1 \\ & 1 & 1 & 1 & & 1 & & & 1 & 2 \\ 1 & & 1 & & & & 1 & & & 3 \\ 1 & & & 1 & 1 & 1 & & & & 4 \\ & 1 & & 1 & & & 1 & & 1 & 5 \end{bmatrix} \begin{array}{c} 1 \\ 2 \\ 3 \ \text{Machines} \\ 4 \\ 5 \end{array} \end{array} \tag{6}$$

Cost c_j 8 7 6.2 5.6 4 5 6 2 3 1

Let

$$x_j = \begin{cases} 1 & \text{if AGV travels on route } j \\ 0 & \text{otherwise} \end{cases}$$

The problem can be formulated as follows:

$$\min Z = \sum_{j=1}^{10} c_j x_j$$

$$\text{s.t. } \sum_{j=1}^{10} a_{ij} x_j \geq 1; \quad i = 1, 2, \ldots, 5$$

$$x_j = 0, 1; \quad j = 1, 2, \ldots, 10$$

An expanded form of the model is:

$$\min Z = 8x_1 + 7x_1 + 6.2x_3 + 5.6x_4 + 4x_5 + 5x_6 + 6x_7 + 2x_8 + 3x_9 + x_{10}$$

$$\begin{aligned}
\text{s.t. } & x_1 + x_2 + x_6 + x_7 & \geq 1 \\
& x_2 + x_3 + x_4 + x_6 + x_{10} & \geq 1 \\
& x_1 + x_3 + x_8 & \geq 1 \\
& x_1 + x_4 + x_5 + x_6 & \geq 1 \\
& x_2 + x_4 + x_7 + x_9 & \geq 1 \\
& x_j = 0, 1; j = 1, 2, \ldots, 10
\end{aligned}$$

The solution of this problem is as follows: select routes 6, 8, and 9 ($x_6 = x_8 = x_9 = 1$; all other variables are equal to 0). The corresponding AGV travelling cost is 10 units.

2.3 Transportation Problem

The transportation problem is one of the most widely used problems in production management. Its discussion will be based on an application to production planning [8].

Assume that a product can be produced in several alternative methods (for example, regular time production, overtime production, or subcontracting). The demand for the product for each period and the cost of manufacturing the product by alternative methods are given. The production capacity of each method is also known. It is required to determine the number of units to be produced by each method so as to meet the demand at a minimum cost. No backorders are allowed. Denote

c_i = cost to produce a product unit by method i
x_{ij} = number of product units to be produced by method i in period j ($i = 1, 2, \ldots, n$; $j = 1, 2, \ldots, m$)
d_j = number of product units required to be produced in period j
s_i = number of product units that can be produced by method i

h_j = cost of holding one product unit from period j to period $j+1$
r_j = number of product units held at the end of period j.

The above problem can be formulated as the following transportation model:

$$\min \sum_{j=1}^{m} \left(\sum_{i=1}^{n} c_i x_{ij} + h_j r_j \right) \tag{7}$$

$$\text{s.t.} \sum_{i=1}^{n} x_{ij} = d_j + r_j - r_{j-1}; \quad j = 1, 2, \ldots, m \tag{8}$$

$$\sum_{j=1}^{m} x_{ij} \leq s_i; \quad i = 1, 2, \ldots, n \tag{9}$$

$$x_{ij} \geq 0; \quad i = 1, 2, \ldots, n; \quad j = 1, 2, \ldots, m \tag{10}$$

As an illustration of the transportation model (7)–(10), consider the following example.

Example 2. A product can be produced in regular time, overtime, or can be subcontracted. It is required to determine the production plan for the next 3 months using the following data:

$c_1 = 70 \quad d_1 = 800 \quad s_1 = 400$

$c_2 = 90 \quad d_2 = 500 \quad s_2 = 200$

$c_3 = 85 \quad d_3 = 900 \quad s_3 = 800$

$h_1 = h_2 = 5$

Table 1. Transportation table for Example 2

	Month 1	Month 2	Month 3	Dummy column	Supply
Regular time production in month 1	70	75	80	0	400
Overtime production in month 1	90	95	100	0	200
Subcontracting in month 1	85	90	95	0	800
Regular time production in month 2	M	70	75	0	400
Overtime production in month 2	M	90	95	0	200
Subcontracting in month 2	M	85	90	0	800
Regular time production in month 3	M	M	70	0	400
Overtime production in month 3	M	M	90	0	200
Subcontracting in month 3	M	M	85	0	800
Demand	800	500	900	2000	4200

Table 2. Solution for Example 2

	Month 1	Month 2	Month 3	Dummy column	Supply
Regular time production in month 1	400				400
Overtime production in month 1				200	200
Subcontracting in month 1	400			400	800
Regular time production in month 2		400			400
Overtime production in month 2				200	200
Subcontracting in month 2		100		700	800
Regular time production in month 3			400		400
Overtime production in month 3				200	200
Subcontracting in month 3			500	300	800
Demand	800	500	900	2000	4200

The transportation table for the problem is shown in Table 1. The production costs which are shown in the cells in Table 1 include production and inventory costs. Since backorders are not allowed, production in a month to meet the demand of a preceding month is not a feasible alternative. Hence, a large value (M) of production cost is assigned to certain cells. Since the number of units available is greater than the demand, a dummy column is introduced to absorb the excess supply. The cost c_i for each cell in the dummy column is zero.

Solving the problem by the transportation algorithm produces the following result (see also Table 2):

1. 400 units are to be produced in regular time in months 1, 2, and 3 ($x_{11} = x_{12} = x_{13} = 400$)
2. 400, 100, and 500 units are to be subcontracted in months 1, 2, and 3, respectively ($x_{31} = 400$, $x_{32} = 100$, $x_{33} = 500$).

2.4 Non-linear Programming

A non-linear programming problem can be formulated as follows:

$$\min Z = f(x_1, x_2, \ldots, x_n)$$
$$\text{s.t. } g(x_1, x_2, \ldots, x_n) (\leq, =, \geq) b_i; \quad i = 1, 2, \ldots, m;$$
$$x_j \geq 0; \quad j = 1, 2, \ldots, n$$

where either $f(X)$ or $g(X)$ or both can be non-linear functions. In general, non-linear programming problems are more difficult to solve than linear programming problems because they do not have certain attractive properties [9]. For example, in a linear programming problem, the extreme-point solutions can be obtained easily and directly, one from another, by linear algebraic transformations. This cannot be done for non-linear programming problems. To solve the non-linear programming problem two classes of methods can be used:

- gradient method
- non-gradient method.

If $f(X)$ is an objective function, then the gradient of $f(X)$ at a point X_0 is denoted by $\nabla f(X_0)$. A necessary condition for X_0 to be an extreme point is:

$$\nabla f(X_0) = 0$$

The gradient methods use this property to determine the optimal point, whereas the non-gradient methods use systematic search procedures to determine the optimal point. For more details on these methods see, for example, McCormick [10] and Bazaraa and Shetty [11].

2.5 Queueing Theory

A queueing system consists of the following four components [12]:

1. A source population of customers
2. Customer arrivals pattern
3. Queue discipline
4. The service mechanism.

The source population of customers refers to the source which generates potential customers. This source can either be finite or infinite depending upon the number of customers the source is capable of generating. For example, the potential customers that can come to a university cafeteria is finite and is equal to the number of staff and students at the university.

The customer arrivals pattern in a queueing system is specified by a probability distribution function. The queue discipline refers to the manner in which the customers will be chosen for service. Some examples of queue disciplines are first-come first-served (FCFS), last-come first-served (LCFS), etc. The service mechanism is specified in terms of a probability distribution function and refers to the service facility which serves the customers.

Depending upon the number of channels and the number of phases, the queueing system can be classified into the following four systems:

1. Single channel, single phase
2. Multi-channel, single phase
3. Single channel, multi-phase
4. Multi-channel, multi-phase.

The number of channels refers to the number of parallel lines of service provided in the service facility, whereas the number of phases indicates the number of service stations a customer is required to go through before service is completed.

The standardized notation (also known as the Kendall notation) for queueing systems has the following format:

$(A/B/C):(D/E/F)$

9. Operations Research Models and Techniques

Table 3. Standard notation for queueing systems

Parameter	Parameter type	Symbol
A and B	constant Poisson arrival and exponential departure Erlangian or Gamma distribution with parameter k general distribution	D M E_k G
C	number of servers	s
D	first-come first-served last-come first-served reservation first emergencies first other	FCFS LCFS
E and F	finite infinite	N ∞

where

A = arrivals distribution
B = service time distribution
C = number of servers in the service facility
D = service discipline
E = maximum capacity of the system
F = size of the calling source.

Table 3 gives a list of symbols which are used to represent parameters A, B, C, D, E, and F.

For example, a queueing system having a Poisson arrival process, exponential distribution of inter-arrival times, 10 servers, a first-come first-served discipline, a maximum queue length of 50 customers with a finite calling source would have the following representation:

$(M/M/10):(FCFS/50/N)$

The following information can be obtained from a queueing model:

1. Steady-state probabilities
2. Expected number of customers in the system, L_s, and expected queue length, L_q
3. Expected waiting time in the system, W_s, and expected waiting time in the queue, W_q.
4. Average utilization of the service facility, ϱ.

An $(M/M/1):(FCFS/\infty/\infty)$ model is presented in Example 3.

Example 3. The arrival of jobs in a machine shop follows an exponential distribution and is at the rate of 12 jobs per day. The processing time is exponentially distributed with a mean service rate of 14 jobs per day. Determine the idle server time per day, mean number of orders, and waiting time per order. Let

λ = arrival rate
μ = service rate
$\varrho = (\lambda/\mu)$.
Expected idle server time per day, $P_0 = 1 - \lambda/\mu$
$= 1 - 12/14 = 0.143$ day
Mean number of orders, L_q $= \varrho^2/(1-\varrho)$
$= 5.13$ orders
Waiting time per order, W_q $= \varrho^2/\lambda(1-\varrho)$
$= 0.043$ day.

2.6 Decision Trees

Decision trees are used to represent decision problems. They make the decision process explicit and clearly state the risks and outcome of a decision. A typical decision tree has the format shown in Fig. 2. In Fig. 2 the square boxes represent decision points and the circles represent outcome points. For example, if decision B is taken and event 2 occurs, then the decision maker has to take one of the following two decisions B_1 or B_2. If decision B_1 is taken, the outcome is b_2.

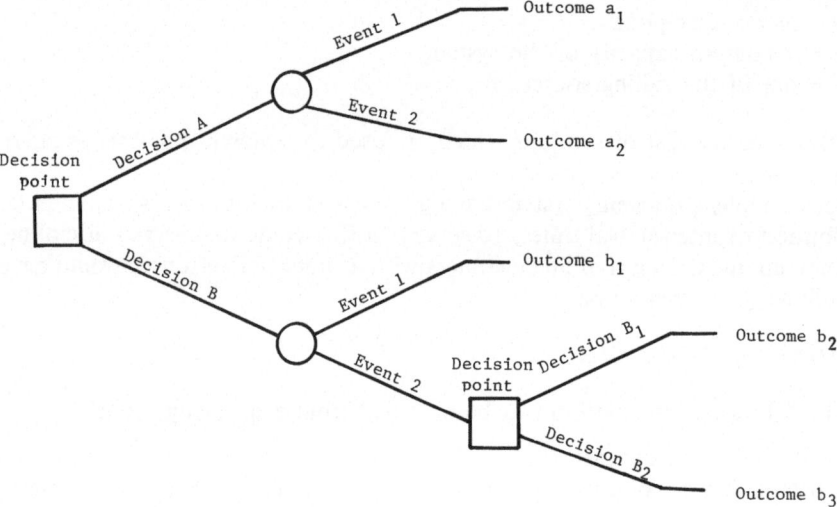

Fig. 2. Example of a decision tree

2.7 Network Models

Network models are often represented by graphs. Some of the most important applications of network models are shortest route problem, minimal spanning tree problem, and maximal flow problem. Phillips and Garcia-Diaz [13] have discussed network models in detail. An application of the network model to the minimal spanning tree problem is given below.

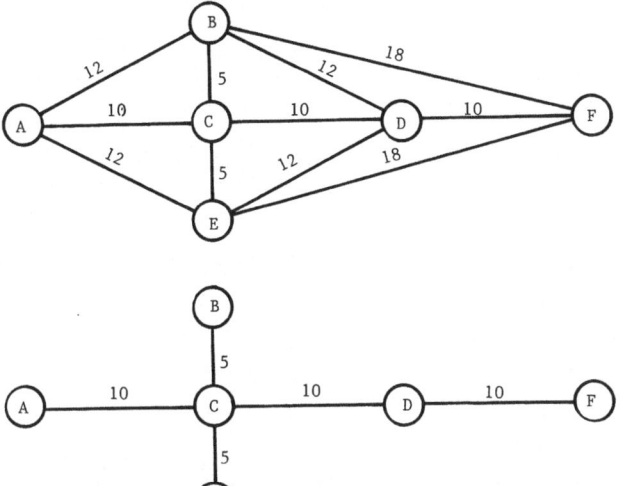

Fig. 3. Distance between centres in Example 4

Fig. 4. Minimal spanning tree solution to Example 4

Example 4. It is desired to connect six centres A, B, C, D, E, and F by a communication cable. The distances between the centres are indicated on the arcs in Fig. 3. The problem is to determine the minimum length of cable required to connect all the six centres.

Solving the above problem using the minimal spanning tree algorithm produces the solution shown in Fig. 4. From Fig. 4 it is clear that the length of cable required to connect all the six centres is $10 + 10 + 10 + 5 + 5 = 40$ units.

3 Application of Operations Research Techniques

There are a number of areas in production management where operations research models have been applied. Table 4 shows a list of sample applications of operations research models.

4 Some Difficulties in Applying Operations Research

To date, a number of operations research models have been applied to production management. However, not all of them are used in practice. The reason is that some of the models, especially the combinatorial models, are difficult to solve. This has forced researchers to develop heuristic methods which require less computational effort (see, for example, Kusiak [33]). However, the solutions provided by such methods are usually suboptimal. Practitioners are willing to ac-

Table 4. Sample operations research models in production management

Model	Application area	Reference
Linear programming	Inventory management and production planning	Gaimon [14]
	Aggregate production planning	Eldredge and Holdman [15]
	Energy utilization planning	Sherali and Staschus [16]
	Detailed scheduling in manufacturing systems	Chang and Sullivan [17]
Integer programming	Loading problem in manufacturing systems	Kusiak [18]
	Aggregate production systems	Randhawa et al. [19]
	Scheduling of material handling systems	Wilhelm and Sarin [20]
	Shop-floor control planning	Egbelu [21]
Transportation problem	Warehouse location	Baumol and Wolfe [22]
	Production Scheduling	Bowman [8]
	Routing of materials handling systems	Chase and Aquilano [2]
	Location of plants in relation to raw materials sources and market areas	Das and Heragu [23]
Non-linear programming	Production planning	Stecke [24]
	Scheduling of parts in manufacturing systems	Kimemia and Gershwin [25]
Queueing theory	Production scheduling	Cheng [26]
	Performance of unreliable manufacturing systems	Vinod and Solberg [27]
	Shop-floor planning	Dar-El and Wysk [28]
	Design of storage systems	Chow [29]
Decision trees	Equipment purchasing	Souder [30]
Network models	Equipment replacement	Bennington [31]
	Production planning	Symthe and Johnson [32]
	Project scheduling	Phillips and Garcia-Diaz [13]
	Fleet scheduling	Phillips and Garcia-Diaz [13]

cept suboptimal solutions if there is a significant reduction in computation effort.

Another difficulty in applying operations research models to production management is that in practice it may not be possible to obtain all the required data, or, in some cases, the data obtained may not be accurate.

In order to make operations research models applicable in production management systems, an attempt should be made to develop more realistic models. Simplifying assumptions make the models less useful.

5 References

1. Buffa, E. S., Basic Production Management. John Wiley and Sons, New York, 1971
2. Chase, R. B. and Aquilano, N. J., Production and Operations Management. Richard D. Irwin, Homewood, Illinois, 1977

3. Shannon, R. E., Long, S. S., and Buckles, B. P., Operations research methodologies in industrial engineering: a survey. AIIE Transactions, Vol. 12, No. 4, 1980, pp. 364–366
4. Gass, S. I., Linear Programming. McGraw-Hill, New York, 1969
5. Budnick, F. S., Mojena, R., and Vollmann, T. E., Principles of Operations Research for Management. Richard D. Irwin, Homewood, Illinois, 1977
6. Lee, S. M., Moore, L. J., and Taylor, B. W., Management Science. Wm. C. Brown, Iowa, 1985
7. Garfinkel, R. S. and Nemhauser, G. L., Integer Programming. John Wiley and Sons, New York, 1972
8. Bowman, E. H., Production planning by the transportation method of linear programming. Operations Research, Vol. 3, No. 1, 1956, pp. 100–103
9. Simmons, D. M., Nonlinear Programming for Operations Research. Prentice-Hall, Englewood Cliffs, New Jersey, 1975
10. McCormick, G. P., Nonlinear Programming: Theory, Algorithms, and Applications. John Wiley and Sons, New York, 1983
11. Bazaraa, M. S. and Shetty, C. M., Nonlinear Programming: Theory and Algorithms. John Wiley and Sons, New York, 1979
12. Gorney, L., Queueing Theory: A Problem Solving Approach. Petrocelli Books, New York, 1981
13. Phillips, D. T. and Garcia-Diaz, A., Fundamentals of Network Analysis. Prentice-Hall, Englewood Cliffs, New Jersey, 1981
14. Gaimon, C., Optimal inventory, backlogging and machine loading in a serial, multi-stage, multi-period production environment. International Journal of Production Research, Vol. 24, No. 3, 1986, pp. 647–662
15. Eldredge, D. L. and Holdman, B., A cost minimization production planning model for continuous process chemical plant operations. Journal of Operations Manegement, Vol. 2, No. 3, 1982, pp. 197–202
16. Sherali, H. D. and Staschus, K., Solar energy in electric utility planning: a linear programming approach. OMEGA, Vol. 12, No. 2, 1984, pp. 165–174
17. Chang, Y. L. and Sullivan, S., Real-time scheduling of flexible manufacturing systems: a conceptual and mathematical foundation. Presented at ORSA/TIMS Meeting, San Francisco, California, 1984
18. Kusiak, A., Loading models in flexible manufacturing systems. In: A. Raouf and S. Ahmad (eds.), Flexible Manufacturing. Elsevier, New York, 1985, pp. 119–132
19. Randhawa, W. V., McDowell, E. D., and Faruqui, S. D., An integer programming application to solve sequencer mix problems in printed circuit board production. International Journal of Production Research, Vol. 23, No. 3, 1985, pp. 543–552
20. Wilhelm, W. E. and Sarin, S. C., A structure for sequencing robot activities in machine tending application. International Journal of Production Research, Vol. 23, No. 1, 1985, pp. 47–64
21. Egbelu, P., Planning for machining in a multi-job multi-machine manufacturing environment. International Journal of Manufacturing Systems, Vol. 5, No. 1, 1986, pp. 1–14
22. Baumol, W. J. and Wolfe, P., A warehouse-location problem. Operations Research, Vol. 6, 1958, pp. 252–263
23. Das, C. S. and Heragu, S., A transportation approach to locating plants in relation to potential markets and raw materials sources. Decision Sciences, Vol. 20, No. 1, Winter 1989
24. Stecke, K. E., Formulation and solution of non-linear integer production planning problems for flexible manufacturing systems. Management Science, Vol. 29, 1983, pp. 273–288
25. Kimemia, J. G. and Gershwin, S. B., Flow optimization in flexible manufacturing systems. International Journal of Production Research, Vol. 23, No. 1, 1985, pp. 81–96
26. Cheng, T. C. E., Optimal due-date assignment in a job shop. International Journal of Production Research, Vol. 24, No. 3, 1986, pp. 503–515
27. Vinod, B. and Solberg, J., Performance models for unreliable flexible manufacturing systems. OMEGA, Vol. 12, No. 3, 1984, pp. 299–308

28. Dar-El, E. M. and Wysk, R. A., Job shop scheduling – systematic approach. Journal of Manufacturing Systems, Vol. 1, No. 1, 1986, pp. 77–88
29. Chow, W. M., Design for line flexibility. IIE Transactions, Vol. 18, No. 1, 1986, pp. 95–103
30. Souder, W. E., Management Decision Methods for Managers of Engineering and Research. Van Nostrand Reinhold, New York, 1980
31. Bennington, G. E., Applying network analysis. Journal of Industrial Engineering, Vol. 6, No. 1, 1974, pp. 17–25
32. Smythe, W. R. and Johnson, L., Introduction to Linear Programming with Applications. Prentice Hall, Englewood Cliffs, New Jersey, 1966
33. Kusiak, A., Scheduling flexible machining and assembly systems. In: K. E. Stecke and R. Suri (eds.), Proceedings of the Second ORSA/TIMS Conference on Flexible Manufacturing Systems: Operations Research Models and Applications. Elsevier, New York, 1986, pp. 521–532
34. Shanthikumar, J. G. and Buzacott, J. A., Open queueing network models of dynamic job shops. International Journal of Production Research, Vol. 19, 1981, pp. 255–266
35. Taha, H. A., Operations Research: an Introduction. Macmillan, New York, 1983

Chapter 10
Artificial Intelligence Approach to Production Planning

Andrew Kusiak[*]

1 Introduction

There are different views on the history or Artificial Intelligence (AI). Some tie the beginning of AI to the year 1842, when Charles Babbage first tinkered with his machines [1]. Others consider the year 1950 when the book *Computing Machinery and Intelligence* by Allan Turing was published as the debut of AI to the academic community [2].

Artificial Intelligence has been applied in several areas. Some of these applications are as follows [3]:

1. Natural language processing
2. Intelligent retrieval from databases
3. Expert systems
4. Theorem proving
5. Robotics
6. Automatic programming
7. Planning and scheduling
8. Perception.

Since expert systems (one of AI's application areas) relate strongly to the area of production management they are emphasized most in this chapter.

Most of the currently operating computer programs (systems) perform tasks according to decision-making logic [4]. These systems do not accommodate a significant amount of knowledge. A typical program consists of an algorithm and data. The algorithm determines how to solve a problem, while the data characterize the parameters. Human knowledge does not fit this model. Since a large portion of human knowledge consists of elementary fragments of know-how, turning this knowledge into application requires new ways to organize decision-making fragments into useful entities.

Expert systems collect these fragments in a knowledge base, then access the knowledge base to reason about specific problems. As a consequence, expert systems differ from conventional programs in the way they are organized, incor-

---

[*] Dr. Andrew Kusiak is also Coauthor of Chapter 9. His biography and photo appear on p. 135.

porate knowledge, execute, and the impression they create through their interactions. An expert system simulates the performance of a human expert.

2 Basic Structure of an Expert System

An expert system consists of the following components (Fig. 1):

1. Knowledge base which contains facts (data) and rules (or other representations) associated with the problem being solved
2. Global database which contains general problem-solving data, provides working space, and keeps track of the problem status
3. Inference engine which solves the problem. It includes an interpreter that decides how to apply the rules to infer new knowledge and a scheduler that decides the order in which the rules should be applied.

A human (domain expert) usually assists in developing the knowledge base. It would be desirable to have a user-friendly natural language interface to facilitate the use of an expert system in three modes: development, problem solving, instruction [5]. In some more sophisticated systems, an explanation module could be included, allowing the user to challenge and examine the reasoning process underlying the expert system's answer.

As an example consider the expert system called DENDRAL developed at Stanford University [5]. DENDRAL's purpose is to generate plausible structural

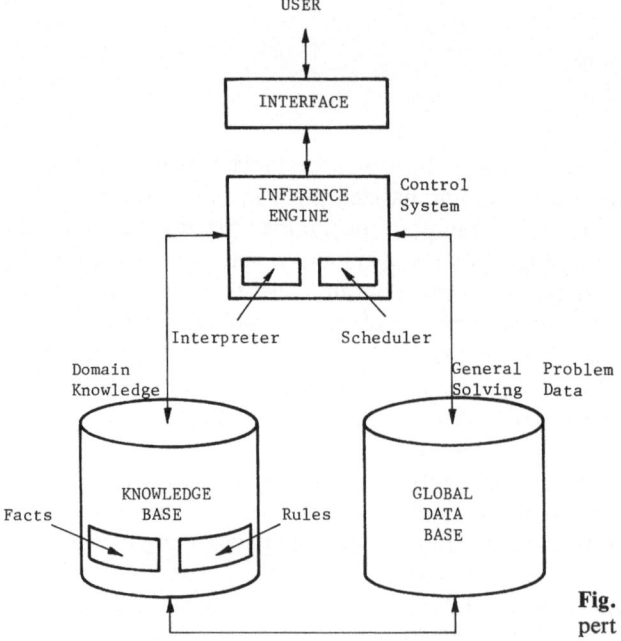

Fig. 1. Basic structure of an expert system

10. Artificial Intelligence Approach to Production Planning   151

representations of organic molecules from mass spectrogram data. The key elements of its knowledge base are:

- Rules for deriving constraints on molecular structure from experimental data
- Procedure for generating candidate structures to satisfy constraints
- Rules for predicting spectographs from structures.

Its global data base contains:

- Mass spectrogram data
- Constraints
- Candidate structures.

The basic functions of its control systems are:

- Forward chaining
- Plan generation and testing.

Since the inference engine is perhaps the most important component of expert systems it is discussed in greater detail below.

2.1 Inference Engine

Many high-level languages for building expert systems, for example EMYCIN, have the inference engine, in some sense, built in as a part of the language [6]. Other lower-level languages, for example LISP, require an expert-system builder (knowledge engineer) to design and implement the inference engine. Both approaches have their advantages and disadvantages. A high-level language with the inference engine built in, means less work for the expert-system builder. However, the builder also has fewer options regarding the organization and access to knowledge. A lower-level language with no inference engine requires a greater development effort, but it provides some basic building blocks so that the system developer can tailor the control scheme to the needs of the problem domain.

3 Knowledge Representation

Knowledge representation is regarded as a core area of AI. Knowledge can be represented by four basic methods:

1. Production rules
2. Semantic nets
3. Frames
4. Predicate calculus.

3.1 Production Rules

Production rules are based on IF *condition* THEN *action* statements. (The term production rule has nothing to do with production management.) When the current problem situation satisfies the IF part of a rule, the action specified by the THEN part of the rule is performed. This action may affect the environment, direct program control, or may instruct the system to reach a conclusion, for example add a new fact or hypothesis to the database [6]. As an example consider the following two production rules:

Rule 1: IF material is available,
THEN print part documentation.
Rule 2: IF the part documentation has been printed,
THEN release part for assembly.

An expert system may involve hundreds or thousands of production rules.

3.2 Semantic Nets

A semantic net consists of:

– Nodes representing objects, concepts, and events

and

– links between the nodes representing their interrelations.

Consider, for example, the simple semantic net presented in Fig. 2.

MANUFACTURING-DEPARTMENT, MACHINING-SYSTEM, MACHINING-CELL and ASSEMBLY-SYSTEM represent sets, and "has-a" and "is-a-component" are names of the links specifying their relationship. Among many possible interpretations of this net is the statement:

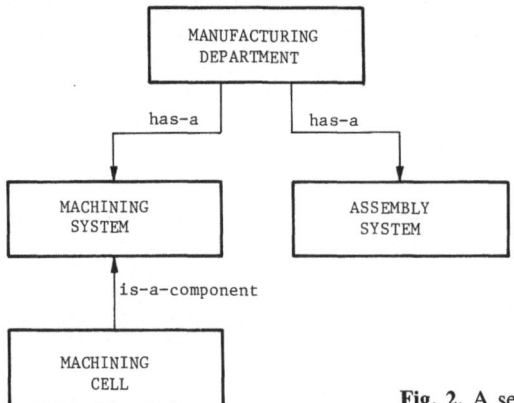

Fig. 2. A semantic net

"A manufacturing department has a machining system".

or

"A machining cell is a component of a machining system.

3.3 Frames

Frames involve a network of nodes connected by relations and organized into a hierarchy. Each node represents a concept that may be described by attributes and values. The topmost nodes represent general concepts and the lower nodes more specific instances of those concepts. In a frame, the concept of a manufacturing department could be organized as shown in Fig. 3.

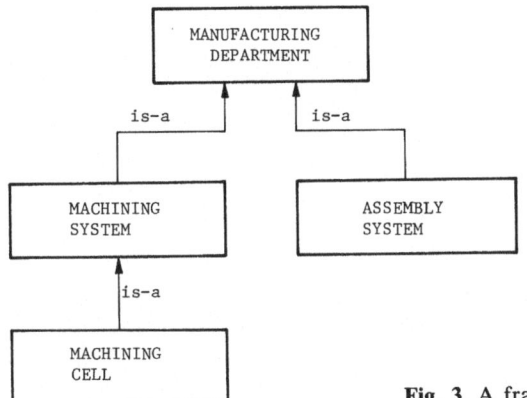

Fig. 3. A frame

So far, the frame representation looks like a semantic net, but in a frame system the concept at each node is defined by a collection of attributes: for example, colour, size, name. These attributes are called slots. There might be a procedure (computer program) associated with each slot. This procedure is executed whenever the information in the slot (the value of the attribute) is changed.

3.4 Predicate Calculus

In order to illustrate application of predicate calculus for a problem formulation consider the following example.

Example 1. Given an initial stack (initial state) presented in Fig. 4 (a), use predicate calculus to construct a stack presented in Fig. 4 (b). The blocks are to be moved by robot R. The initial state presented in Fig. 4 (a) can be described as follows:

CLEAR(B),   CLEAR(C),   ON (C, A),   GRIPPEREMPTY,
ONTABLE(A),   ONTABLE(B).

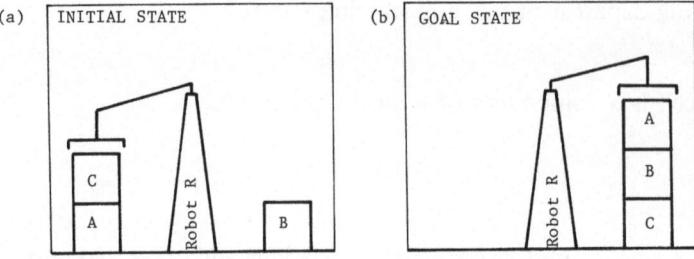

Fig. 4a,b. Representation of initial and goal states

The goal state can be formulated as follows:

ON(B, C) & ON(A, B)

For more details on knowledge representation see, for example, Nilsson [3], Waterman [6], and Winston [7].

4 Hierarchy of Expert Systems

Developments in computer hardware have led to their application in many different areas. To provide managers with information, software in the form of Management Information Systems (MIS) has been developed. Since MIS does not take any decision it may be referred to as a passive system. Based on information provided by MIS, Decision Support Systems (DSS), Optimization (OPT) systems, and, more recently, Expert Systems (ES) have been developed. The DSS is appropriate for rather loosely structured problems, while the OPT system is characterized as being appropriate for well-structured problems.

Expert systems can be loosely divided into three categories:

1. Expert Systems (ES). Note that the term "expert system" can mean either a specific category or a class of categories (see Fig. 5)
2. Knowledge-based expert systems of, for short, Knowledge-Based Systems (KBS)
3. Expert Support Systems (ESS).

All the above mentioned three categories of expert systems are compared in Fig. 5.

The expert system and knowledge-based system categories are where most of the current research and development work is concentrated.

An expert system incorporates skills of a domain expert and is typically not able to perform very broad analysis and justification of the decision. This is due to limited knowledge included in the problem domain.

A knowledge-based system employs human knowledge to solve problems that ordinarily require human intelligence [8]. The difference between an expert

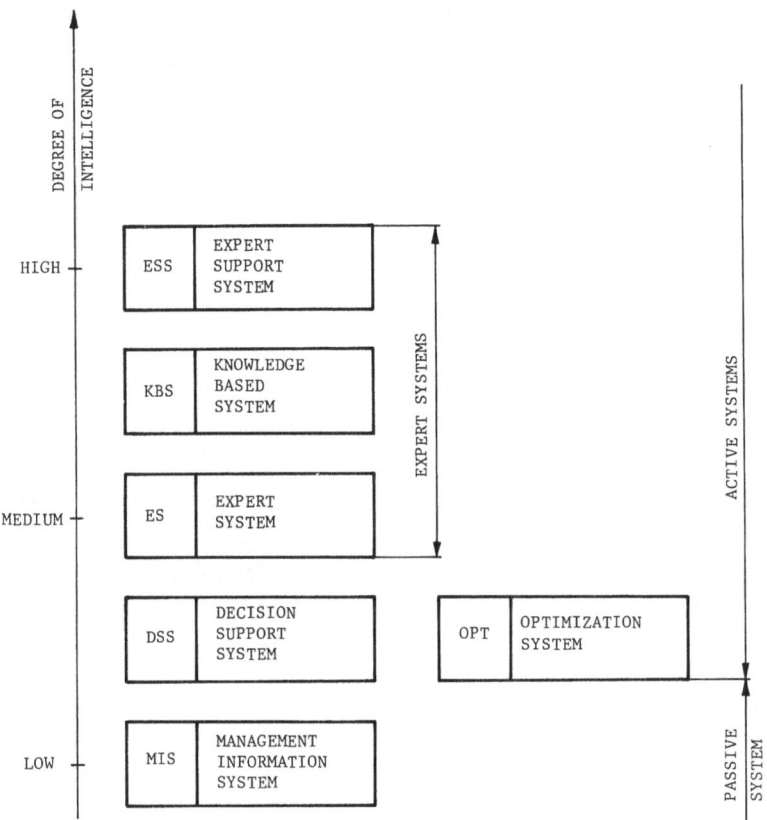

Fig. 5. A relationship among various categories of systems

system and a knowledge-based system is in the breadth of the problem domain. Knowledge-based systems possess a broad knowledge of the problem domain and perform much deeper analysis than expert systems. Expert support systems is a category of expert systems to be developed in the future.

Since DSS, OPT, ES, KBS and ESS take decisions, they have been classified as active systems, as opposed to MIS which is a passive system (see Fig. 5).

The difference between the DSS and expert systems is in the degree of intelligence. DSS can be briefly characterized as a system being able to perform "what-if" analysis. Nowadays, expert systems are able to reason, postulate, learn, justify their advice, and relate interactively with the user [9]. In the future, expert support systems are expected to comprehend, assess, create, and innovate.

5 Artificial Intelligence – Planning and Scheduling

One of the aspects of AI study is planning, i.e. automatic generation of a sequence of actions that transform the world from a given initial state to a desired goal state. Planning occurs in a variety of different environments including production. Traditionally, optimization methods have been used most to solve production planning problems. Due to the high computational complexity of these problems, new approaches have always been appreciated. Typically, in production management literature the term "planning" is associated with a long horizon. For short-horizon planning, the term "scheduling" is used. In this section a scheduling problem in a particular type of a manufacturing system, called a Flexible Manufacturing System (FMS), is discussed. For more details on FMSs see, for example, Kusiak [10–12].

5.1 Scheduling Flexible Manufacturing Systems

The scheduling problem in FMSs has been an intriguing subject for many researchers and practitioners. There are two basic approaches to this problem:

1. Optimization (based on scheduling theory and mathematical programming)
2. Artificial intelligence.

The scheduling papers published so far are discussed in relation to an FMS which consists of machining and assembly subsystems (Fig. 6).

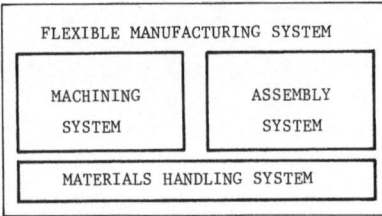

Fig. 6. Structure of a sample FMS

Optimization Approach

Using the optimization approach the FMS scheduling problem can be formulated as follows:
Schedule resources such as: parts, machines, tools, fixtures, and pallets in a way that optimizes the objective function (e.g. the rate of machine utilization) subject to constraints related to the limited resources, precedences, due dates, etc.

Most scheduling papers treat the machining and assembly subsystems independently. One of the first reports on scheduling FMSs has been published hy Hitz [13]. He developed and solved a mathematical programming formulation for the specific type of manufacturing system called a flexible machining line. His periodic release strategy was explored by Erschler et al. [14]. Stecke and Solberg [15] applied sixteen heuristic rules for scheduling an existing flexible machining system. Carrie and Petsopoulos [16] reported on computational results with seven heuristic rules for an existing flexible machining cell.

Chang and Sullivan [17] presented a binary formulation of the scheduling problem for a flexible machining system. To solve this problem they developed

10. Artificial Intelligence Approach to Production Planning

a two-phase suboptimal algorithm. Afentakis [18] showed that under certain assumptions the flexible machining system can be transformed into a flow line type system. Perhaps the first approach, linking scheduling the machining system with the assembly system, is presented by Kusiak [19].

Artificial Intelligence Approach

Most AI scheduling papers refer to the machining system only. Bullers et al. [20] discussed an evaluation-net (E-net) approach to controlling a flexible machining system. The evaluation-net was used to represent and describe the information flow and system control. Fox [21] and Bourne and Fox [22] presented an actual AI scheduling system named ISIS. The system has been tested in an industrial machining environment and is still under development. It employs constraint-directed reasoning and an advanced knowledge representation scheme. Morton and Smunt [23] proposed the PATRIARCH system for a machining system. PATRIARCH incorporates the following four levels: (1) strategic planning, (2) capacity planning, (3) scheduling, and (4) dispatching. The last two levels 3 and 4 have been discussed in greater detail. Shaw and Whinston [24] developed and implemented in LISP a planning and scheduling system for a flexible machining system. The system is able to determine routes and schedule jobs in a stochastic machining evironment.

Ben-Arieh [25] designed an expert system for scheduling a machining and assembly system, where some of the machined parts were designated for assembly. He reported on simulation results for seven different scheduling policies.

5.2 AI Formulation of the Scheduling Problem

An AI-based approach linking scheduling of the assembly system with the machining system is presented. Assume that the aggregated schedule for the assembly system has been specified (see Fig. 7). This aggregated schedule can be disaggregated into the schedule for parts and operations.

As an example consider an assembling and machining structure of the product P_1 presented in Fig. 8. Assembling a single product P_1 requires 2 parts p_1, 1 part p_2, 3 parts p_3, 1 part p_4, and 1 part p_5. Each part consists of 1 to 3 operations. For example there are 3 machining operations o_7, o_8, and o_9 to be performed on part p_5. Operations o_7 and o_8 can be performed in any sequence, but both prior to the operation o_9. Based on the above considerations one can present the AI formulation of the FMS scheduling problem. In this formulation the

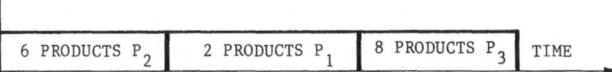

Fig. 7. An aggregated assembly schedule

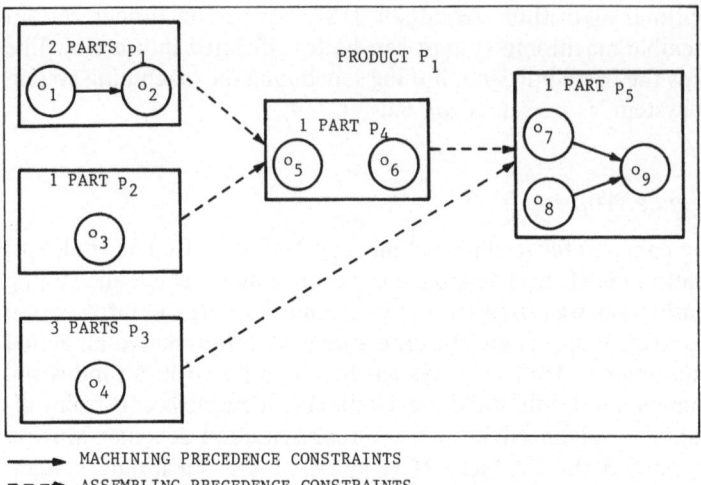

Fig. 8. An assembling and machining structure of product P_1

schedule for the assembly system will be the goal schedule for the machining system.

Given the goal schedule, schedule resources of the machining system such as: parts, machines, tools, fixtures, and pallets in a way that the deviation between the goal schedule and the current schedule is minimized. Of course, while scheduling these resources, the constraints related to their availability, precedences, etc. have to be considered.

5.3 Artificial Intelligence Versus Optimization Approach in Planning

It is worthwhile to exploit some differences between the AI approach and the optimization approach to the planning (scheduling) problem. The first approach involves optimization of the objective function subject to a set of constraints. All decision variables as well as the value of the objective function are determined from the optimization. Some algorithms may require initial (starting) solution.

In the AI approach the goal state as well as the initial state are known in advance (the goal corresponds to the value of the objective function in the optimization approach), while the decision variables result from the search process. The two approaches are compared in Fig. 9.

5.4 Example of an Expert Scheduling System

One of the most known expert scheduling systems is ISIS [21, 26].

The system uses a variety of representation and search techniques for reasoning with constraints. ISIS provides a general solution to the manufacturing

10. Artificial Intelligence Approach to Production Planning

OPTIMIZATION APPROACH	
OBJECTIVE FUNCTION	Determined by optimization algorithm
PARAMETERS AND CONSTRAINTS	Specified prior to optimization
DECISION VARIABLES	Determined by optimization algorithm
INITIAL* SOLUTION	Given prior to optimization

ARTIFICIAL INTELLIGENCE APPROACH	
GOAL STATE	Given prior to search
PARAMETERS AND CONSTRAINTS	Specified prior to search or modified by the search procedure
DECISION VARIABLES	Determined by search procedure
INITIAL STATE	Given prior to search

*required by some algorithms

Fig. 9. A comparison of two approaches for solving the FMS scheduling problem

system scheduling problem and has been developed within the environment of the Westinghouse turbine component plant in Winston-Salem, North Carolina, USA.

Fox [21] has viewed the task of constructing schedules as constraint-directed activity. He makes the point that scheduling is a function of two basic activities, both of which are required to enable the other to be carried out:

1. The selection of a sequence of operations
2. The assignment of start and finish times per operation and the allocation of resources.

Scheduling is constrained by activities and information from all elements of the factory. These constraint sources and their effect upon the schedule have been investigated by Fox and Smith [26]. This study has identified five broad categories of constraints; they are as follows:

1. Organizational goals comprising
   - due dates
   - work in progress
   - shop stability
   - shifts
   - cost
   - productivity goals
   - quality
2. Physical constraints comprising
   - machine physical constraints
   - setup times
   - processing time
   - quality

3. Causal restrictions (i.e. conditions that must be satisfied before a job can start) comprising
   - operation alternatives
   - machine alternatives
   - tool requirements
   - material requirements
   - personnel requirements
   - inter-operation transfer times

4. Availability constraints comprising
   - resource reservations
   - machine downtime
   - shifts

5. Preference constraints comprising
   - operation preferences
   - machine preferences
   - sequencing preferences.

The manner in which ISIS constructs a schedule is by conducting a hierarchical, constraint-directed search in the space of all possible schedules. In this way ISIS uses the wide range of constraints that typically influence scheduling decisions.

The search is conducted at four different levels. At each level the scheduling problem is represented by multiple abstractions each of which in turn is a function of the specific constraint types considered at that level. Communication between levels is achieved by constraint exchange.

The objectives of the search, on the four levels respectively, are:

1. To select an order to be scheduled according to a priority algorithm based upon the category of the order and its due date
2. To perform a capacity analysis of the plant to determine the availability of the machines required by the selected order, using a critical path method
3. To perform detailed scheduling of all resources necessary to produce the order
4. To select and assign reservations for resources required in the schedule.

At each of these levels processing proceeds in three phases:

1. Pre-analysis to determine the bounds of the level's search space
2. Use of search to perform the actual problem-solving in this space
3. Assessment of the quality of the results produced by the search in phase 2. The results are in fact coded as constraints to be used by the next level.

ISIS is able to generate realistic job schedules because it uses a variety of representation and search techniques. The system is a general methodology for incorporating constraint knowledge to enable automatic schedule generation.

6 Tools for Building Expert Systems

Tools for building expert system can be divided into four groups [6] as follows:

1. They can be programming languages, which can be either problem-oriented languages such as FORTRAN and PASCAL or symbol manipulation languages such as LISP and PROLOG.
2. They can be knowledge engineering languages. A knowledge engineering language consists of an expert-system-building language integrated into an extensive support environment. There are two classes of knowledge engineering languages: skeletal systems, for example EMYCIN, KAS; and general-purpose systems, for example PROSPECTOR, MYCIN.
3. System-building tools consist of programs that help acquire and represent the domain expert's knowledge, and programs that help design the expert system, for example AGE, TEIRESIAS, and ART.
4. Support tools consist of facilities which help with programming, such as debugging tools, knowledge-based editors, and facilities that enhance the capabilities of the finished system, such as built-in input/output and explanation mechanisms. These facilities come usually as a part of an expert-system language and are designed to work with that language.

6.1 Automated Reasoning Tool (ART)

Automated Reasoning Tool (ART) is one of the recently developed automated tools which can be used by knowledge engineers to build expert systems. ART's features include [27]:

1. Rule-based programming, featuring the interaction of forward and backward chaining rules with a powerful pattern-matching language
2. Schema-based knowledge representation, which allows the program to reason about objects and their relations to one another
3. A mechanism called Viewpoint for providing a powerful means of modelling hypothetical alternatives and situations that change dynamically with time
4. An interactive development environment called ART Studio which provides monitoring and debugging aids for system development
5. A graphics interface package called ARTIST that enables the user to create a custom interface using animation graphics and mouse-sensitive menus.

In order to build an expert system with ART, a knowledge engineer must learn appropriate information from a human expert and code it in a form acceptable by ART. As soon as the program is generated, its behaviour is evaluated by the expert, who also helps the knowledge engineer expand and refine the ART program. This cycle is typically repeated several times until the program is acceptable (see Fig. 10). Each time the cycle is completed the program becomes more powerful and more intelligent.

A knowledge engineer builds an expert according to the following five-stage methodology [4]:

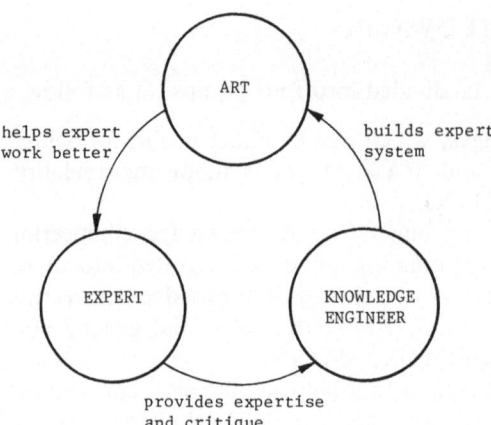

Fig. 10. A development cycle for a typical expert system using ART

Stage 1. Identification: problem characteristics are determined and requirements for the conceptualization phase are specified.
Stage 2. Conceptualization: concepts for knowledge representation are determined.
Stage 3. Formalization: design structure for knowledge organization is defined.
Stage 4. Implementation: rules for other knowledge representation schema for embodying knowledge are formulated.
Stage 5. Validation rules for other knowledge representation schema for organizing knowledge are validated.

All or some of these stages are repeated in each cycle represented in Fig. 10.

7 Future Outlook

The future economic imperatives will pull artificial intelligence in many new directions. Almost all industries and services will be affected. Progress in AI theory will lead to many efficient applications. New automated tools for building AI systems as well as new hardware technology will emerge. The number and variety of AI computers and especially microcomputers will increase. Parallel architectures are now being considered for future AI machines. Expert systems will allow the distribution of scarce expertise, reduce the costs of low human performance, and provide help to humans trying to access information and apply computers [4]. The primary factor limiting the growth of AI is lack of trained people, software tools and easily understood application categories. Many countries have increased their involvement in AI research, notable among these are Japan through MITI, the UK through the Alvey Commission, the European Community through ESPRIT, and the USA through the Strategic Computing Program.

In the 1990s, Japan's fifth-generation computer project is expected to yield an extremely powerful AI parallel machine running PROLOG at one billion logical inferences per second.

8 References

1. Winston, P. H., Perspective. In P. H. Winston and K. A. Prendergast (eds.), The AI in Business: Commercial Uses of Artificial Intelligence. MIT Press, Boston, Mass., 1983, pp. 1–13
2. Ringle, M., Philosophy of artificial intelligence. In M. Ringle (ed.), Philosophical Perspectives in Artificial Intelligence. Humanities Press, 1979, pp. 1–20
3. Nilsson, N. J., Principles of Artificial Intelligence. Tioga Publishing Co., Palo Alto, Calif., 1980
4. Hayes-Roth, F., The knowledge-based expert system: a tutorial. IEEE Computer, September 1984, pp. 11–28
5. Gevarter, W. B., Artificial Intelligence, Expert Systems, Computer Vision and Natural Language Processing. Noyes Publications, Park Ridge, N.J., 1984
6. Waterman, D. A., A Guide to Expert Systems. Addison-Wesley, Reading, Mass., 1986
7. Winston, R. J., Artificial Intelligence. Addison-Wesley, Reading, Mass., 1977
8. Barr, A. and Feigenbaum, E. A., The Handbook of Artificial Intelligence. William Kaufman, Menlo Park, Calif., 1981
9. Chorafas, D. N., Fourth and Fifth Generation Programming Languages, Vol. 1. McGraw-Hill, New York, 1986
10. Kusiak, A., Flexible manufacturing system: structural approach. International Journal of Production Research, Vol. 23, No. 6, 1985, pp. 1057–1073
11. Kusiak, A., Application of operational research models and techniques in flexible manufacturing systems. European Journal of Operational Research, Vol. 24, No. 3, 1986, pp. 336–345
12. Kusiak, A., Modelling and Design of Flexible Manufacturing Systems. Elsevier, New York, 1986
13. Hitz, K. L., Scheduling of flexible flow shops, Report No. LIDS-R-879, Laboratory for Information and Decision Systems, MIT, Cambridge, Mass., 1979
14. Erschler, J., Roubellat, F., and Thuriot, C., Periodic release strategies for FMS. Presented at the TIMS/ORSA Meeting, San Francisco, 1984
15. Stecke, K. E. and Solberg, J. J., Loading and control of item flow in a flexible manufacturing system. International Journal of Production Research, Vol. 19, No. 5, 1981, pp. 481–490
16. Carrie, A. S. and Petsopoulos, A. C., Operations sequencing in an FMS. Robotica, Vol. 3, No. 4, 1985, pp. 259–264
17. Chang, Y. L. and Sullivan, R. S., Real time scheduling of FMS. Presented at TIMS/ORSA Meeting, San Francisco, 1984
18. Afentakis, P., An optimal scheduling strategy for flexible manufacturing systems, Working Paper No. 85-012, Dept of Industrial Engineering and Operations Research, Syracuse University, Syracuse, N.Y., 1985
19. Kusiak, A., FMS Scheduling: a crucial element in an expert system control structure. In Proceedings of the 1986 IEEE Conference on Robotics and Automation, 7–10 April 1986, San Francisco, pp. 653–658
20. Bullers, W. I., Nof, S., and Whinston, A. B., Artificial intelligence in manufacturing planning and control. IIE Transactions, December 1980, pp. 351–363
21. Fox, M. S., Constraint directed search: a case study of job-shop scheduling, Ph.D. Thesis, Dept. of Computer Science, Carnegie-Mellon University, Pittsburgh, Pa., 1983
22. Bourne, D. A. and Fox, M. S., Autonomous manufacturing: automating the job-shop. IEEE Computer, September 1984, pp. 76–86
23. Morton, T. E. and Smunt, T. L., A planning and scheduling system for flexible manufacturing. In Proceedings of the First ORSA/TIMS Special Interest Conference on Flexible Manufacturing System, Ann Arbor, Mich., 1984, pp. 313–326
24. Shaw, M. J. and Whinston, A. B., Applications of artificial intelligence to planning and scheduling in flexible manufacturing. In A. Kusiak (ed.), Flexible Manufacturing Systems: Methods and Studies. North-Holland and Elsevier, New York, 1986, pp. 223–242

25. Ben-Arieh, D., Knowledge based control system for automated production and assembly. In A. Kusiak, (ed.), Modelling and Design of Flexible Manufacturing Systems. Elsevier, N.Y., 1986, pp. 347–368
26. Fox, M. S. and Smith, S. F., ISIS – A knowledge based system for factory scheduling. International Journal of Knowledge Engineering, Vol. 1, No. 1, 1984
27. Clayton, B. D., ART Programming Primer. Inference Corporation, Los Angeles, 1985
28. Adiga, S., Artificial intelligence and its relevance to industrial engineers. In Proceedings of the Annual International Industrial Engineering Conference, Chicago, Ill., 1986, pp. 309–314
29. Bennett, J. L., Building Decision Support Systems. Addison-Wesley, Reading, Mass., 1983
30. Finke, G. and Kusiak, A., Network approach to modelling of flexible manufacturing modules and cells. Revue Française d'Automatique, d'Informatique et de Recherche Opérationnelle (RAIRO-APII), Vol. 19, No. 4, 1985, pp. 359–370
31. Kusiak, A., Flexible Manufacturing Systems: Methods and Studies. North-Holland and Elsevier, New York, 1986
32. Stecke, K. E. and Suri, R., Flexible Manufacturing Systems: Operations Research Models and Applications. J. C. Balzer, Basel, Switzerland, 1985

9 Appendix: Glossary of Expert System Terms

Artificial intelligence. The area of science concerned with developing intelligent computer programs. This includes programs that can solve problems, learn from experience, understand language, interpret visual scenes, and, in general, behave in a way that would be considered intelligent if observed by a human.

Backward chaining. An inference method where the system starts with what it wants to prove, e.g. Z, and tries to establish the facts it needs to prove Z. The facts needed to prove a conjecture (Z) are typically given in rule form; e.g. IF A & B, THEN Z. If A and B are not known (are not available as data), the system will try to prove A and B by establishing any additional facts (as specified by other rules) needed to prove them. The additional facts are established in the same way that A and B were established, and the process continues until all needed facts are established or the system gives up in defeat.

Domain expert. A person who, through years of training and experience, has become extremely proficient at problem solving in a particular domain.

Domain knowledge. Knowledge about the problem domain, e.g. knowledge about geology in an expert system for finding mineral deposits.

Expert system. A computer program that uses expert knowledge to attain high levels of performance in a narrow problem area. These programs typically represent knowledge symbolically, examine and explain their reasoning processes, and address problem areas that require years of special training and education for humans to master.

Expert-system-building tool. The programming language and support package used to build the expert system.

Explanation facility. That part of an expert system that explains how solutions were reached and justifies the steps used to reach them.

10. Artificial Intelligence Approach to Production Planning 165

Forward chaining. An inference method where the IF-portion of rules are matched against facts to establish new facts.

Frame. A knowledge representation method that associates features with nodes representing concepts or objects. The features are described in terms of attributes (called slots) and their values. The nodes form a network connected by relations and organized into a hierarchy. Each node's slots can be filled with values to help describe the concept that the node represents. The process of adding or removing values from the slots can activate procedures (self-contained pieces of code) attached to the slots. These procedures may then modify values in other slots, continuing the process until the desired goal is achieved.

Heuristic. A rule of thumb or simplification that limits the search for solutions in domains that are difficult and poorly understood.

Inference chain. The sequence of steps or rule applications used by a rule-based system to reach a conclusion.

Inferene engine. That part of a knowledge-based system or expert system that contains the general problem-solving knowledge. The inference engine processes the domain knowledge (located in the knowledge base) to reach new conclusions.

Inference method. The technique used by the inference engine to access and apply the domain knowledge, e.g. forward chaining and backward chaining.

Knowledge. The information a computer program must have to behave intelligently.

Knowledge base. The portion of a knowledge-based system or expert system that contains the domain knowledge.

Knowledge-based system. A program in which the domain knowledge is explicit and separate from the program's other knowledge.

Knowledge engineer. The person who designs and builds the expert system. This person is, for example, a computer scientist experienced in applied artificial intelligence methods.

Knowledge engineering. The process of building expert systems.

Knowledge representation. The process of structuring knowledge about a problem in a way that makes the problem easier to solve.

Object-oriented methods. Programming methods based on the use of items called objects that communicate with one another via messages in the form of global broadcasts.

Predicate calculus. A formal language of classical logic that uses functions and predicates to describe relations between individual entities.

Problem-oriented language. A computer language designed for a particular class of problems: e.g. FORTRAN designed for efficiently performing algebrain computations, and COBOL with features for business record keeping.

Production rule. The type or rule used in a production system, usually expressed as IF *condition* THEN *action*.

Production system. A type of rule-based system containing IF – THEN statements with conditions that may be satisfied in a database and actions that a change the database.

Representation. The process of formulating or viewing a problem so that it will be easy to solve.

Rule. A formal way of specifying a recommendation, directive, or strategy, expressed as IF *premise* THEN *conclusion* of IF *condition* THEN *action*.

Rule-based methods. Programming methods using IF-THEN rules to perform forward or backward chaining.

Scheduler. The part of the inference engine that decides when and in what order to apply different pieces of domain knowledge.

Schema. A frame-like representation formalism in a knowledge-engineering language (e.g. SRL).

Search. The process of looking through the set of possible solutions to a problem in order to find an acceptable solution.

Search space. The set of all possible solutions to a problem.

Semantic net. A knowledge representation method consisting of a network of nodes, standing for concepts or objects, connected by arcs describing the relations between the nodes.

Slot. An attribute associated with a node in a frame system. The node may stand for an object, concept, or event: e.g. a node representing the object employee might have a slot for the attribute name and one for the attribute address. These slots would then be filled with the employee's actual name and address.

Tool. A shorthand notation for expert-system-building tool.

Tool builder. The person who designs and builds the expert-system-building tool.

Tools for knowledge engineering. Programming systems that simplify expert system development. They include languages, programs, and facilities that assist the knowledge engineer.

Part IV

The Computerized Production Management System

Chapter 11
Databases

Johan C. Wortmann

Johan C. Wortmann studied industrial engineering and management science at Eindhoven University of Technology, The Netherlands. He has been active in the development of information systems for production/inventory control since 1973 and wrote a doctoral thesis on the subject. He has been involved in a number of practical application, both in component manufacturing and in assembly operations. He worked in the first half of 1985 as visiting professor at Rutgers University, New Jersey, and is currently employed at Eindhoven University of Technology. J. C. Wortmann is a member of IFIP WG 5.7.

1 Introduction

Effective management of production organizations relies heavily on timely and accurate information. The continuous pressure to increase flexibility and to reduce lead times in the production stage requires a quick reaction by many decision centres. Such a quick reaction is only possible if decisions are based on *up-to-date* operational information and on quick exchange of information with related decision centers. The continuous pressure to decrease stocks and to increase turnover rates requires also *accurate* information. This is immediately evident and widely accepted for such approaches as MRP II and OPT, but on second thought it is also true for an approach such as JIT. Although JIT does not require accurate information on floor stocks it does require accurate bills-of-materials, customer orders, etc.

Many operational and tactical decisions require availability of a high volume of operational data. It takes considerable effort to create and maintain these data. Therefore, such a collection of data, a database, represents an important asset. It is more and more recognized that this asset has to be managed – by an organizational function, by adequate procedures, and by technical means. The organizational function is often called the Data Base Administration (DBA). The procedures are concerned with the definition of the concepts (entity types, relationship types, attributes) and with the authorization of users of the database. The technical means embodies a collection of software, designed to manage the database as a recource and to support the DBA. Such a collection of software is called a Data Base Management System (DBMS).

In the remainder of this chapter, we concentrate on the nature of a DBMS in Sect. 2, on the architecture of a DBMS in Sect. 3, and on the definition of the concepts used in the database in Sect. 4. The latter point will be illustrated by examples from the production field. In passing, we will introduce the so-called *relational* approach to databases. Sect. 5 closes this chapter with a number of remarks on other issues related to databases, and with literature references.

2 The Nature of Database Management Systems

The above introduction to databases might wrongly suggest that any collection of data is considered to be a databases. This idea is certainly incorrect. The term "database" has a much more specific meaning, involving at least the following characteristics.

Shared data. Data in the database are shared by various applications. This point is best illustrated by its opposite: the exclusive control and usage of private files by each application. This situation has the disadvantage that uncontrolled redundancy of data may occur. Such an uncontrolled redundancy is often expensive, because multiple data-entry and multiple update is required. Moreover, uncontrolled redundancy leads to inconsistency between various versions of the same data collections. Finally, uncontrolled redundancy leads to ambiguity with respect to the meaning of the concepts used in each application: if data on inventories, operation times, etc. are different, then it is not obvious whether there is a difference in meaning or an error in content.

Data independence. If the physical storage structure of files and the corresponding access methods are hidden for the application programs, the architecture provides data independence. This means that the individual application programs employ standardized commands or subroutines to access data. The advantage of data independence lies in the fact that storage structures and access methods can be changed without affecting existing applications. Therefore, conflicting requirements with respect to the physical storage structure can be balanced while designing the database. Furthermore, a separation between the logical and the technical design is made possible.

Apart from shared data and data independence, databases usually exhibit some other features, such as the following.

Integrity. Maintaining the integrity of the database aims at avoiding update transactions which are probably or certainly incorrect. The most obvious violation of integrity occurs if two contradicting facts are stored. This violation of integrity is sometimes called inconsistency. Much consistency is avoided already by controlling redundancy. Another obvious violation occurs if a value is assigned to an attribute, which is not concordant with the set of values allowed for the attribute (e.g. a value XII for a numerical field, such as weight). Rules of this type are sometimes called *domain integrity rules*. Still another type of integrity rule is called *referential integrity*: this type of rule specifies that an attribute A which refers to an entity type E can only have values occurring as the key of some actual occurrence of type E. For example, if the entity type OPERATION has an attribute CAPACITY TYPE, a referential integrity rule could require that any allowed value for this attribute occurs as the key of an actual occurrence of the entity type FACILITY GROUP. These examples of integrity rules are certainly not exhaustive. Integrity rules are sometimes called *constraints*, which is a more general notion.

Security. While integrity rule protect the database against invalid updates, security rules aim at protection against deliberate misuse. Security is provided by means of authorization measures, which restrict the update and query facilities for certain users. The DBMS provides the DBA with tools to grant users authorization for updates and queries.

Recovery. While integrity and security protect the database against accidental or deliberate user errors, the recovery procedures of a DBMS provide a safeguard against system errors. Such system errors occur due to failures in the hardware or operating system, or due to anomalies in the DBMS itself (such as deadlock). A modern DBMS is equipped with routines to restore the database to a previous state and to redo automatically all transactions from this state onward.

Support of multiple views. Different applications may require a different view of the same data. For example, calendar data may be presented in the MM-DD-YY (month-day-year) format or in the DD-MM-YY format. This is a fairly trivial example, but much more complicated differences in views can be supported by a DBMS.

Performance monitoring. The DBMS provides the DBA with a number of utilities to measure the performance of the database system and to reorganize the internal structure of the database.

The above list of features shows that a DBMS is much more than a file management system, and that a database is much more than an arbitrary collection of data. The architecture of the current DBMS is usually in concordance with the so-called three-level architecture (ANSI/SPARC [1]), which is presented below.

3 The Three-Level Architecture

The three-level architecture distinguishes three different views on the database: the internal view, the external view, and the conceptual view. The internal view describes the internal construction of the database. The external view describes the way in which a specific user sees the database. The conceptul view describes the meaning in the real world of the entities and relationships to be stored in the database. Each view is specified in a so-called *schema* by using a specific language. This language is called the *data-definition language (DDL)*.

The *internal view* is closest to the physical storage structure of the data. The internal view is described formally in the internal schema, by means of the internal DDL. This internal schema specifies the nature of files, the formats of fields in stored records within files, the physical sequence, indexes, etc. The internal schema does not necessarily deal with the physical properties of the storage medium (such as blocks, cylinders, overflow areas). Therefore, the internal schema often relies on a file management system within the operating system in order to cope with the physical properties of the storage medium. In addition to the internal schema, a *mapping* has to be specified, which defines the correspondence between the internal schema and the conceptual schema.

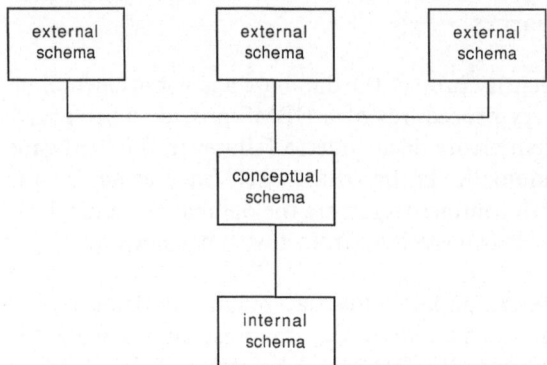

Fig. 1. Three-level architecture

The *external view* is closest to the users of the database. The external view specifies how the user sees the database, in terms of e.g. entity types, attributes of entity types, and relationships between entity types. For example, PART, SUPPLIER, and PROJECT could be entity types, and STOCK-ON-HAND could be an attribute of PART; and CONTRACT could be a relationship among PART, SUPPLIER, and PROJECT. The users of a database are either application programs or end-users interacting "directly" from a terminal with the database.

If the user is some application program, this program is written in a programming language, e.g. COBOL. Such an application programming language is called a *host language*. For accessing the database, the host language is extended with statements for retrieval and update of data from the database; these statements constitute the *data manipulation language (DML)* which is embedded in the host language. In many current implementations, the DML takes the form of a call to subroutines. The DBMS provides the application programmer with a library of subroutines for several host languages.

If the user is an end-user interacting "directly" with the database, the user enters commands on his screen in a so-called *data sublanguage*, such as SQL (see

Chamberlain and Boyce [2]). Many modern DBMSs provide powerful data sublanguages, enabling end-users to retrieve and update data in a non-procedural way. From a technical point of view, however, the user interacts with another application, which interprets his (sometimes rather complex) commands and operates on the database. This application, the data sublanguage interpreter, is a part of the DBMS software collection and may be written in any suitable language. As has been indicated in the previous section, different users may employ *different external views* on the database. Several factors contribute to the need to support these different view:

- *Partial view*. Many users are only interested in a (small) part of the database. For reasons of simplicity and security, these users are only aware of such a part, which is sometimes called a Subschema.
- *Derived view*. Many users are not interested in the data as stored in the database, but in some derived collection of data. This derived form is often an abstraction, in which certain details of the data are omitted. For example, if the database contains inventory data on parts in different storage locations, a specific external view may contain an attribute STOCK-ON-HAND which equals the sum of available stock on all locations.
- *User-specific syntax*. An external view should always specify in what format and in which units of measure specific data should be provided by the DBMS to the end-user or application. Our earlier example on the format of calendar dates is concerned with this issue. The way in which the end-user or application receives and provides the data is called the syntax of the data.

The external views are specified as external schemas by means of an external data definition language (external DDL). In addition to each external schema, a *mapping* should be specified, which defines the correspondence between the external schema and the conceptual schema.

The *conceptual view* describes the entire structure of the database, and not just a part. Its focus is not on the internal storage, and not on the external syntax but on the meaning of the concepts used in structuring data. This is sometimes called the *semantics* of the data. The syntax of the data is rather uninteresting at the conceptual level. It should be noted that a database can be considered to represent the state of affairs in a part of the real world. This part of the real world is sometimes called the *universe of discourse* [3]. The conceptual view describes the pieces of data which can be obtained on the universe of discourse. It also specifies the integrity rules which apply to any actual state of affairs in the universe of discourse.

In a production environment, the universe of discourse consists of entity types such as products, capacities, routings, suppliers, customers, and orders, together with relationships among these entity types, and integrity constraints. The conceptual view serves to define all these concepts in a precise way. This precision greatly improves the communication between different information systems design teams, because it provides a common language. Many authorities feel that the conceptual view should not only specify the semantics, but also the *pragmatics*. In other words, it should not only specify the meaning of the complete set of operational data, but also the usage of these data in applications and the exchange of information between users. This would ultimately lead to a complete enterprise model with well-defined boundaries between applications [4]. Although the future will perhaps bring such an enterprise model, we will not speculate on this issue here.

The conceptual view is described formally in the conceptual schema, by means of a conceptual data-definition language (conceptual DDL). The nature of the conceptual schema and of the conceptual DDL are discussed in more detail in the next section.

Before concluding this section, some attention has to be paid to the way in which all the above schemas and rules are stored. We have introduced many ways to describe data, at the internal, external, and conceptual level, and to relate these descriptions to each other through mappings. Together, this collection of descriptions and rules are called *metadata*. The metadata are stored in a so-called *data dictionary/directory* (DD/D) which is a considerable help for the DBA (Data Base Administrator). It is fascinating that the DD/D is itself often stored in a (separate) database, enabling the DBA and other users to address it through the normal facilities of the DBMS. However, the conceptual schema of this DD/D-database is usually predefined. This schema may be called the *metaschema*. It is an interesting exercise for the reader to think about the structure of this conceptual schema of the DD/D-database (perhaps after reading the next section).

4 Conceptual Modelling

The conceptual view of a database describes the universe of discourse. It tells us what kind of entities do exist in the real world. It tells us about the properties of these entities, and their allowed values. It specifies relationships and constraints. Thus, the conceptual view describes all possible states of the universe of discourse. In order to model the universe of discourse into a conceptual view, we need a language. Many such languages have been proposed (for an overview, see Tsichritzis and Lochovsky [5]). For the purpose of illustration, we shall use a fairly simple language and diagramming technique here. Our aim is to demonstrate the process of conceptual modelling, together with an explanation of a simplified version of a conceptual model often encountered in practice.

Anything of interest in the universe of discourse is called an *entity*. Entities can have properties, which will be called *attributes*. We assume that entities can be clustered into *entity types*; an entity type consists of all entities with the same set of attributes. Examples of entity types are products, customers, suppliers, orders, etc. Each entity belongs to precisely one entity type.

As mentioned, entity types can participate in *relationships*. Often, such relationships restrict the allowed set of values in any actual database which corresponds to the conceptual view. Consider for example the following situation:

- The conceptual schema contains the entity types OPERATION and FACILITY GROUP.
- The entity type OPERATION has an attribute CAPACITY TYPE.
- The entity type FACILITY GROUP has a key attribute GROUP CODE.
- A relationship may specify that operations cannot be defined for non-existent facility groups; more formally, each value of the attribute CAPACITY TYPE for any actual occurrence of an OPERATION-entity should occur as a value of GROUP CODE in an actual occurrence of a FACILITY GROUP-entity.

This type of relationship is called an *N-to-1* relationship (or a *functional dependency*). More generally, there exists an N-to-1 relationship from entity type *A* to entity type *B*, if each *A*-entity is associated with exactly one *B*-entity; conversely, each *B*-entity may be associated with an arbitrary number of *A*-entities. The above example can be characterized therefore as an N-to-1 relationship from OPERATION to FACILITY GROUP. Alternatively, it is said that FACILITY GROUP is functionally dependent on OPERATION.

Entity types and their N-to-1 relationships can be depicted by so-called *data-structure diagrams* [6]. In such diagrams, entity types are represented by boxes (rectangles). The N-to-1 relationship from entity type *A* to entity type *B* is represented by an arrow from box *B* to box *A*. The reader is warned that other conventions are elsewhere in use, but this convention seems to be most generally accepted. Figure 2 shows a simplified data structure for a production situation.

Figure 2 shows an entity type PRODUCT ITEM. Each item defined in the company is represented in the data structure by an entity of this type. In order to produce items, the company employs internal work orders. These real-world entities are represented in Fig. 2 by the entity type WORK ORDER. Each work order is released for the manufacture of a lot size of a specific item; therefore, each work order entity is associated with one product item entity. Conversely, a product item entity may be associated with an arbitrary number of work order entities. In other words, there is an N-to-1 relationship from WORK ORDER to PRODUCT ITEM.

For manufacturing of an item, specific operations are required. In order to emphasize that these operations exist independently of the existence of any work orders, such operations are called normative operations. The corresponding entity type is shown in Fig. 2 as NORMATIVE OPERATION. Again, there is an N-to-1 relationship from NORMATIVE OPERATION to PRODUCT ITEM. It is left to the reader to explain in detail the meaning of this relationship. Fig. 2 shows also an N-to-1 relationship from NORMATIVE OPERATION to FACIL-

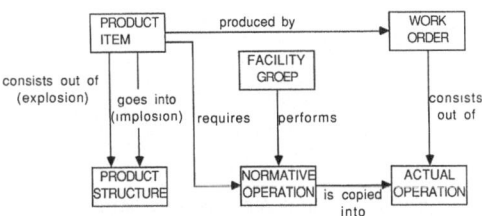

Fig. 2. A simplified data structure diagram

ITY GROUPS. This relationship has been explained above while introducing the N-to-1 relationship. Whenever a work order is created, the normative operations of the associated item are copied as actual operations and attached to the work order. This is shown in Fig. 2 by the entity type ACTUAL OPERATION. The entity type ACTUAL OPERATION has an N-to-1 relationship towards NORMATIVE OPERATION, and another one to WORK ORDER.

Before discussing the entity type PRODUCT STRUCTURE in Fig. 2, a number of new concepts have to be introduced. Until now, we have been assuming implicitly that the only relationships between entity types are N-to-1 relationships. Generally speaking, this is incorrect. Many other relationships can be defined, depending upon the modelling language employed. The most important relationship is the so-called *N-to-M* relationship between two entity types A and B. This relationship represents the fact that any entity of type A can be associated with an arbitrary number of entities of type B and vice versa. Such a relationship is sometimes represented by a line with an arc at both ends. However, in most cases the need is felt to associate attributes with such a relationship – which makes it difficult to distinguish this type of relationship from a genuine entity type. For example, the entity type ACTUAL OPERATION in Fig. 2 could be depicted as an N-to-M relationship between NORMATIVE OPERATION and WORK ORDER. Due to the fact that it has attributes of its own, such as ACTUAL DURATION and YIELD, we prefer to model this N-to-M relationship as a new entity type with two N-to-1 relationships. As mentioned, this is often the case.

We are now in a position to take a closer look at product structure data, or *bills-of-material* as it is commonly called. Other terms are "goes-into" data, "parent – component" relations, etc. Each product item, as a parent, is assembled out of an arbitrary number of other product items, called the components. Conversely, each product item – as a component – is consumed by an arbitrary number of parents. This situation can be modelled as an N-to-M relationship between the entity type PRODUCT ITEM and *itself*.

For a more detailed study of this point, we depict a simple product structure in Fig. 3. The capital letters represent product items, and the arrows represent goes-into relations. This product structure could be made visible at the external level by means of a display as in Fig. 4. This representation is called an *extension* of the database, whereas Fig. 2 shows an *intension*.

Upon closer examination of Fig. 4, it becomes clear that each bill-of-material relation (i.e. each product structure entity) is associated with two product item entities: one parent item and one component item. On the other hand, each prod-

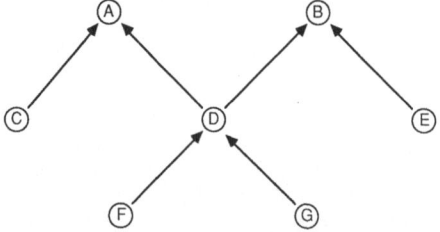

Fig. 3. Example of a product structure

Entity type	Attributes			Entity type	Attributes			
Product item	code	stock-on-hand	etc.	Product structure	code parent	code cmpnt	quant.	etc.
	A	25			A	C	1	
	B	40			A	D	3	
	C	13			B	D	1	
	D	2			B	E	10	
	E	88			D	F	4	
	F	0			D	G	0.5	
	G	7							

Fig. 4. Extended data structure of Fig. 3

uct item entity can act as a parent item in an arbitrary number of product structure entities; similarly, it can act several times as a component item. This reveals why Fig. 2 shows two N-to-1 relationships from PRODUCT STRUCTURE to PRODUCT ITEM. These two relationships are often called the *explosion* and the *implosion* relationship.

This concludes our discussion of conceptual modelling in relation to Fig. 2. It is worthwhile to note that the structure shown in Fig. 2 is at the heart of the vast majority of current databases in the production area.

5 Concluding Remarks

5.1 The Three Dominant Approaches

It is often stated that there are three dominant approaches to the design of a DBMS: the *relational* approach, the *hierarchical* approach, and the *network* approach (see Date [7] for an excellent treatment of all three). The above statement suggests that the three approaches are completely comparable, which should be doubted. More specifically, the relational approach is nowadays increasingly accepted as a proper foundation for an end-user data sublanguage. Accordingly, it is more and more used as an external DDL and a conceptual DDL. Consequently, the vendors of DBMS-packages have developed a relational "layer" around an internal structure which is based on another approach. We shall shortly characterize the relational approach, which is largely based on the work of Codd [8].

In the relational approach, the database is seen as a collection of tables, such as in Fig. 4. These tables are called *relations*, which explains the term "relational approach". We will avoid this term, however, in order to prevent confusion with the term "relationship" used above. Each table represents an entity type, and the columns of a table correspond to the attributes of the entity type involved. The ordering of the columns is irrelevant. The rows of the table represent the individual entities; the ordering of the rows is, again, irrelevant. The interesting point is that *all* information is stored in such tables. Relationships between entities can only be stored by attributes with related values, and never by addresses of data. For example, even if an N-to-M relationship does not have any additional attributes, such a relationship is stored in a table as the right-hand table in Fig. 4. In the other approaches, the first two columns of this table would probably contain address-information, such as pointers, but in the relational approach this is impossible.

The fact that all information is stored in tables, and that the relationships are restricted to value-relations, enables the relational model to employ high-level data-sublanguages. Codd defined two "pure" types of languages: the so-called relational algebra, and relational calculus. Many currently available languages to operate "directly" on databases are a mixture of these types.

5.2 Normalization

Another topic to be discussed briefly is normalization. This is a process followed in the conceptual design phase in order to avoid certain anomalies. First of all, a conceptual design should eliminate all "repeating groups". For example, if work orders in Fig. 2 were not considered to be separate entities, they would have been represented as part of the PRODUCT ITEM entity type. In this case they would constitute a repeating group. Such a situation cannot be represented easily in the form of tables. Therefore, such a repeating group is described as a separate entity type in a separate table. Next, the conceptual design should aim at tables, in which all non-key attributes are fully functionally dependent on the key-attribute. Sometimes, a conceptual schema contains partial or transitive functional dependency within one entity type. It can be shown that this has leads to (conceptual) problems in updating. The process of normalization can be interpreted as a process of decreasing the size of tables until only fully functional dependencies remain.

5.3 Distributed Databases

Finally, a few words seem to be justified about distributed databases. Usually, the term "distributed databases" refers to the situation where the physical storage media are remote from each other, and where there is a network connecting these media. The key idea is that a distributed database should ideally look like a non-distributed one. In other words, the fact that the database is distributed should be coped with at the internal level and not at the conceptual level or the external level.

Especially in a manufacturing environment it can be doubted whether this approach is always justified. Some researchers would argue that a deliberate partitioning of the database which is recognized at the conceptual level is to be preferred, both from an organizational point of view [9] and from a design point of view [10].

6 References

1. ANSI/X3/SPARC Study Group on Data Base Management Systems, Interim Report. FDT (ACM SIGMOD bulletin), Vol. 7, No. 2, 1975
2. Chamberlain, D. D. and Boyce, R. F., SEQUEL: a structured English query language. In Proc. ACM SIGMOD Workshop on Data Description, Access and Control. Available from ACM
3. Van Griethuysen, J. J. (ed.), Concepts and Terminology for the Conceptual Schema and the Information Base, ISO/TC97/SC21 – N 197. ANSI, New York, 1985
4. Flavin, M., Fundamental Concepts of Information Modelling. Yourdon, New York, 1981
5. Tsichritzis, D. H. and Lochovsky, F. H., Data Models. Prentice-Hall, Englewood Cliffs, N.J., 1982
6. Bachman, C. W., Data structure diagrams. Data Base (journal of ACM SIGBDP), Vol. 1, No. 2, summer 1969. Available from ACM
7. Date, C. J., An Introduction to Database Systems. Addison-Wesley, Reading, Mass., Vol. I 1985 and Vol. II 1983
8. Codd, E. F., A relational model of data for large shared data banks. CACM, Vol. 13, No. 6, June 1970
9. Pels, H.-J. and Wortmann, J. C., Decomposition of information systems for production management. Computers in Industry, Vol. 6, No. 6, December 1985, pp. 435–452
10. Yeomans, R. W., Choudry, A. and ten Hagen, P. J. W., Design Rules for a CIM System. Elsevier, Amsterdam, 1985
11. Bertrand, J. W. M. and Wortmann, J. C., Production Control and Information Systems for Component Manufacturing Shops. Elsevier, Amsterdam, 1981
12. Ullman, J. D., Principles of Database Systems. Computer Science Press, Rockville, Md., 1982
13. Weldon, J.-L., Data Base Administration. Plenum Press, New York, 1981

Chapter 12
User Interface

Eero Eloranta

Dr. Eero Eloranta has a M.S. from Helsinki University of Technology in Industrial Engineering. He received the degree of Ph.D. in 1981 at Helsinki University of Technology, the Laboratory of Information Processing Science. His major fields of interest are production management and distributed information systems. He has previously spent one year at Eindhoven University of Technology (The Netherlands). At the moment he is working as a project leader of "Distributed Production Management Systems" in the national Finprit development program in Finland. He is also working as a lecturer in industrial management and in information processing science at Helsinki University of Technology. Dr. Eloranta is in IFIP WG 5.7 and is representing Helsinki University of Technology in the CAM–I factory management program.

1 Interface Problems

Any production management system (PMS) can be divided into three interconnected parts:

- User interface
- Application (with the necessary systems software connections)
- Data (or knowledge) base.

Traditionally, the major attention has been paid to the application software, i.e. all the application-oriented production management operations. A drawback of such focusing was inefficient and inconsistent data organization and user interface. Gradually, with the development of database management systems, some of the problems in data management were eliminated, even though distributed parallel processing and knowledge-based data representation still cause a lot of problems, even at the theoretical level.

User interface problems are still relevant for the effective use of a PMS. Clumsiness in interaction, inconsistent communication conventions even within a single application system, and inefficient communication procedures are more or less common in practice. This is not to say that the present character-oriented user interfaces are inferior in their domain, but more to emphasise the inadequacy of such a limited domain in general.

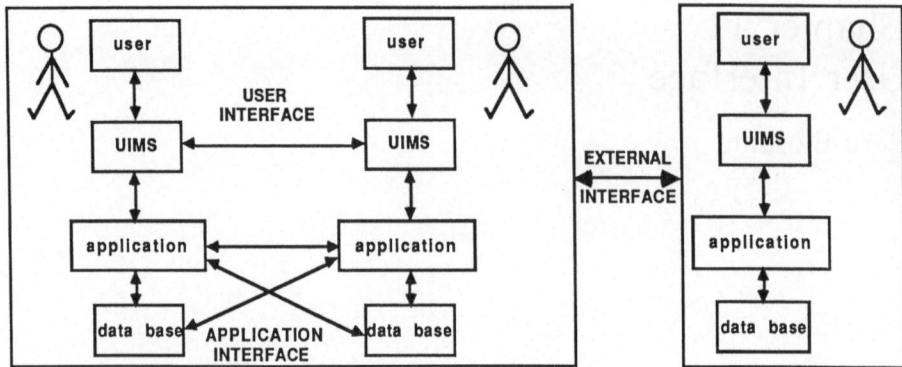

Fig. 1. Interface problems: user interface, application interface, and external interface

In addition to the user interface problem, the state-of-the-art in PMS interfaces covers other related interface problems: application interface and external interface. These three types of problems are illustrated schematically in Fig. 1.

User interface problems relate primarily to the boundary surfaces between the application software and the user on one hand, and between the database and the user on the other. In the state-of-the-art, direct communication with the database is seldom possible from the user level, which emphasizes the potential for further development. Quite recently, a discussion about a uniform user interface management system (UIMS) has originated, not only among CAD systems (see e.g. Takala, [1]) but also in the PMS area.

Application interface problems relate to the boundary surfaces between application software and database on one hand and between different applications on the other. The latter problem has been subjected to reconsideration lately, partly as a by-product of multi-function application generation software, partly as an implication of the rise of distributed production management.

External interface problems are concomitant with modern data technology, i.e. increased machine intelligence on factory floors and distributed processing in planning and monitoring applications. External interface problems relate to the interfaces between different machinery, including different computers. This problem area is tackled by the efforts of international standardization among users (e.g. MAP [2] and TOP [3]), standardization organizations (e.g. ISO, [4]) and computer, robot, and machinery manufacturers, but commercial domination based on the *de facto* standardization is vital as well (e.g. SNA and the related systems – program to program connections, among other things).

Our emphasis in this paper lies in the *user interface* problem, leaving aside the problems in the application interface and external interface.

12. User Interface

2 The Evolution of the User Interface

2.1 Logical Design

Traditionally, before the era of computers in production management, the user interface was based on various archives, manual or poorly mechanized, copiers, and manually maintained wall charts of schedules (networks or Gantt charts). Most activities in production management bureaus were labour-intensive. However, the advantages of those systems were the *visuality* of the wall charts and the close access and flexibility of the manual archives (any measurable data could be recorded without any major update efforts in the "information system").

The very first computer-assisted production management systems, appearing in markets primarily in the 1960s, were provided with user interfaces that were not even equal to the contemporary manual systems, especially in their input procedures: instead of any rationalization in data collection, some additional tasks were originated. This together with slowly responding *batch*-oriented *processing* and lack of capabilities in efficient data management are among the most important explanations for the masses of failures of PMS projects in those days.

The outputs were represented as *printer listings,* one example of which is presented in Fig. 2.

The second stage in evolution, commencing approximately in the 1970s, can primarily, be characterized as the origin of transaction-based *real-time* systems. As far as the user interface is concerned, the input/output facilities were expanded with display terminals. In a short run, this advance was nothing but a replacement of a batch-processed printer listing by a display panel by reducing information to match the limited size of a CRT (e.g. from $n \times 132$ to 24×80 characters). The progress from the batch-oriented mode of data processing into some really transaction-oriented processing was very slow, partly because of the inertia

```
09/02/67                          REQUIREMENTS GENERATION                                    PAGE 01

ITEM 03867   DESC    BEARING ASSEMBLY   TYPE A   U/M A   BALOH 15   SFSTK 05   ALLOC 05   SHRKF 05
OPCD E       ORDQY         MINOQ 10   MAXOQ 100   MLTOQ 05   LTCD M   LDTM 10

              09/11/67   09/25/67   10/09/67   10/23/67   11/06/67   11/21/67   12/07/67   12/20/67
GROSS            30         25         40         35         25         40         40         45
OPEN ORD         50
NET                                    42         37         27         42         42         48
PLAN ORD                    85                               70                    90
OFFSET           85                                70                    90


ITEM 4965H   DESC    FLAME WASHER       TYPE P   U/M A   BALOH 165  SFSTK 40   ALLOC      SHRKF
OPCD D       ORDQY 180  MINOQ       MAXOQ        MLTOQ        LTCD P   LDTM 20

              09/11/67   09/25/67   10/09/67   10/23/67   11/06/67   11/21/67   12/07/67   12/20/67
GROSS            65         55         85         65         75         85         70         80
OPEN ORD
NET                                    80         65         75         85         70         80
PLAN ORD                               180                   180                   180
OFFSET          180                    180                   180
```

Fig. 2. An example of the outputs of the early production management systems: printer listing of a materials requirements planning system [5]

```
        MLI VS FORECAST/ORDERS   ITEM TYPES: EXPLICIT        PLANNER: 00901      AMM351 J4
        -   ITEM    - - ENG/DRAW NO - -      DESCRIPTION     - UM   VENDOR    AVAILABLE
        99001                              SPRAY UNIT              EA            202
        START DATE: 10/17/78   CURRENT DATE: 11/07/78 ·          SAFETY STOCK:    21
                       REQUIREMENTS PLAN              ANTICIPATED DEMAND          EXPECTED
        SEQ#   DATE    S   QTY    VS DEMAND  GREATER FORCAST BACKLOG REFERENCE   INVNTORY
               6/09/78                20-        20               20   C003148      182
               6/09/78                35-        35               15   C003148      167
        0020  11/07/78       300      49       216     216                           43-
        0010  11/08/78       200     249                                             43-
              11/14/78                33       216     215                           85
              11/21/78               183-      216     216                          131-
        0010  11/22/78       200      17                                             44
              11/29/78               199-      216     216                          172-
        0010  11/30/78       225      26                                             53
              12/06/78               190-      216     216                          163-
        0010  12/07/78       225      35                                             52
              12/10/78                35                         40   C003149        62
              12/10/78                35                         60   C003149        62
              12/10/78                11       240                   140 C003149     38
              12/13/78               205-      216     216                          178-
                                                     CK01 RESTART-PLANNER  CK05 CHG/DELETE
                    CONTINUED     ENTER DATE  000000 CK02 RESTART-ITEM     CK06 NEXT ITEM
                                                     CK04 ADJ              CK24 END OF JOB
```

Fig. 3. An input/output form of a transaction-processing-oriented PMS [6]

among software vendors caused by the heavy investments in the previous generation of application software.

In transaction processing the display terminals were considered not only as output media but also as input and even input/output panels. An example from this era is presented in Fig. 3. In short, the appearance of the form in Fig. 3 can fairly be characterized as the *state-of-the-art in the user interface* of production management systems. As can be seen on the basis of Fig. 3, the target of ergonomic efficiency of display terminal forms was difficult to achieve. It was also customary that just one subject at a time could be displayed. In slightly more sophisticated versions the screen area was divided into sections, e.g. the primary operation at the top and a secondary (often an inquiry) at the bottom.

The very first display panels were designed and programmed simply with the aid of formating rules provided in the host application programming language. Such a working method was highly labour-intensive and error-prone. As the development of appropriate systems software for display from programming proceeded, better methods – first *subroutine packages* and thereafter more and more advanced *display generators* – were adopted.

2.2 Input/Output Devices

The development of data capturing in PMSs has not been linear. In the very first production management systems no special equipment was adopted. In most cases data were first written on special forms, thereafter submitted to a punch card preparation department and finally fed into a computer via a punch card

12. User Interface

reader. This tedious procedure implied an inevitable recession in the development from the previous manual counterparts of production management systems, both in costs and in processing times — even in data reliability, because each of the multiple steps in the data input procedure was a potential source of errors. The throughput time of an application run was dominated by frequent sort and merge operations, necessary because of the serial data organization methods inherent in magnetic tape data storage. This in turn together with infrequent periodic batch runs inflated the response time and thus the quality of the user interface.

Line printers were applied as the physical means of data output in the early production management systems. Compared with the previous manual alternative (type- or handwritten lists) such a change could really be considered as an advance. Yet, it provided no substitute for the previous, manually produced graphics.

With the enhancement of transaction processing, specially designed factory data terminals were developed. Today, this equipment can be characterized as follows:

- General-purpose terminals
- General-purpose terminals for heavy conditions (humidity, dust, temperature, electric disturbances, etc.)
- On-line/off-line (factory) terminals with
    - keyboard and alphanumeric display (potentially limited character set)
    - card readers (cartoon, plastic, metal, etc.)
    - optical character recognition (OCR) equipment
    - bar code equipment
    - speech recognition equipment
    - combinations of the options above
- On-line/off-line portable factory terminals with options described above
- Event-driven automatic data-capturing equipment.

PMS data input in process industries dramatically differs from that in discrete manufacturing. Ever since the 1970s, the masses of input data in process industries have been collected by means of (automatic) sensors for the various types of physical and chemical variables. This in turn has facilitated the automatic output of control variables, by such as mechanical relays, electronic switches, etc. Thus, the level of algorithmic control of production systems can be considered to be more ambitious in process industries than in the highly heuristically controlled discrete processes — partly because of the interface problems with the physical production system.

Event-driven *automatic data-capturing* equipment has begun to be adopted in discrete manufacturing together with the exploitation of direct numerically controlled (DNC) machinery. However, so far, this potential for direct computer input/output between administrative data-processing systems and e.g. welding, machining, and assembly systems is extremely rare. It can be expected that a wider use of event-driven automatic data capturing and control will be correlated with increasing/widespread applications of robotics in discrete manufacturing.

Factory terminals have in many cases evolved as a by-product of some other product development, such as cashier, bank, and security systems. In the 1970s one of the primary obstacles before widespread use of dedicated terminals was cost. However, modern hardware technology (optical and laser code readers) is rapidly changing the situation. Additionally, the development of LANs (*local area networks*) will reduce the total costs of such a communication system in future. As for the future in the factory environment, optical fibres in data communication will improve transmission reliability together with extensions in channel capacity.

Bar code technology is in a stage of commercial breakthrough. A reasonable level of *de facto* industrial standardization (Code 39) in the different codes has been reached [7]. Bar code data-capturing technology has proven extremely beneficial in the fields of consumer products. The logistics chain from a manufacturer to a consumer could be controlled in all the intermediate stages (operations at the manufacturing site, each step of transportation, wholesaler, retailer) by means of a single code. However, in a pure manufacturing environment there is no such tag on the side of a product capable of increasing the value added in a product. Thus, even though the market for bar code technology can be expected to grow extensively, it will never beat solutions where any product marking has been totally avoided in the spirit of just-in-time manufacturing.

Portable data-collection equipment has its merit in some dedicated tasks, such as the physical count of inventories. However, such instruments, binding a person in the data collection task, can be considered ideal only in a few cases, because of costs and reliability.

2.3 Representation Capabilities

Traditionally, the user interface of a PMS has been capable of carrying information presented in *textual* form, i.e.

- Strings
- Alpha-Numeric fields.

In the last few years lots of tools for the generation of displays and reports containing textual information have been developed. This development has proven beneficial both in the quality of input/output forms and in design efficiency. However, much further progress cannot be expected in this area because of the shortcomings of the textual data representation itself, which lacks most of the variety of cognitive association mechanisms.

Computer *graphics* in production management can be exploited in two different modes:

- Visualisation graphics
- Interactive graphics.

Visualization graphics stands for the display of data objects and attributes as graphical symbols (histograms, bar, line, and pie charts, curves of different shapes, icons, miniatures, etc.). *Interactive* graphics is complementary to visualization graphics. It covers not only the display of data objects in graphical form but also the *manipulation* of objects in their graphical representation.

12. User Interface

In the state-of-the-art in production management there is not much of any form of graphics applied in practice. Some rough visualization graphics, e.g. histograms of machine loading, made by line printers, have been used for years. After the launching of spreadsheet software, more advanced visualization graphics have been adopted, primarily in a mainframe-microcomputer environment.

The role of interactive graphics is modest in the state-of-the-art. Yet, there are some symptoms of development in that direction, in project planning systems (e.g. MacProject, MsProject, Quicknet), flow shop scheduling [8] and graphical loading [9], to mention some application areas.

The application of *voice* in the input and output of data is in the experimental stage. Yet, so far, the price/benefit ratio is far from acceptable in most industrial applications. We even believe that this application field is perhaps not one of the most beneficial for such information technology, primarily because of the shortcomings of the nature of verbal information representation. Instead, for text-processing applications, voice input and output will apparently offer tremendous opportunities – also in a factory environment.

In the following section we focus on the applications of computer graphics in PMS, which we consider as the most promising direction in the future development of representation capability, at least in the short run.

3 Graphics and Object Orientation – a Potential for the Future

3.1 Representation Capability

The application of graphics, both for visualization and interaction, requires a wide variety of features in the user interface, e.g.

– Icons
– Unstructured figures
– Structured pictures
– Miniatures
– Windows, subwindows
– Fixed and arbitrary multi-windows
– Transient subwindows
– Pop-up menus
– Intelligent fields.

Production management systems themselves give rise to some *application-oriented features* in representation capability, such as

– Forms
– Intelligent forms
– Networks (e.g. CPM and Gantt charts)
– Rules, regulators, and associated yardsticks, etc. (in knowledge-based systems).

These features are beyond the commercial state-of-the-art in the user interface of PMS. However, research laboratories are experimenting with all the features described above – and their factory management applications.

3.2 Information Value of Graphical Data Representation and Manipulation

Research results are far from unidirectional concerning the merits of computer graphics in decision support applications. According to an extensive literature study by DeSanctis [10] there is no consensus on

- Superiority of tabular vs. graphical representation
- Implications of graphics in decision efficiency and effectiveness
- Relative merits of different visualization forms (histograms vs. pie-charts vs. line charts, etc)
- Benefits of colour vs. monochromatic representation.

The conflicting research results have been interpreted to reflect that the most appropriate data representation

- is based on the *personal* cognitive and decision-making style and reasoning procedures among individuals
- is *task* dependent, i.e. information representation should match the particular decision-making problem type

However, the research results gained so far concern the application of visualization graphics. Extensive empirical evidence of the merits/demerits of interactive graphics is still lacking. Our experience, although completely insufficient to serve as statistically acceptable evidence, will favour graphical interaction instead of textual manipulation in several production-planning tasks.

Visualization Graphics

In the light of the lack of clear evidence of the superiority of data visualization in decision making, non-interactive computer graphics ought to be considered more as a complement to, rather than a substitute for traditional tabular data representation. Modern multi-function spreadsheet software, for example, provides a uniform multi-view opportunity for the focusing of numerical decision-making problems (Fig. 4).

An example where visualization graphics is applied with success is given in Fig. 5. Fig. 5 (a), about the ABC analysis of materials items, shows clearly the problems met in the particular company in the inventory turn of class B. Fig. 5 (b), in turn, shows the abnormal behaviour of cost accumulation of a make-to-order product. Figure 5 shows nicely the facility for *two-dimensionality* of graphics in data representation: e.g. in Fig. 5 (b) both the abnormal S-shape and the timing (= the second month in the project) of the problems can easily be identified.

12. User Interface

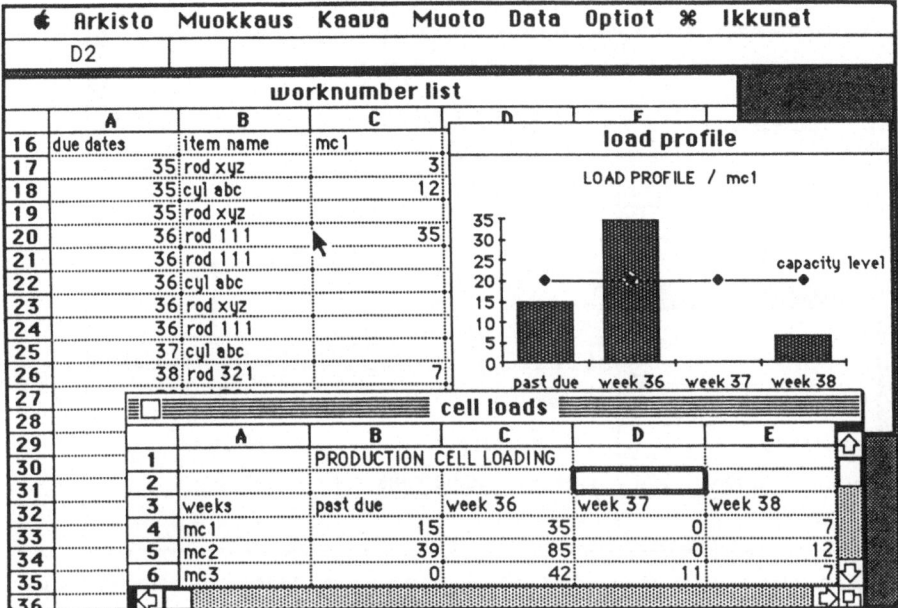

Fig. 4. A multi-view of the same data elements — a sample session on Excel [11]

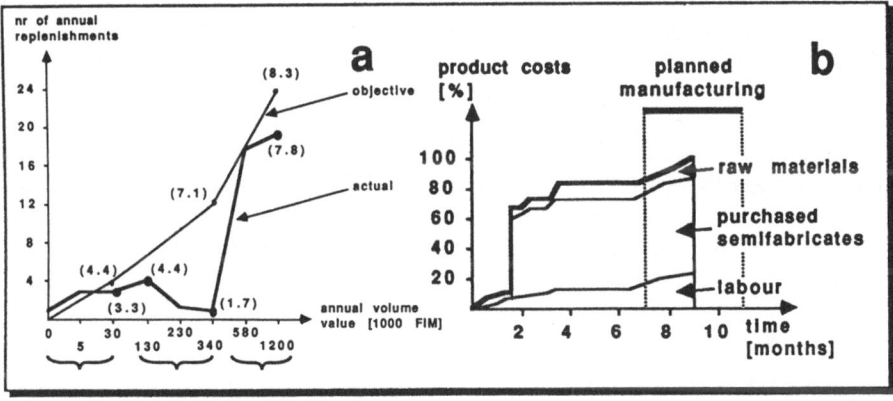

Fig. 5. Applications of visualization graphics in production management: (**a**) ABC-analysis of materials items (average inventory turn in each volume value interval presented in parentheses); (**b**) a cost accumulation curve of a make-to-order product, showing abnormal double-S shape

Interactive Graphics

There are a variety of production management tasks most naturally represented in graphical forms. One of the most favourable examples is any timetable, such as a Gantt chart of parts manufacturing and assembly of a product. The manipulation of such a timetable was performed very often as a graphical "lego-play" before the era of computers. This procedure was logically simple and the

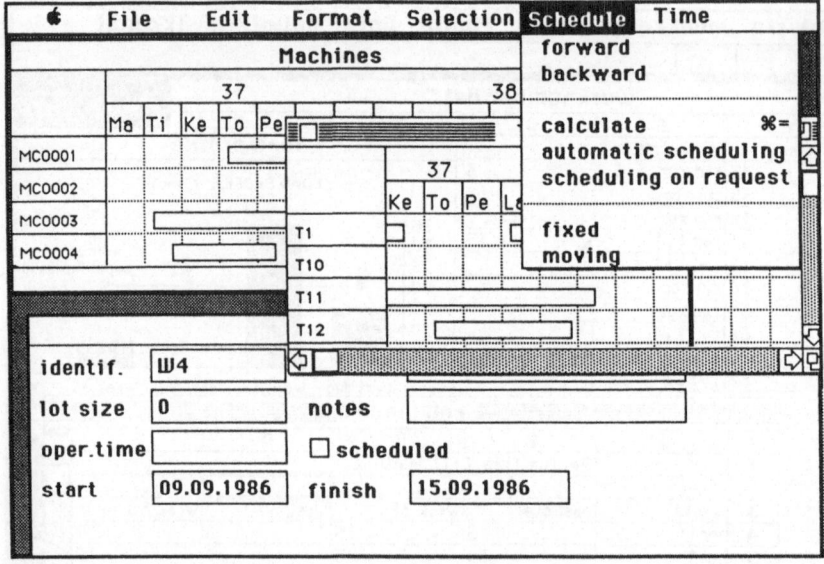

Fig. 6. A sample session of a loading and sequencing system. LOADtool, based on the idea of graphical interaction in the manipulation of scheduling objects

result easy to understand compared with their later computer-aided(?) counterparts, in which e.g. the rescheduling of a work operation requires the typing of different codes (work number, machine code, operation code, etc.) and perhaps even two compound operations (delete the previous work order, create a new one) from the user side.

Our limited experience of graphical interaction in planning tasks confirms the intuitive idea of the easiness of learning graphical interaction instead of some textual interaction. Our primary workbench has been a graphical loading and scheduling system, LOADtool [9], a session of which is presented in Fig. 6.

Visualization and Interactive Graphics Combined

Even though there are not enough theoretically justified guidelines for the best possible option in graphical data representation, it can be expected that combination of options will also be oriented towards multiple decision-making styles. Figure 7 gives an example where visualization graphics and interactive graphics are combined. Figure 7 represents a layout simulator, LOSIM [9], where the materials flow of a factory for a simulated production plan is *animated* (the background window). In addition to the heuristic results of simulation, a more analytical orientation in the statistics of the simulated materials flow is provided, represented in a textual form in this case. It is possible to reach even further and represent the simulated event data. In this example the foreground window shows the distribution of the behaviour of a particular simulated variable — queue length of a particular machine centre in this case.

12. User Interface

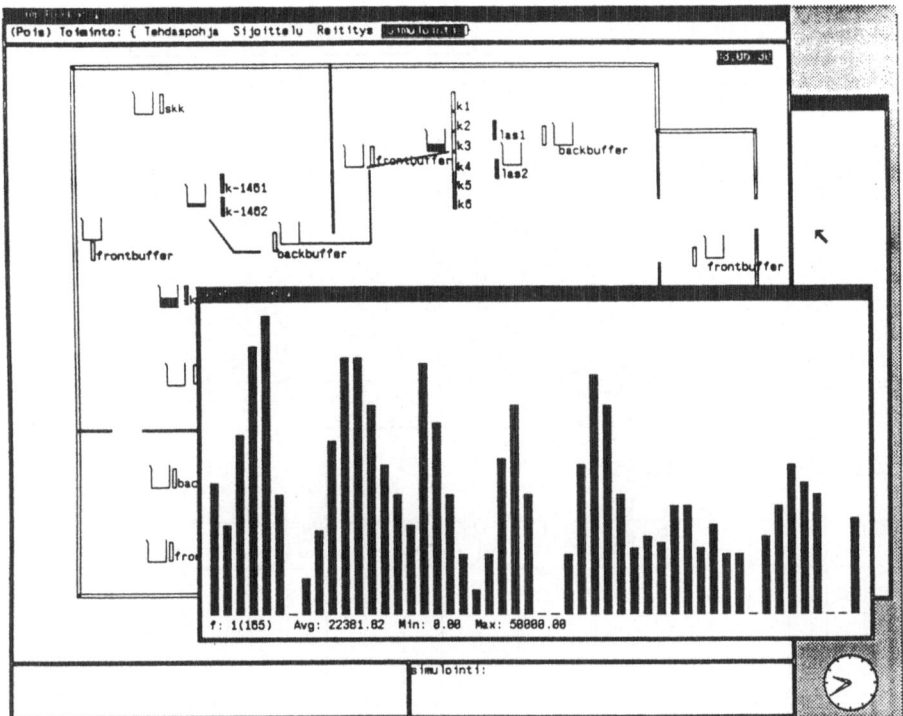

Fig. 7. Visualization graphics, interactive graphics, and character-based statistics combined: a sample session of a layout-simulator, LOSIMtool

3.3 Graphical Manipulation

Interactive graphics can be applied for different functions in production management:

- Graphical transfer
- Graphical editing
- Graphical arithmetic
- Graphical modelling.

Transfer functions cover such operations as

- Move
- Point and position
- Cut and paste

on graphically represented objects. An example of a system based merely on transfer functions is a graphical tender analysis system of electric distributor panels (Fig. 8).

In addition to move, cut and paste, *graphical editing* covers functions for some elementary

- Reshaping

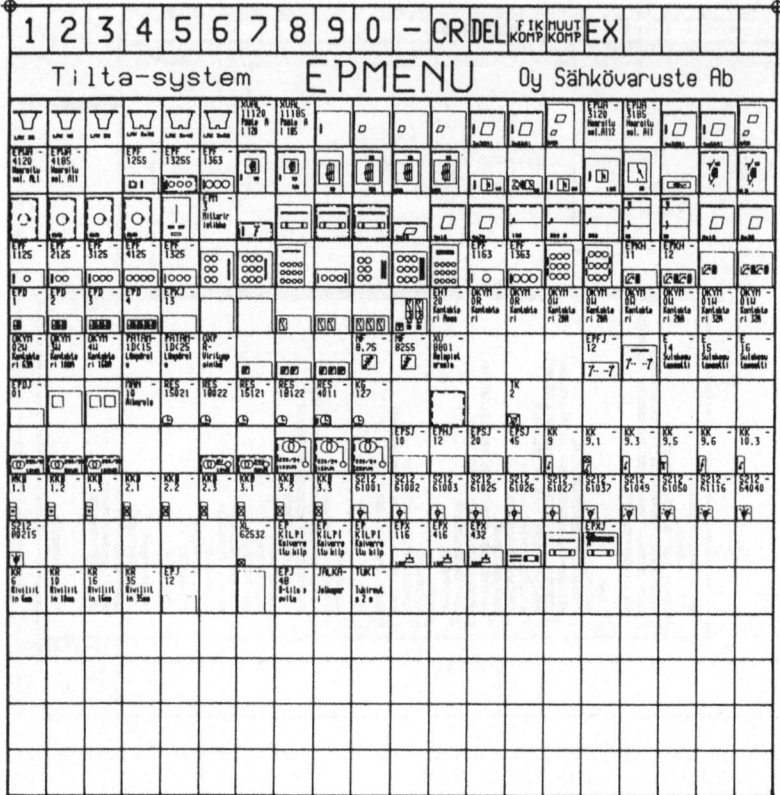

Fig. 8. A graphical tender analysis form for electric distributor panels

of graphically represented objects. Reshaping provides an opportunity to update some attributes of data objects. A loading system for single-item production should support the presence of different load profiles [12] for different stages of production or for loading of different groups of specialists. If these profiles can be reshaped via graphical interaction, we can talk about interaction according to some graphical editing functions.

Graphical arithmetics extends the palette of graphical interaction capabilities by

– Calculation

functions. For example, graphical arithmetic could make it possible to apply some algebra for operations between load profiles (also unary operations). Operations, such as the calculation of the total costs or manpower of load profiles could be applied.

Graphical modelling is today an essential feature of an CAD system. Graphical modelling, applied in the PMS framework is less clear. However, in network planning systems (see, e.g. MacProject [13]) or graphical product structuring systems (e.g. STRUCTtool [14]) some elementary graphical modelling is

12. User Interface

of major value. In layout planning tasks some more advanced graphical modelling capability is a necessity (see e.g. Fig. 7).

3.4 Working Environments

In the state-of-the-art in production management the *working environment* of a production planner is a heterogeneous composition of computer listings, a character-oriented terminal, a note pad and a pen for storing intermediate results, a desktop (or pocket) calculator, perhaps some manual archives, etc. In the most unfortunate cases one terminal has proved insufficient, and different terminals to communicate with different systems have been required.

We believe that some of these data and computational objects can be represented in a more homogeneous working environment. This idea is a direct quotation of working environments introduced first in office automation applications (e.g. [15]).

There are different roles bound to different production management tasks. Some of these roles are located even outside the organization of the production plant. It is quite obvious that the different roles should be supported by different working environments, because the relevant operations and data sets are different in each role. Among the most important roles, we should mention:

- Production planner
- Buyer
- Designer
- Process planner

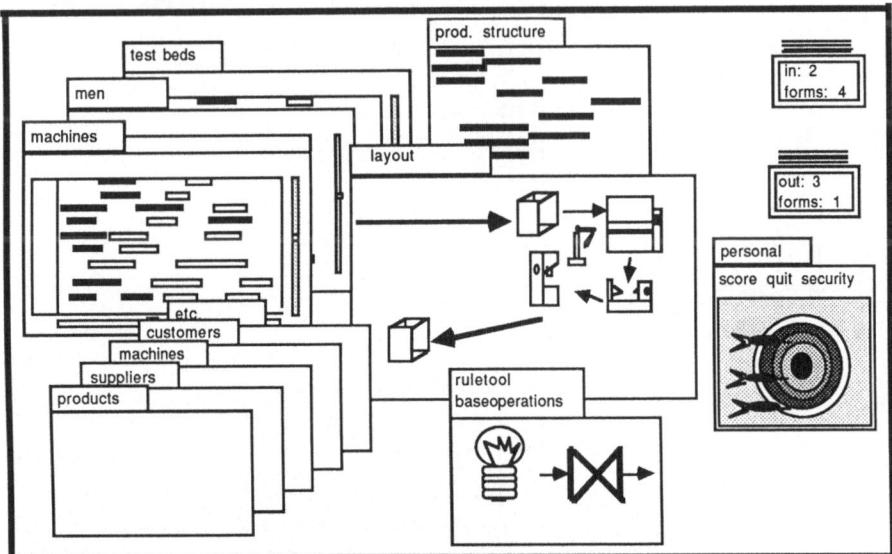

Fig. 9. An imaginary working environment of a production planner

- Production manager
- Plant supervisor
- Subcontractor
- Customer.

The last two are examples of organization external interest groups!

Figure 9 is a fictitious illustration of some of the central ideas related to the working environments of future factory management systems — the working environment of a production planner in this case. In the figure both elementary building blocks and basic design tools as well as some applications can be identified in a uniform but versatile working environment.

3.5 Hierarchical Representation of Objects

Reconsidering Fig. 9 — or any complicated information system — it would be unrealistic even to try to represent all the objects to be manipulated and operated as a single-level model. Thus a multi-level representation is a necessity. Factory description, for example, is a composition of machine and assembly shops,

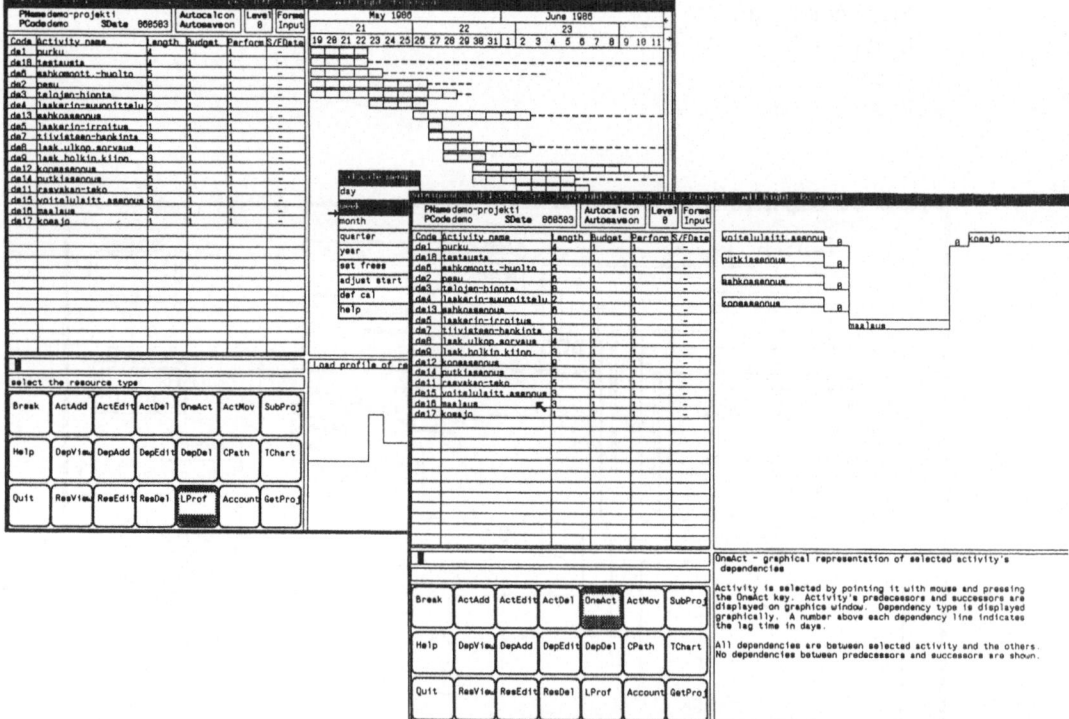

Fig. 10. A project planning and monitoring system, PRETTY, supporting the hierarchical representation, of project activities

which in turn represent compositions of machine groups, product lines or production cells. Each group is a composition of machines, etc.

It is not the task of the user interface software to make any consideration on the depth or context of representation hierarchy. Instead, the interface software should provide the user with tools for the specification of the desired structure.

Figure 10 is an example of the means to implement the ideas of hierarchical representation and manipulation of objects in the graphical working environment. This example deals with a project planning and monitoring system, PRETTY [16]. The foreground window represents a subproject of the project in the background window − an activity explosion with the necessary interconnections among the schedule, budget, and resources.

3.6 Object Orientation

The decomposition and composition of *object* representations should be supported by the user interface of PMS. In addition to the direct vertical hierarchy, horizontal data representation should be supported as well. In this context horizontal representation refers to the *aspects* associated with a certain object. For example, a production *cell* has an association with

- *Data aspects* (machine data, layout illustration, production plan (graphical Gantt chart or CPM representation), customers and suppliers, materials items (raw mtrl., semifabr., prod.), inventory, product structures, order book, materials requisitions, manning, tooling, etc.)
- *Application-related base operations* (loading, bill-of-materials explosion and implosion, due date assignment, etc.)
- *General base operations* (intelligent-form specification, (form) mail functions, *ad hoc* database operations, etc.)
- *Application software related to the object* (any piece of software code with cross references with the object under consideration)
- *Representation at the user interface,* e.g. as an intelligent form.

It is possible to specify object-related windows according to the concept of an intelligent form (see Section 3.7). An object-related window represents the object as a whole with the corresponding subobjects and the related aspects. Some of the aspects can be objects themselves.

3.7 Intelligent Form and Intelligent Mail

Most of the factory management systems today have been constructed with the aid of some application screen-formatting tools. Unfortunately, such forms are relatively dull, which means that their specifications are equipped with just minor field validation means.

On the other hand, intelligent form-based tools are capable e.g. of forming flow specifications, form- and field-related data manipulation, and database interface. Flow code could be implemented via object-oriented pointing operations

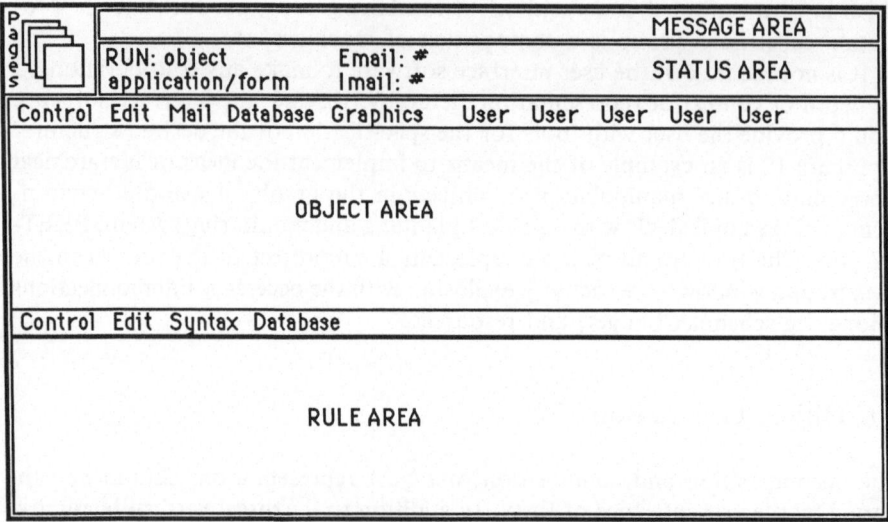

Fig. 11. An illustration of the intelligent-form mail system MAILtool

with a mouse. Multi-windowing capabilities make it possible to handle inter-form communications thus facilitating even multi-form application generation without actual programming. In terms of programming conventions, intelligent forms can be considered as one example of object-oriented programming style.

Given the capabilities of the intelligent form it seems natural to enlarge its application area for communication purposes. This idea is a consequence of organizationally distributed production management. We have a strong belief that a major part in future factory data processing is bare *communication*.

The more advanced electronic mail facilities can be provided with form specifications, with the respective form intelligence. Electronic *intelligent-form mail* is a means to implement distributed production management. The "electronic secretary" may handle quite a lot of routine repetitive message-related operations. She can even initiate some processes based on knowledge at her disposal about some regularly appearing forms. In fact, there are several types of mail objects associated with different treatments in a distributed system, clarified e.g. by Tsichritzis [17].

Figure 11 is an illustration of a form mail system, MAILtool [18], based primarily on the intelligent-form generator PAGEStool [19]. It is at laboratory experimentation stage at Helsinki University of Technology.

3.8 Simulation and Animation

One of the common characteristics of the present R & D efforts in production management systems is the orientation towards simulation and knowledge engineering. Both of these topics, but especially simulation, rely on the visualization capabilities of

12. User Interface 197

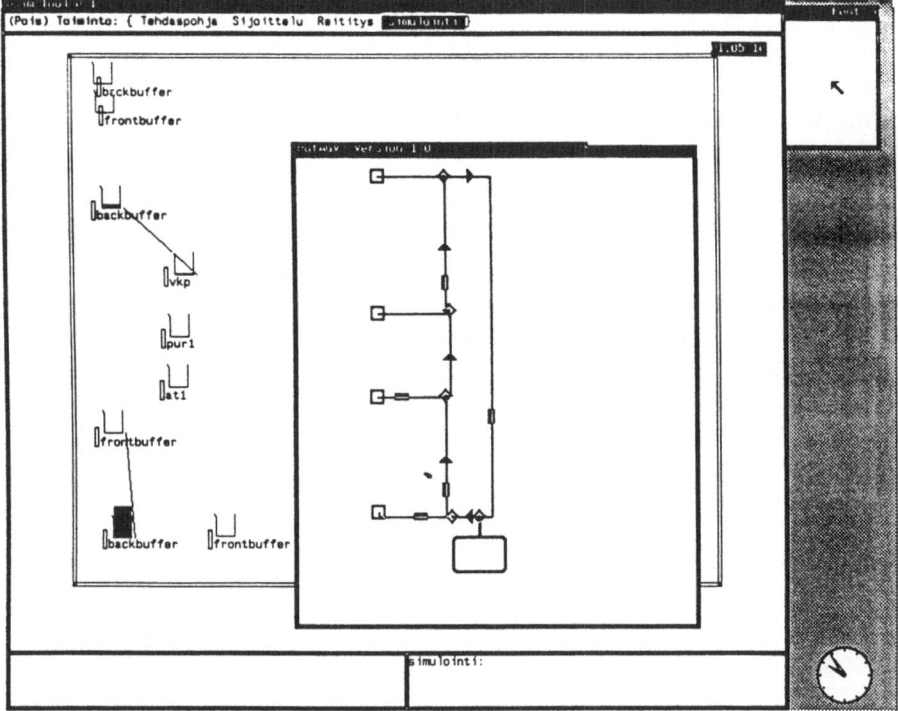

Fig. 12. Two simulators connected online: a layout simulator, LOSIM (background window), and an AGV transportation system simulator (foreground window)

- The object system under study (e.g. factory layout)
- Simulation run
- Simulation results analysis

in the user interface of PMS. The visualization of simulation runs has brought along an opportunity of simple *animation* of the production process phenomena under study. This approach should be encouraged.

Inoue [8] has developed a flow shop simulation system with animation features. Figure 12, in turn, covers a tandem simulation with animation! A general job shop simulation is visualized in a layout simulator, LOSIM [9], in the background window. Interestingly, this simulation has an online process connection with a transportation system simulator [20] representing the behaviour of an AGV transportation system.

4 Conclusions

This paper has dealt with the state-of-the-art and the near future of the user interface of production management systems. The state-of-the-art could be characterized as a logical design and physical devices dedicated to string and alphanumeric field manipulation.

The background to the future process of change is electric. It has its origin in manufacturing technology, management conventions, organizational arrangements, and, finally, in information technology. Interactive raster graphics with multi-windowing capabilities constitute a technical prerequisite for advanced user interfaces – radically more versatile than those familiar in the state-of-the-art. However, the problems behind the improved user-friendliness and visuality are technically complicated. Additionally, lots of standardization efforts in all the aspects of interfaces of production management are required.

History provides, however, a lesson for the evolution progress. The heavy investment in the systems representing the state-of-the-art is a source of inertia opposing the drive for change. Thus there is apparently time enough to overcome the technical problems posed to the development of more human-oriented user interfaces of production management systems.

6 References

1. Takala, T., User interface management system with geometric modeling capability: a CAD system's framework. IEEE CG&A, April 1985, pp. 43–50
2. Kaminski, M., Protocols for communicating in the factory. IEEE Spectrum, April 1986, pp. 56–62
3. Farowich, S., Communicating in the technical office. IEEE Spectrum, April 1986, pp. 63–67
4. International Standardization Organization (no author). Open Systems Interconnection – Basic Reference Model, ISO 7498, ISO, 1984
5. IBM (no author). System/360 Requirements Planning Application Description Manual, Publication GH20-0487-3, IBM, 1st edn., 1968
6. IBM (no author). Introducing Advanced Manufacturing Applications; Manufacturing Accounting and Production Information System (MAPICS), Publication GH30-9002-0, IBM, 1983
7. Hurwitz, D., Data capture for MRP II systems. In Autofact 1985 Conference Proceedings, North-Holland, 1985, pp. 17–36
8. Inoue, I., Yamada, Y., and Adachi, T., A tools system in decentralised production management systems. Computers in Industry, Vol. 6, No. 6, 1985, pp. 465–476
9. Eloranta, E., Hynynen, J., Hämmäinen, H., Jahkola, J., Kyhälä, A., and Räisänen, J., A workbench for distributed production management systems. Computers in Industry, Vol. 6, No. 6, 1985, pp. 413–425
10. DeSanctis, G., Computer graphics as decision aids – directions for research. Decision Science, Vol. 15, No. 4, 1984, pp. 465–487
11. Microsoft (no author). Microsoft Excel User's Guide. Microsoft Corp., 1985
12. Buffa, E., Modern Production/Operations Management, 6th. edn., Wiley, 1980
13. Kaehler, C., MacProject. Apple Computer Inc., 1984
14. Eloranta, E., Mankki, J., and Raunio, R., Graphical visualisation and manipulation of product structures. APMS '87 Conference, North-Holland, 1987 (to appear)
15. Smith, D., Irby, C., Kimball, R., Verplank, B., and Harslem, E., Designing the Star user interface. BYTE, April 1982, pp. 242–282
16. Opas, J., PRETTY – A Workstation Based Project Planning and Monitoring System, Reference Manual. Helsinki Univ. of Technol., 1986 (in Finnish)

17. Tsichritzis, D., Mail Objects, Working Paper. Université de Genève, 1985
18. Hämmäinen, H., Intelligent Mail and its Application in Factory Management Systems, Lic. Tech. Thesis, Helsinki Univ. of Technol., 1986
19. Hämmäinen, H., Jahkola, J., and Kyhälä, A., Intelligent form as a user interface for end-user oriented application generator. In The Future of Command Languages, IFIP W.G. 2.7 Working Conference, North-Holland, 1985
20. Maamies, T., An Interactive Planning and Simulation System for AGVs, Ms. Sci. (Tech.) Thesis, Helsinki Univ. of Technol., 1986 (in Finnish)

Chapter 13
Systems Analysis Techniques

Guy Doumeingts*

Guy Doumeingts is Professor in the field of CIM (Computer Integrated Manufacturing) at Bordeaux Institute of Technology, France. He obtained a Ph.D. in 1964 and a higher doctoral thesis in 1984 at the University of Bordeaux. His research topics deal with the GRAI method, a design method for CIM and with techniques to control a manufacturing unit. He is presently Deputy Director at the GRAI Laboratory of the University of Bordeaux (1). He has managed over 20 industrial research contracts and has published over 80 papers in the field of CIM. Professor Doumeingts is a member of IFIP TC 5 and WG 5.7.

1 Introduction

In today's challenging industrial climate, the search for efficiency in production cost, quality, flexibility, and due dates compels us to design, implement and operate manufacturing systems in which increasingly sophisticated techniques are involved.

For an industrial user, the right choice of equipment and software components for an advanced manufacturing system becomes a major issue. In the past, to invest in manufacturing equipment was a difficult but controllable exercise for managers. Nowadays, to invest in an advanced manufacturing system means a cost in the range $ 1 – 45 million ($ 45 million for the latest, most ambitious project in France), which can make one hesitate. One cannot afford a mistake, as a single mistake may lead one's company to bankruptcy.

In our field, the difficulties increase with the specific conditions:

- Knowledge about manufacturing system
- Capability of design methods
- Process of designing a concrete system initially based on concepts.

---

* The author wishes to thank Eric Poumeyrol, Didier Darricau, and Michel Roboam for their kind cooperation on this paper.

As far as manufacturing is concerned, the knowledge does not exist in one mind, as the above-mentioned manufacturing systems have numerous functions which require skills from various sectors. We have then the new situation of manufacturing system design: the close relationship between the products to be manufactured, the facilities and means necessary to manufacture them, and the management of the whole system. We can sum up this first point and say that the knowledge is not in one mind, but in several: we must collect and organize it.

Our experience in the field of methods for designing manufacturing systems leads us to search for methods to describe manufacturing systems and also to formalize them. The difficulties arise from the very features of the manufacturing systems:

- *The need to model:* from a global viewpoint through a top-down approach (because it is an organizational system with several aspects), and in detail through a bottom-up approach (in order to obtain accurate running)
- *The need to study,* not only the static but also the dynamic behaviour, particularly when a disruption occurs. This comment emphasizes the role of simulation in the design process. We think it is necessary first to design the model from a static point of view (*structuring*) then to study the behaviour over time from a dynamic point of view (*simulation*).

This paper tries to show an overview of the methods and tools and conceptual model used to design production management systems. The paper is divided into two parts:

- State-of-the-art
- Attempt at a classification.

2 State-Of-The-Art in Design/Specification Methods and Tools for Manufacturing Systems

A method consists of three basic elements:

- A conceptual model: an abstract scheme which gives an invariant representation of a system for a problem-solving class
- An application methodology: represents the practical aspects of the method (implementation steps of the concept, human intervention, etc.)
- Tools and representation rules.

Below, we present the state-of-the-art of conceptual models, then methodologies and tools.

2.1 Conceptual Models

We need models to describe and formalize manufacturing systems in the design process. These models can be used as a reference and a guide all along the design process. Such models are called conceptual models because they describe the manufacturing systems in terms of concepts and relationships. We present three significant cases: CAM−I, NBS, GRAI.

The CAM−I Conceptual model

According to the starting procedures of the other projects, CAM−I set up in 1976 an interest group on job shop control. This group adopted in 1978 an approach for developing functional specifications for advanced job shop control systems. The functional specifications for an advanced factory management system were proposed in March 1979. The methodology to design an advanced factory management system described in these specifications is divided into nine steps:

1. Development of a conceptual model of factory activities and their management
2. Definition of a hierarchical structure for factory management
3. Description of the events and activities to be managed
4. Determination of the basic algorithmic solutions and information flow
5. Definition of interface requirements to other high-level systems
6. Evaluation of available hardware components which could be used for data collection and processing
7. Description of user/system interfaces and requirements for a factory command language
8. Assessment of the economic feasibility of implementation
9. Identification of design factors from a human point of view.

There are two basic modes of discrete parts manufacturing:

− Batch production
− Line production.

Batch production involves the non-continuous processing of a variety of parts in small groups or lots. Typically, a common set of production resources is used to manufacture a large variety of part types. Line production involves a continuous flow of similar parts through dedicated work-stations and is generally associated with a high volume of production.

In step 2 (definition of a hierarchical structure for factory management) the CM− divide factory management for controlling batch production into four levels.

Factory level: entire production facilities for one or more end-products are planned and checked at this level.

Job shop level: entire production facilities for manufacturing detailed parts or components are planned and checked at this level.

Work-center level: the use of production facilities for performing a particular type of process or processes on a variety of part batches is planned and checked at this level.

Resource level (or unit level, or shop floor level): tools, materials, and other items necessary for the performance of specific processes or tasks are checked at this level.

Because of the hiearchical structure of the control system, the planning horizon becomes smaller at lower levels of control. In step 1, CAM− has developed a conceptual model of factory activities and their management, then for each level described in step 2 we have some basic activities. For each level there are three basic types of decisions required to perform the managerial function:

− To decide that something should be done
− To decide what is to be done
− To decide how and when to do it.

The results of these three basic decisions lead to a management directive to those doing the work. There are four fundamental steps required to perform the management directive once it has been received:

− To acquire the necessary resources and information to do the job
− To execute the plan
− To compare the actual results with what was requested
− To deliver the results.

In fact, CAM−I gives seven activities at each level:

Decide: a decision of action to be taken based on higher-level requirement and constraint
Design: development of a conceptual end-product ordered by the decision
Plan: a sequential ordered set of instructions to achieve the conceptual product
Acquire: monitoring and control of the actual receipt of items, both hand goods and information, used to manufacture the product according to the sequential ordered set of production
Make: monitoring and control of the execution of the instructions, an action that consumes time and energy to manufacture a product
Verify: to make the comparisons of the actual products obtained from executing a planned, sequential, ordered set of instructions with the intended concept, and report the difference
Deliver: monitoring and control of the distribution of the completed products.

The resulting model is a generic conceptual representation of the information and activities required to manufacture a product (Fig. 1). We have this representation of management activities for each level defined by CAM−I. Then they defined a hierarchical structure for the factory management (Fig. 2). CAM−I has studied separately each level and defined a conceptual model for every one:

− Dacom model for work-center level
− Pritsker model for shop floor level.

13. Systems Analysis Techniques

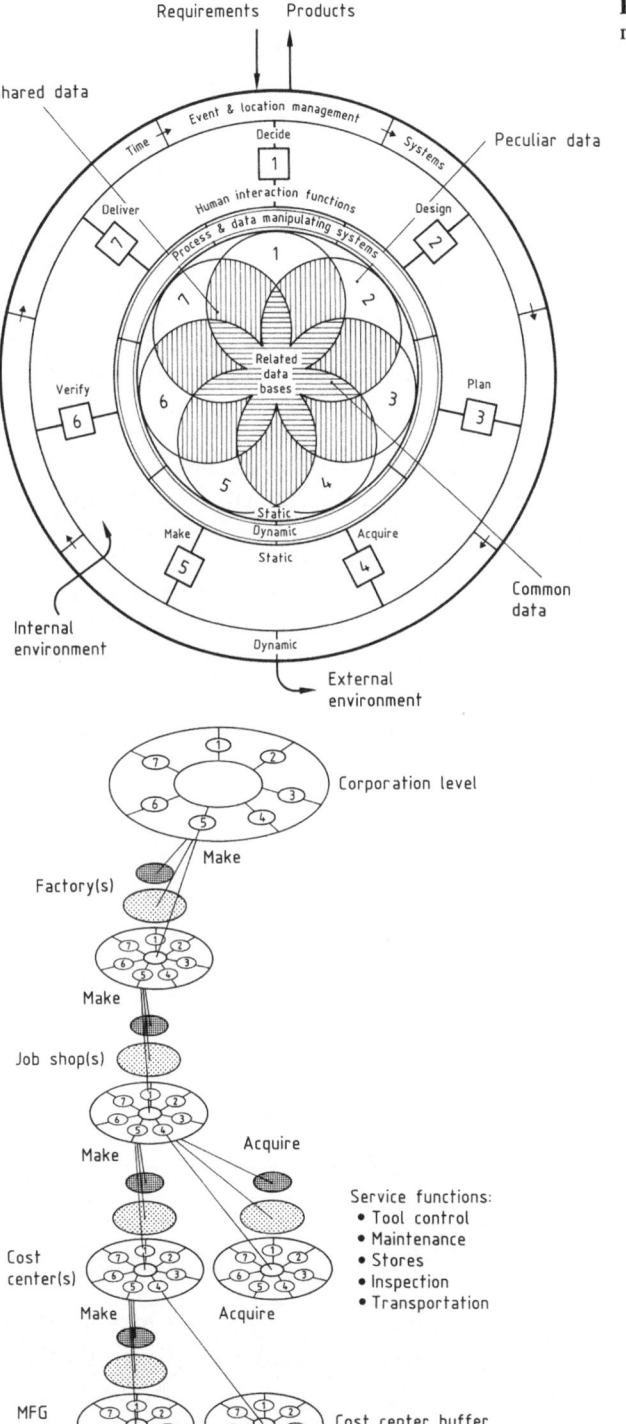

Fig. 1. CAM-I conceptual model

Fig. 2. CAM-I hierarchical model

The NBS Conceptual Model

The National Bureau of Standards (NBS) is establishing a model which will be used mainly for research on interface standards and metrology in an automated environment. NBS has developed a manufacturing systems architecture in order to support standardization of interfaces.

The manufacturing system is partitioned into a five-level hierarchical structure (Fig. 3), as follows.

- *Facility*. This highest level of control comprises three major subsystems: manufacturing engineering, information management, and production management.
- *Shop*. This level is responsible for the real-time management of jobs and resources on the shop floor through two major modules: task management and resource management.
- *Cell*. Controllers at this level manage the sequencing of batch jobs of similar parts or subassemblies, such as materials handling or calibration.
- *Work-station*. This level directs and coordinates the groupings of equipment on the shop floor. A typical test-bed work-station consists of a robot, a machine tool, a material storage buffer, and a control computer.
- *Equipment*. These controllers are linked directly to all automated pieces of equipment of the shop floor, be they robots, numerically controlled machine tools, coordinate measuring machines, delivery systems, or various storage retrieval devices.

Complementary guidelines for the control architecture are as follows.

1. Each level should have a unique control language, developed for the particular level of abstraction and the needs of that level.

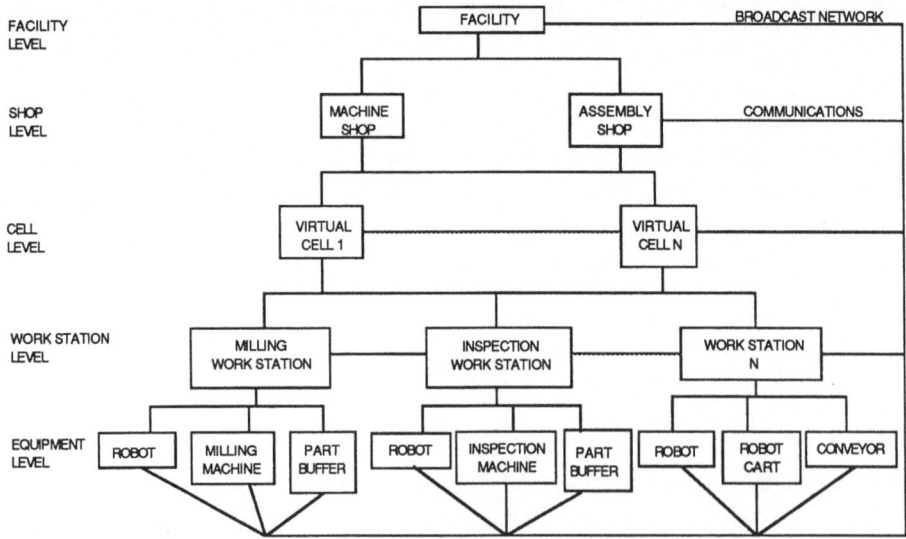

Fig. 3. NBS hierarchical model

13. Systems Analysis Techniques

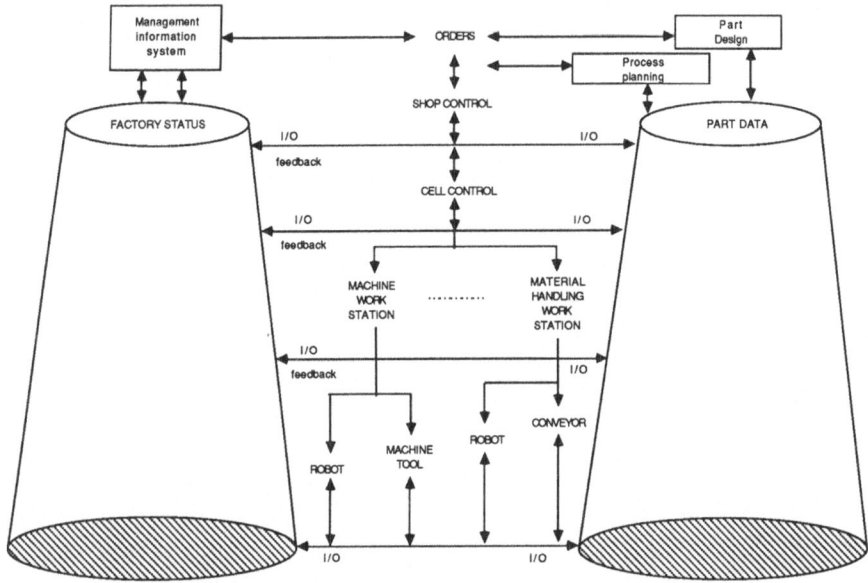

Fig. 4. NBS conceptual model

2. Commands are given in a top-down mode with successive decomposition at each level. At each level, the decomposition shall be based on procedures, functions, or rules providing a string of lower-level commands. The decomposition shall be made taking into account the "feedback information from the sensor processing hierarchy".
3. The commands should be decomposed using a finite-state machine approach with (interactive) learning facilities, i.e. updating of problem-solving rules in the operational phases.
4. The processing shall be made in globally synchronized cycles.
5. Data are grouped into two structures, each leveled according to the five-layer model (Fig. 4 – only four layers are shown). One contains part data (part design, process planning), the other holds the more dynamic type of data – factory status.
6. The system has distributed data administration. Data are exchanged between the processes through data administration systems, using mail-boxes.
7. Planning horizon is defined on a level-by-level basis.
8. Data communication takes place via a "Broadcast Network". The lower level is called "common local data path". It is (maybe only schematically) indicated that there is a common node for each of the elements on the shop level, another for cell level, and one for each group of equipment which constitutes a work-station.

The GRAI Conceptual Model

The conceptual model of the GRAI method has been developed from theories on complex systems and organization systems; it comprises two parts: the first one

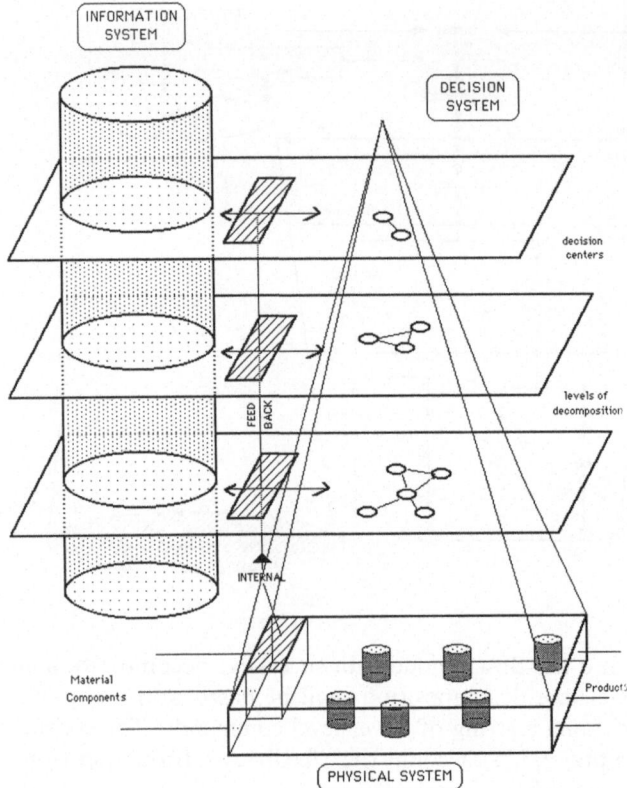

Fig. 5. GRAI conceptual model

outlines the organization of a Production System, the other one details the activities of a Decision Center (DC).

Structure of a Production System

A production system may be divided into three subsystems (Fig. 5):

- *The physical subsystem* is composed of machines, materials, workers, techniques. It is divided into work-centers. These centers are structured units or manufacturing cells built on group technology principles.
- *The decision subsystem* shows a structure organized into a hierarchy; each level is determined according to a horizon (the interval during which a decision is valid) and a period (the interval at the end of which a decision has to be reviewed). This subsystem is divided into decision-making levels, each one composed of one or several decision centers. These DCs receive a decision frame from the upper level and give decision frames to the DC of the lower levels or the same level.
- *The information subsystem* provides the decision subsystem with relevant information. It is a link between decision and operating systems. Decision and information subsystems combine together to make the Production Manage-

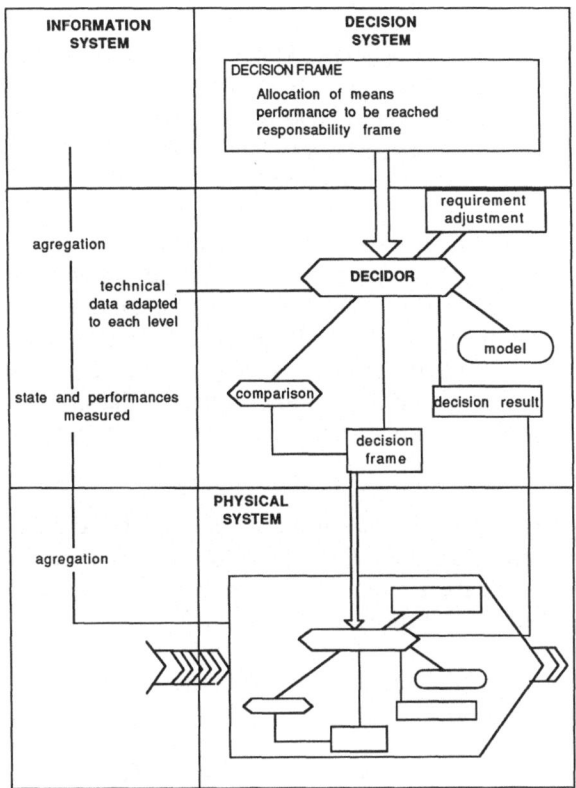

Fig. 6. GRAI decision centre conceptual model

ment System. The GRAI method is to study the consistency of the decision structure of this system.

Structure of a Decision Center (Fig. 6)

This model aims to conceptualize the operation of a DC, in a steady or altered state. We can therefore define:

— The various activities of a decision center
— Its decision frame (variables and decision limits)
— Decisions made by the DC
— Information used by the DC.

Both models represent various concepts enabling the construction of a consistent, valid, and adapted representation — a set of design and analysis rules.

2.2 Tools and Methods to Represent the Models

Tools help to represent very clearly the elements of the model which are the reference all along the design process. We present below several tool examples.

SSAD Concepts

Basic concepts. Structured System Analysis Design (SSAD) is a tool used to build a model of a system. We build a logical model (non-physical) through graphical techniques. We draw a data flow diagram which represents only data flow or information between processes (which represent activities). The representation has a static characteristic. We have four symbol conventions, as follows.

External entities are more often logical classes of things or people that represent a source or destination of transactions. The system we are considering accepts data from another system or provides data to it. An external entity can be symbolized by a solid square with the upper and left sides in double thickness to make the symbol stand out from the rest of the diagram (Fig. 7). By designing some thing or some system as an external entity, we are implicitly stating that it is outside the boundary of the system we are considering. The external entities are the sources or the destinations of data.

Data flow is symbolized by an arrow, preferably horizontal and/or vertical, with an arrowhead showing the direction of the flow. Each data flow is to be thought of as a pipe down which parcels of data are sent. We have data flow between external entity and process, and between process and data store.

Process can be symbolized by an upright rectangle, with the corner rounded, optionally divided into three areas (Fig. 8).

Data store can be symbolized by a pair of horizontal parallel lines, closed at one end, preferably just wide enough to hold the name. Each store can be identified by a "D" and an arbitrary number in a box at the left-hand end, for easy reference (Fig. 9).

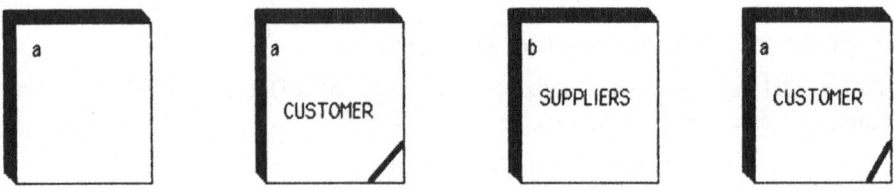

Fig. 7. SSAD symbols for external entities

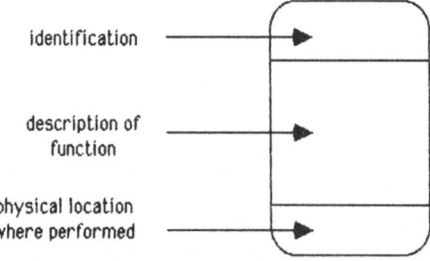

Fig. 8. SSAD symbol for process

Fig. 9. SSAD symbol for data store

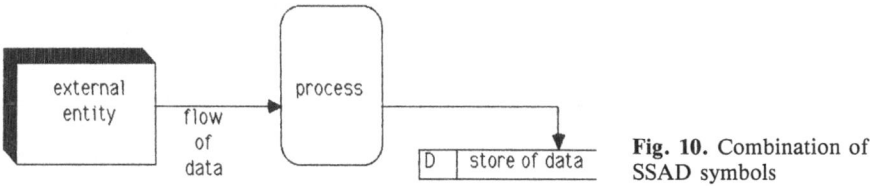

Fig. 10. Combination of SSAD symbols

These symbols and the concepts they stand for are at logical level. A flow or data may physically be contained anywhere that data passes from one entity or process to another. A process may physically be a combination of activities (Fig. 10). A data store is a file which contains information.

The three steps of the analysis are as follows:

- To identify the information
- To structure collected information with the aid of the activity diagram
- To ask the users to check the results.

In these activity diagrams, the boxes represent activities or functions and the arrows represent data (either information or object). There is a breaking down procedure which enables us to perform the analysis according to a top-down approach: the diagrams at the high levels encompass a wide range of details so that the words in the boxes and on the arrows indicate an aggregation of concepts. Successive diagrams at the lower levels disclose the details with more specific terms.

GRAI Concepts

Graphic tools (Fig. 11). When applying the GRAI method, two graphic tools are used: structuring and dialogue tools:

- The GRAI grid gives a hierarchic representation of the whole structure of Decision Centers of the Production Management System.
- The GRAI graphs (Graphs with Results and Activities Interrelated) enable the representation of activities and decisions made at the level of a DC, and of links between various centers. Links between these graphs allow us to draw up the activity networks.

Application of GRAI method (Fig. 12). Its application must be structured and abide strictly within well-defined procedures. Several actors must be involved in this approach:

- The synthesis group leads the project and validates the results. It is composed of the main executives of the manufacturing system.
- The analysis group collects the information from the manufacturing systems and is a true sensor embedded in the system.

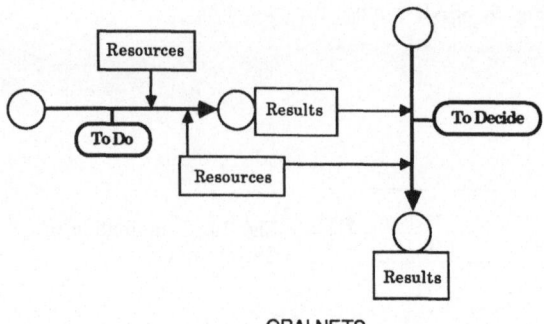

Fig. 11. GRAI tools

- The specialist group, comprising specialists in the GRAI method, helps the analysts in gathering information in GRAI form.
- The group of interviewed people is composed of executives who have a decision-making function at any level of the structure.

In the analysis phase we collect all information from the system. We thus gather the expert knowledge of the factory management people.

In the top-down analysis we determine, with the synthesis group, the content of the Grid (Fig. 11): Horizon/Period, Decision Center (DCs), links between DCs (the relationship between two DCs is represented by a double arrow, and the main exchanges of information by a single arrow).

In the bottom-up analysis, we start with the lower levels of the grid and analyse through the GRAI nets the detailed activities of each DC. We thus collect:

- The data necessary for the decision making
- The data related to decision frames:
    - decision variables
    - decision operators
    - decision rules
    - constraints

13. Systems Analysis Techniques

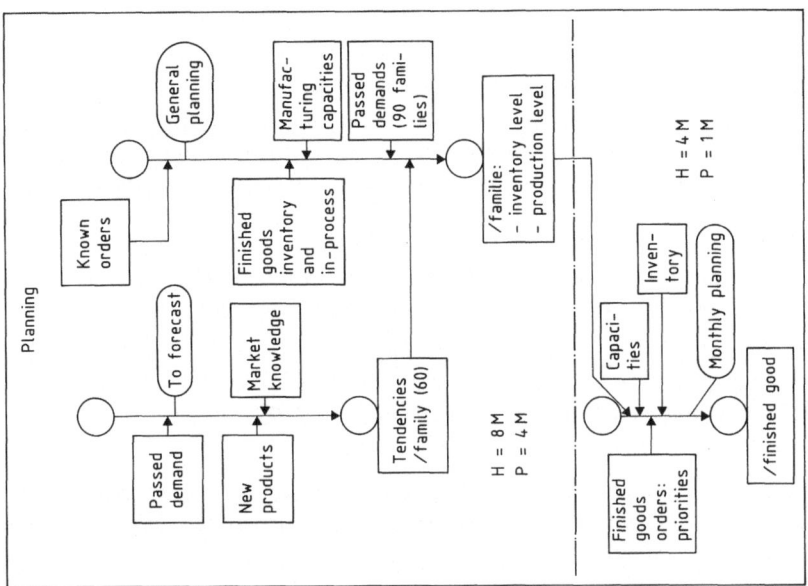

Fig. 12. Sample result of application of GRAI method

- criteria
- objectives.

This analysis is performed in a static way (when all information and supports are present) and in a dynamic way (in an altered state when a support is missing).

In such a study the executives within the manufacturing systems must be involved in the synthesis group. We are aware that these executives do not have a lot of time; that is why the structured approach and the method have been specially designed in order to take only two or three hours of their time every two weeks during a period of four to six months. The result of this analysis phase is detailed in an analysis book in which we collect:

- One or several grids, according to the complexity of the systems. We usually have an aggregated grid and several detailed grids.
- Several GRAI nets; each square in the Grid has a GRAI net which shows the detailed decision activities of this decision center.

During the analysis phase, and thanks to the conceptual model and the formal rules we have set, we are in a position to detect any inconsistencies. These inconsistencies deal with:

- information
- structure of decision centers
- coordination between decision centers
- dynamic behaviour.

Examples of inconsistencies relate to:

- The information systems:
    - lack of updating of the information
    - lack of provisional information
    - delay in the transfer of information
    - redundancies of information
    - inconsistencies between units of measure
- The structure of the decision centers:
    - inconsistency between the period of information updating and the period of decision making
    - lack of decision frame
    - lack of performance evaluation
- The coordination between decision centers:
    - when one decision center must comply with two different decision frames
    - when the decision frames skip several levels
    - inconsistencies between the horizon/period of each level
- The dynamic behaviour:
    - analysis of decisions in an altered situation
    - search for missing elements.

Once the inconsistencies have been detected and the objectives have been given by the top management (strategic objectives, provisional development of manufacturing systems), we start the design phase.

The design approach follows also a top-down and bottom-up design:

- During the top-down design we design a new grid in order ro resolve the inconsistencies; we use the design rules given by the GRAI conceptual model.
- During the bottom-up design we follow two steps to design GRAI nets:
  - First, the MACRO-GRAI gives us the overall relationship between decision centres.
  - Second, the detailed GRAI gives us the final design.

Once the design phase has been carried out, we are in a position to write the specification book.

So the GRAI conceptual model may be considered as the foundation of knowledge representation, and the GRAI method as the technique to acquire the knowledge from experts.

$IDEF_0$ Concepts

Basic concepts are a combination of structured analysis concepts. Together, they form a discipline that can be applied to systems generally, from planning to design. The basic concepts are:
1. To understand a system by creating a model that graphically shows things (objects or information) and activities (performed by men or machines). The model must properly relate both aspects.
2. To distinguish what functions a system must perform, from how the system is built, in order to perform those functions. The distinction must be clearly evident in a model.
3. To structure a model as a hierarchy with major functions as the top and successive levels revealing well-bounded details. Each model must be internally consistent.
4. To establish an informal review cycle to proofread the developing model and record all decisions in writing. This insures that a model reflects the best efforts of a committed team.

The $IDEF_0$ activity diagram comprises the following:

1. *Activity diagrams.* A model is a series of diagrams which break down a complex subject into its component parts. The initial diagram is the most general or abstract description of the whole system. This diagram shows each major component as a box. The details of every component (that is, the insides of each box) are shown on other digrams. These boxes can be broken down into further diagrams, until the system is described at any desired level of detail (Fig. 13).
2. *Boxes.* When the boxes on a diagram represent activities, they are each described by an active verb phrase, written inside the box. Each box on a diagram is numbered in its lower right corner, in order from 1 to, at most, 6.
3. *Arrows.* Arrows connected to a box represent real objects, or information needed by or produced by the activity. They are each labelled with a noun phrase, written beside the arrow. Data may be information objects, or any-

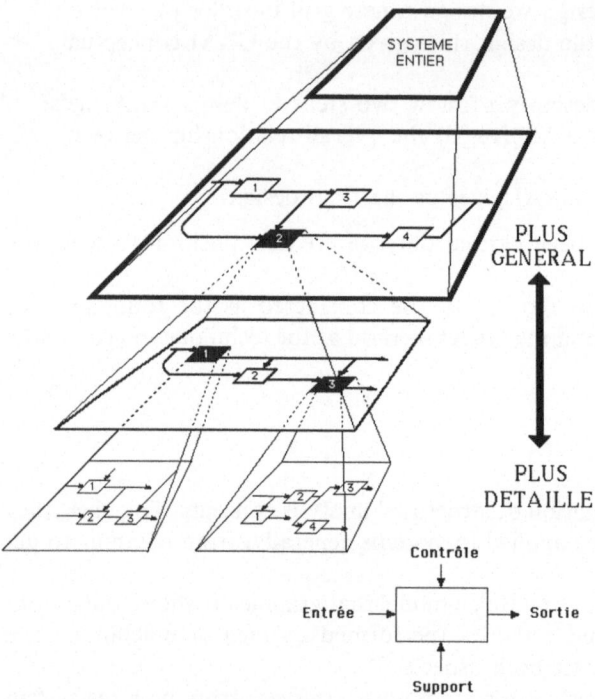

Fig. 13. IDEF$_0$ model

thing that can be described with a noun phrase. The side of the box at which an arrow enters or leaves shows the arrow role as an input, a control, or an output. Incoming arrows (left and top of box) show the data needed to perform the activity. Outgoing arrows (right of box) show the data created when the activity is performed. From left to right (input to output), an activity transforms data.

A control describes the conditions or circumstances that govern the transformation. The bottom of a box is reserved to indicate a mechanism, which may be the person or device which carries out the activity. A mechanism indicates the means by which the activity is performed. A mechanism may be a person or a committee or a machine or a process. In short, the output or input shows *what* is done by the activity, the control shows *why* it is done, and the mechanism shows *how* it is done.

The arrows on an activity diagram represent data constraints. They do not represent flow or sequence. Connecting the output of one box to the input or control of another box shows a constraint. The box receiving the data is constrained, since the activity cannot be performed until the data are made available by the box that produces it. The arrows entering a box show all the data that is needed for the activity to be performed.

IDEF$_0$ can certainly used at various levels of abstraction. The ICAM project has developed several models with IDEF$_0$, for example one of the most commonly used is "ICAM – Manufacturing Zero Architecture: MFG$_0$". IDEF$_0$ profits from the same computer support as SADT: SPECIF which is a software developed by the French company IGL.

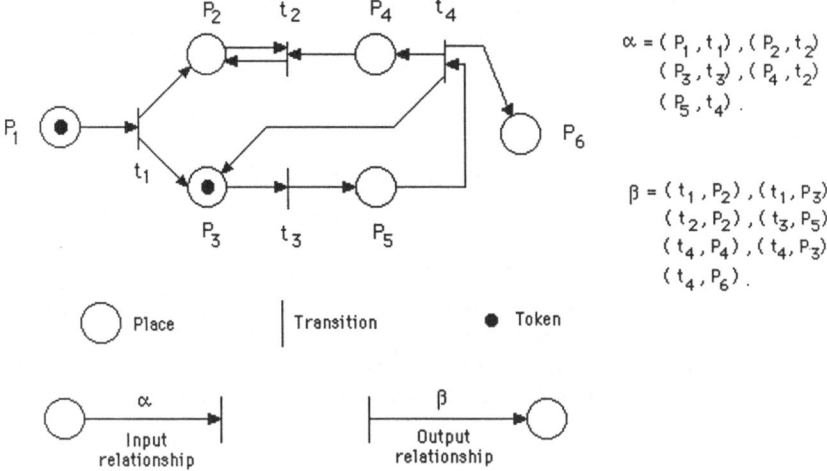

Fig. 14. Marked PETRI net

PETRI Concepts

As far as the physical resources are concerned, PETRI nets [1] seem well adapted to study the dynamic development of the system if we are in a position to describe the system with a linear model. A PETRI net (Fig. 14) is represented by a bipartite directed graph. This bipartite directed graph R includes two kinds of nodes, the P places and T transitions, represented respectively by circles and bars. These nodes are linked by directed arcs from places to transition and from transition to places. A Petri net is thus a four-triple $R = (P, T, a, b)$ such that:

P is a finite and non-empty set of places $P = \{P_i\}$

T is a finite and non-empty set of transitions $T = \{t_j\}$

$a \subseteq P \times T$: Input relationship $(P_i, t_j) \subset a \Leftrightarrow \$$ directed arc from P_i to t_j

$b \subseteq P \times T$: Output relationship from P_j to t_j $(t_k, p_1) \subset b \Leftrightarrow \$$ directed arc from t_k to P_1

Moreover, we define a marking M or R as being an application of P into N (set of positive numbers and zero).

GRAFCET Concepts

GRAFCET [2, 3] which can be defined from a particular case of PETRI nets, is also a tool which enables us to study synchronization between events and activities. We mention for this level GEMMA [4], which is not a tool but a method which enables us to study the running of a machine in a normal and altered state.

MFM Concepts

Another method can be used: MFM (Multilevel Flow Modelling) [5]. MFM has been developed at Risö Laboratory, Denmark. The method was primarily aimed at describing complex (power) production plants. The modelling concepts are based on thermodynamics: the model oriented to analyse physical systems deals with three types of flow functions: mass, energy, information. MFM could be used in manufacturing after adaptation. We wish to mention the role of Group Technology techniques which allow us to structure in a cell according to several criteria. We may consider that Group Technology is a technique for analysing and designing Physical subsystems.

In this field we find the greatest amount of research work, methods, tools, and computer support, because of the connection with information processing and the need for software engineering.

MERISE Concepts

MERISE [6, 7] has been developed in the late 1970s. It is a design and development method for information systems. It is based on system theory and is supported by three cycles:

- Life cycle of the information system
- Abstraction cycle
- Decision cycle.

MERISE is divided into several steps: master plan, preliminary study, detailed study, completion, implementation, maintenance. To describe the dynamics of information, MERISE uses PETRI nets. The MERISE method has been used for several projects which have validated its underlying assumptions. There are several computer supports. Lately the SEMA company has developed a new product combining MERISE and GRAI: GRAI – MERISE.

AXIAL Concepts

AXIAL is a method developed by IBM France [8]. AXIAL looks like MERISE with specific features.

$IDEF_1$ Concepts

$IDEF_1$ is the dual method of $IDEF_0$ for describing information systems IDEF, is based on the entity classes and relationship model. The diagrammatic notation of $IDEF_1$ (Fig. 15) consists of boxes which represent entity classes, lines representing links, and possibly line terminators indicating the type of class ratio

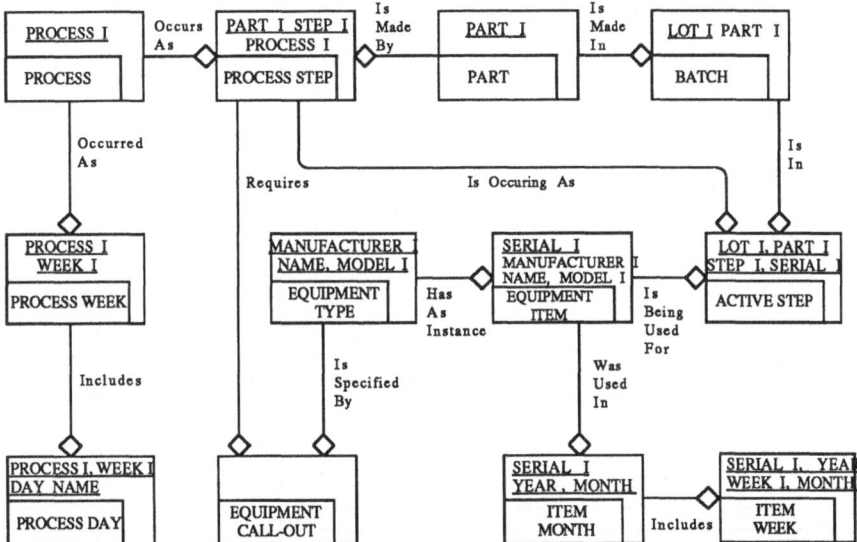

Fig. 15. Example of IDEF$_1$ diagrammatic notation

according to each class. The third aspect of entity modelling, data attribute, is supported by textual and tabular definitions, rather than by direct diagrammatic representations. There is an analysis process and the model may be mapped onto the corresponding IDEF$_0$ model.

IDA Concepts

IDA (Interactive Design Approach) [9] is based on computer-aided tools and is used mainly during the specification and design phases of a software project. The resulting model consists of three submodels, each one corresponding to a specific view of the information system: the process event, the process data, and the process resource. The process concept achieves the integration of the three different views. IDA encompasses a specification language called DSL (Dynamic Specification Language) which provides documentation, simulation, and prototyping.

GRAICO Concepts

The GRAICO method follows the same approach as the GRAI method. There is a conceptual model which enables us to describe the operating, physical, and informational subsystems. The conceptual model for the operating subsystem has a different structure than the decisional system in the GRAI method because the type of running is quite different. The various steps of the method are:

- To analyse and organize into a hierarchy the objectives of the control
- To describe the static running of the physical process

PROCESSUS DECOMPOSITION \ FUNCTIONS	CONTROL			FOLLOW UP			SECURITY	
	ELABORATION OF OBJECTIVES	COORDINATION	INDISPENSIAL	PROCESS	EQUIPMENT	MEMORIES	ELABORATION OF ALARMS	PROTECTIONS
P0								
P1								
P11								
P12								
P13								
P14								
P15								
P16								
P2								
P21								
P22								
P23								
P3								
P31								
P32								
P33								
P34								
P4								
P41								
P42								

Fig. 16. GRAICO grid

- To describe the dynamic running of the physical process
- To specify the needs of the control
- To specify in detail the functions of the control.

Figure 16 shows an example of the grid which represents the conceptual model.

IMMS Concepts

IBM has developed an Integrated Manufacturing Modelling System (IMMS) [10]. The goal of IMMS was to build a generic software package for manufacturing modelling. The package makes interactive models possible as well as building

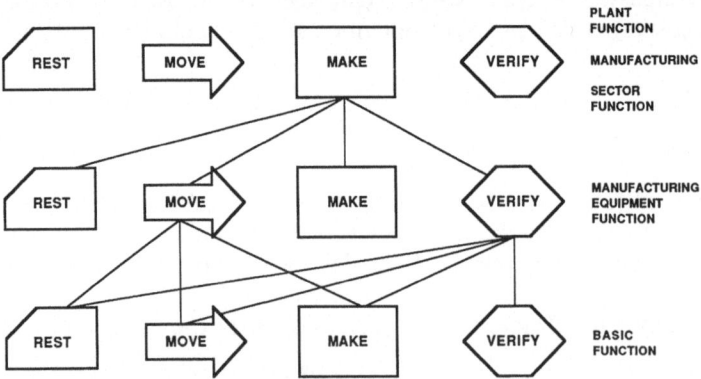

Fig. 17. IMMS materials flow hierarchy

and testing. IMMS integrates the fundamental components of modelling, namely system description, data acquisition and management, graphic input and output, and performance simulation. The main idea is to describe the flow of information, materials, and resources by using parametric standard model-building blocks. The input is done by placing function symbols on a graphic screen and by filling in blanks for parameter data. The output includes all common performance measurements such as utilization and work-in-process (WIP) in graphic form. The simulation language used to implement the model building blocks is RESQ2 (RESearch Queuing Package).

This example looks very interesting: we think it is well adapted to the physical and perhaps decisional level in limited functions (the control functions are REST, MOVE, MAKE, VERIFY: Fig. 17).

Simulation Concepts

Finally, we want to mention the role of simulation, as it is becoming a necessary tool. At present there are several approaches and a lot of simulation tools.

The first approach consists in developing from a language (FORTRAN, PASCAL) a simulation program which is well adapted to the problems involved, but this requires a large amount of development work without any possibility of using it again in other studies.

The second approach uses a simulation language such as GPSS, SLAM, SIMAN, ECSL-CAPS, Q NAP 2 or SIMGRAI (developed in collaboration with our Laboratory), which is more user-friendly and allows us to develop the programs faster. The evolution of such programs is to have three disconnected sections in the language: one to enter information, one to process it, and one to use results.

The third approach has appeared lately; it uses object-oriented language which enables us to get a description of a problem in a user language. In France the SGN company, in cooperation with CERT-DERA, is developing such a language: the project is called OASYS. Most people tend to forget the part of structuring before simulating. A good relationship between both these tools seems necessary today. This will be the topic of the following section. We have shown in this part that the various methods and tools can be used for designing a manufacturing system.

3 Classification Attempt

From our point of view, the design phase deals with:
- The determination of the functional specifications of the system to be built
- The global architecture
- The detailed specifications of the components.

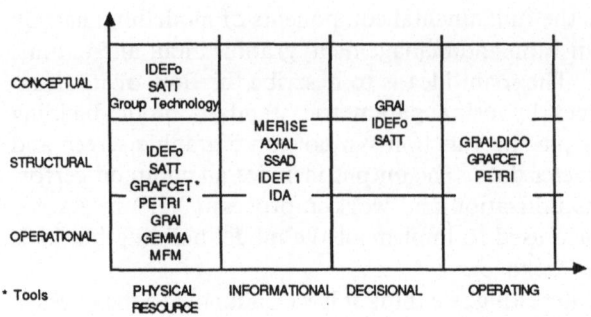

Fig. 18. Classification of tools and methods for Analysis/Design/Specifications phase

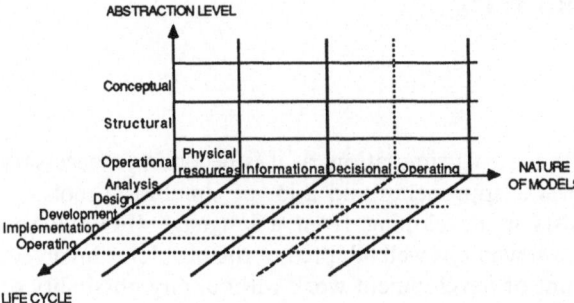

Fig. 19. Various kinds of models in the life cycle of manufacturing systems

In the state-of-the-art, which we have achieved recently through the ESPRIT Project 418 "Open CAM Systems" [11], completed by our own work, we have tried to classify current methods and tools. The difficulties of the task lay in the novelty of the field, and in the lack of formalism and research work on the subject. We will limit our classification to several methods which include tools and to some tools without methods. We have represented the results of our investigation in Fig. 18, limiting the matrix proposed in Fig. 19 to the first phase of the life cycle: Analysis/Method/Specification. (We have grouped the first two steps because usually the same tools and the same methods are used.)

The three steps of the design phase may be linked with three abstraction levels of the design process, enabling us to work from the conceptual level towards the concrete level:

- The conceptual level, which enables us to represent the system mainly with concepts and functions
- The structural level, which enables us to group the functional specifications and define an architecture: for example, at this level, we have to determine the manual and automatic procedures
- The operating level, which, from the detailed specifications of the hard components, enables us to determine the hard components, the software packages, and the specific software. This step must take into account the realizational constraints.

We believe that the design must follow the three previous steps if we want to be able to start from a conceptual point of view and reach a concrete realization.

Finally, further to our research in this field (GRAI conceptual model) a manufacturing system can be divided into three subsystems [12]:

- Physical resource subsystem: this subsystem includes men, machines, materials flow, techniques; its role is to transform raw materials into end-products
- Decisional subsystem: this aims to control the physical subsystem in order to reach to economic and social targets while taking the constraints into account. We have improved our model lately by introducing an operating level which links the decisional and physical parts (this level includes the control of machines and all procedures to be controlled: security, quality maintenance)
- Informational subsystem which links the two previous subsystems.

To sum up, we can determine three axes of subdivision:

- One axis for the life cycle
- One axis for the level of abstraction
- One axis for the various subsystems.

If we combine these three subdivisions (Fig. 19), we can determine a model for each case (for instance the decisional model at the conceptual level in the analysis phase). The various boxes are linked and we must determine the transformations ($\mathcal{T}$) to go from one model of a subsystem to another (decisional model at conceptual level towards decisional model at structural level, or decisional model in the analysis phase towards decisional model in the design phase) and interfaces (I) between the subsystems).

4 References

1. Favrel, Oh, and Campagne, Graphic Modelling by PETRI Nets for the production planning. In Proc. IFIP conference, APMS 82, Bordeaux, France, August 1982, North-Holland
2. Blanchard, M., Comprendre Maîtriser et Appliquer le GRAFCET. Publishing Cepadues, Paris, 1979
3. Gagi, Ladet., Grafcet synthetised computer system and procedure description in FMS. In Proc. IFIP conference, APMS 82, Bordeaux, France, August 1982, North Holland
4. GEMMA: Guide d'Etude des Modes de Marches et d'Arrêts. ADEPA Publishing, Paris
5. Lind. Morten, A system modelling framework for the design of integrated process control systems. Documentation Risö – M 2409, December 1983
6. Tardieu, Rochfeld, and Coletti. La méthode MERISE: Principes et Outils. Les Editions d'Organisation, 1984
7. Rochfeld and Tardieu. MERISE, an information system design and development methodology. Information and Management, Vol. 6. North-Holland, 1983
8. Pellaumail., AXIAL, une méthode de conception de système d'information proposée par IBM France. Informatique et Gestion, No. 118, Paris, October 1980
9. Bodart, Pigneur, Hennebert, and Leheureux., The experimentation of information system requirements by simulation and prototyping in the IDA project. ISDOS-IDA Prise Meeting, Paris, January 1985
10. Engelke et al., Integrated Manufacturing Modelling System. IBM Journal of Research and Development, Vol. 4, July 1985

11. Research work carried out within the project 418 "Open CAM System" of the ESPRIT Programme Intermediary report. Commission of the European Communities, October 1985
12. Doumeingts, G., Methode GRAI: méthode de conception des systèmes en productique, Higher Doctoral Thesis, University of Bordeaux (1), 13 November 1984
13. CAM – I., Conceptual information model for an advanced factory management system – Work center level – Final report, R-83-FM-01. CAM – I, Arlington, Texas, June 1983
14. Doumeingts, G. Methodology to design computer integrated manufacturing and control of manufacturing unit. Advanced Course on Computer Aided Manufacturing – CAM 83, Karlsruhe, FRG, September 1983
15. Doumeingts, G., How to decentralize decisions through GRAI model in production management. IFIP W.G. 5.7 International Working Conference, Munich, FRG, March 1985

Chapter 14
Fourth Generation Languages

Jarle Aaram

Jarle Aaram holds an M.Sc. in Mechanical Engineering from the Norwegian Institute of Technology (NTH). During 1974–1976 he worked in industry on development and maintenance of computer systems for finite element analysis. As from 1976 he has worked as a scientist at the Production Management Section of Production Engineering Laboratory NTH-SINTEF. His main area of work has been computerized production management systems.

1 Four Generations of Languages

Programming languages are usually divided into four generations. The first generation is the machine languages, where programmers have to write each binary-coded machine instruction. Binary coding is voluminous and laborious to write, and the programs are difficult to read and maintain, and highly machine dependent.

The second generation is the assembly languages, which introduced a major improvement by symbolic coding using mnemonics. Assembly programs are also voluminous, difficult to maintain, and machine dependent.

The third generation is the high-level languages like FORTRAN and COBOL. Using these languages, the volume of programming is reduced significantly, as one statement represents several assembly instructions. In addition, programming is done in a more "natural" language, and the programs are to some extent machine independent.

The fourth generation is the very high-level or non-procedural languages, where the volume of the programs and the time for programming may be reduced to only 10% of that for third generation languages. For the first three generations, programming is a detailed specification of *how* to process data, whereas programming using a fourth generation language is specifying *what* to do. Although programs written in a fourth generation language are basically machine independent, very few fourth generation languages are implemented on several types of computers.

2 Characteristics of Fourth Generation Languages

A distinction should be made between fourth generation tools and fourth generation languages. Fourth generation tools comprise

− Query languages
− Report generators
− Application generators
− Fourth generation languages.

Query languages are designed for simple, interactive query on selected data. Report generators are designed for easy specification of reports, including programming if necessary. Application generators are programs that generate application program code from specified screen layout, data structure, program logics, etc., whereas fourth generation languages are very high-level programming languages that do not produce any additional code. This chapter is limited to fourth generation languages only.

Fourth generation languages are characterized by high efficiency in program development, and most fourth generation languages use non-procedural or result-oriented coding (i.e. specifying *what* to do, not *how* to do it). Simplified, it might be said that these languages contain procedural logics as preprogrammed macros, so programming is reduced to specifying which macros to use and appropriate parameters. Some languages combine non-procedural and procedural coding to handle special logics not covered by the non-procedural part.

Several languages utilize data dictionaries to define data elements and structures. This may be done in two ways. Passive data dictionaries are used at compile time only. Thus, changes in a data dictionary require recompiling some programs. Active data dictionaries are utilized at run time. Changes may be performed more easily, as no recompiling is required. On the other hand, computer resource consumption is increased. Thus, active data dictionaries are preferred during systems development, whereas passive data dictionaries may be preferred when development is finished.

Fourth generation languages contain high-level description of screen layout, input/output control, and database interface. Default options are used to a large extent to minimize the volume of code required for standard applications. These options may be overridden by specifying appropriate values. In this way, first versions of application programs (prototypes) may be created quickly, and may also be extended and modified easily. Most fourth generation languages are designed for online applications, and report functions often include graphics.

Some fourth generation languages are integrated in a systems development environment comprising editor, compiler, etc. This will ease the systems development as the users may easily switch between editing, compiling, testing, and so on. Some fourth generation languages are designed for interpretative use, whereas others have to be compiled. There are also languages for optional interpretation or compiled usage. Application programs developed by a fourth generation language need at least a run-time version of the language implemented even when compiled and linked.

As for interface to other systems, fourth generation languages vary from "open" to "closed". Some languages store data on files using internal formats, which are not accessible from outside. These languages contain special export and import functions to transfer data to and from external files using a standard format. Some languages, however, may utilize general databases also accessible from other systems.

Using fourth generation languages impacts the systems development process heavily. The volume of code written, as well as development time, is reduced drastically. A more important impact, however, is that these languages allow a quite different way of systems development. The stepwise, sequential procedure of specification, programming, and testing may be replaced by experimental systems development. Prototypes may be used as parts of specifications, starting with some central parts and expanding successively. Using prototypes will make possible a more active user participation in the systems development. End-users may test the prototypes and suggest improvements, thus assuring that the final system will meet end-user requirements.

Maintenance of application systems is normally a large task. Programs written in a fourth generation language are expected to be more easy to maintain since the code is more compact. As development time is reduced, the number of application programs developed, as well as the diversity of the programs, will increase. The intensity of change in environment and systems requirements will also increase. This will all cause increased need for maintenance. If this is not taken into account at systems design and development stage, the maintenance backlog may be enormous even if fourth generation languages are used.

Fourth generation languages are often regarded as self-documenting. To some extent this may be true, depending on what they are compared with. This does not mean, however, that documentation and comments should be omitted. For third generation languages the comments specify *what* to do, and the code specifies *how*. For fourth generation languages, the code specifies *what* to do, and the comments should specify *why*.

3 Overview of Languages

As most fourth generation languages are mainly non-procedural, this means that procedural logics must be intrinsic to the languages. Many fourth generation languages are not general purpose, but are limited to some specific areas of application. Normally there is a trade-off between generality and flexibility of the language and easy systems development for the application areas selected. Different users and application areas cause different requirements for the programming language. Therefore, no "best" language can be chosen in general. The language must be chosen according to application areas and application environment.

Table 1 shows some properties of a few, selected fourth generation languages. The list of languages, as well as their description, is far from complete, and is included for the sole purpose of showing some of the diversity.

Table 1. Some selected fourth generation languages

Language	Supplier	Computers	Database	Proc. coding avail.	Screen def.	Report funct.	Compiled/ interpret.	Data struct. indep. use	Run-time avail.	Graphics	Data dict.	Remarks
ADMINS	Admins Inc.	VAX, PDP	Internal files	Yes	Yes	Yes	C	Yes	No	REGIS interf.	No	
CONSENSUS (UFO)	Martin Marietta Data Systems	IBM mainfr.	IBM external databases and files	Yes	Yes	Yes	Report I Proc. C Nonproc. I	Yes	Yes	RAMIS II report gen.	Active	
FOCUS	Information Builders Inc.	IBM mainfr. and PC, VAX, WANG, etc.	Internal files	Yes	Yes	Yes	C/I	Yes	Yes	Simple graphics incl.	Passive	
IDEAL	Applied Data Research Inc.	IBM mainfr. and PC	Datacom, open	Yes	Yes	Yes	C	Yes	Yes	GDDM incl.	Active	
INFO	Henco Inc.	HB, PRIME, VAX, IBM mainfr. and PC, etc.	Relational, open	Yes	Yes	Yes	C	Yes	No	Graphics incl.	Active	
INGRES/OSL	Relational Technology Inc.	IBM mainfr. and PC, VAX, sev. UNIX-computers	Relational, open	Yes	Yes	Yes	I, partly C	Yes	Yes	Graphics incl.	Active	
MANTIS	Cincom Systems Int.	IBM mainfr., VAX, HB	Rel., open, interf. to external db	Yes	Yes	No	C/I	Yes	Yes	GDDM interf.	No	
MAPPER	Sperry Corp.	UNIVAC 1100, Sperry PC, some UNIX-comp.	Relational, open	Yes	Yes	Yes	I	Yes	No	Graphics incl.	No	

14. Fourth Generation Languages

Name	Company	Hardware	Database							Graphics	Active/Passive	Notes
MITROL (MIMS)	Mitrol Inc.	IBM mainfr.	Internal	Yes	Yes	Yes	C/I	Yes	Yes	GDDM interf.	Partly active	Special PM procedures incl.
NATURAL	Software AG	IBM mainfr. VAX, Siemens	Relational, open	Yes	Yes	Yes	C/I	Yes	Yes	GDDM interf.	Passive	
ORACLE	Oracle Corp.	IBM mainfr. and PC, PDP, VAX, HP, DG, sev. UNIX-comp.	Relational, open	Yes	Yes	Yes	I, partly C	Yes	No	SAS-GRAPH interf., SQL* GRAPH, EASY* SQL	Active	
PROGRESS	Data Language Corp.	IBM PC and comp., most UNIX-5 computers	Internal files	Yes	Yes	Yes	C/I	Yes	Yes	No	Parly active	
RAMIS II	Martin Marietta Data Systems	IBM mainfr. and PC	Network, open	Yes	Yes	Yes	I	Yes	No	GDDM interf.	Active	

Although no language can be chosen as the best in general, some general requirements for fourth generation languages can be given. A fourth generation language should be flexible and allow for experimental systems development, thus covering all activities from early specifications until implementation and maintenance. Changes in program logics as well as data structures should be easy to perform and not require rewriting of existing programs. The language should also be flexible regarding user interface like screen layouts and reports, and choosing between menus or commands. Use of data elements must be allowed in screen presentations or reports independent on the underlying database structure. Data elements as well as data sets must be protected against unauthorized use. In addition, it is important that the language is sufficiently general and flexible to cover all transactions in a system.

A fourth generation language should cover most of the application by non-procedural (result-oriented) code to make it easy to describe exactly what is to be performed. However, procedural code should also be available for special logics. Describing the results in a fourth generation language means primarily describing the dialogues, reports, controls, calculations, and database interactions.

A fourth generation language should also be fairly independent of computer type, operating system, database (if external), peripheral devices, etc. This will make the language more flexible and ease the transferral of application programs to different computers. The number of users of a particular language is also important. A large number of users provides a better base for sharing experience and will also ensure maintenance and further development of the language.

To shorten response time, compiled code should be available. A run-time system for the language should be available for users not doing development. A flexible interface to external systems is also important. Graphic features or interface to graphic systems should be available.

The language or the database system used should also provide logging of database updates and automatic recovery after breakdowns. To many production management systems this is a vital property.

4 Application Areas Within Production Management

Fourth generation languages are applicable to prototyping as well as to ordinary systems development and *ad hoc* programming. As previously described, prototyping is a very convenient method for developing systems specifications. Fourth generation languages support this way of development since a first version of transactions may be operational in a short time and refined successively.

The main area of application for fourth generation languages is, however, development of final systems based on ordinary systems specifications or prototypes. Today's fourth generation languages are well suited for a large variety of management systems where procedural logics are well established. The strength of fourth generation languages to these applications is mainly the reduction of time and effort in systems development as well as systems maintenance.

Due to the small volume of code written and the flexibility of program logics, fourth generation languages are also well suited for *ad hoc* programming when special requirements arise.

Non-procedural languages can, in general, only be used for problems where procedural logics are intrinsic. To some extent, this may limit the area of application for some languages. However, a large majority of production management transactions are general information handling, like storing, retrieving, and updating data, screen handling, reporting, etc., which do not require any special production management logics. Therefore, a large number of fourth generation languages will be appropriate for less advanced production management applications.

Production management also includes some special procedures, like bill-of-materials explosion, requirements calculation, scheduling, and loading. The logics for these functions may be difficult to program in a non-procedural language unless the language contains special macros for these logics. As many production management systems will include such functions, the problems of programming these will limit the use of some fourth generation languages within production management.

Some interesting applications have been developed, where fourth generation languages are used for production management systems interfacing process control systems. Stock control transactions written in fourth generation languages are interfaced to control systems for transporters of automatic stocks. The stock control system monitors quantity on stock for each part, where stored, free stock positions, etc. When parts are to be stored or retrieved, the stock control system sends and receives appropriate data to or from the transporters' control system. Similarly, fourth generation language systems may be interfaced to CNC and DNC systems, providing data like parts to manufacture and quantities, and receiving feedback data like quantity completed and time used.

5 The Need for Fourth Generation Languages

The computer industry is growing rapidly, and the price/performance ratio for hardware is decreasing. This is not the case for software development, and development of application systems has become a bottleneck in the growth of computer applications. As the cost of computers decreases, still more application systems will be required. In addition, the rapidly changing environment will also increase the need for shortening the time for development and maintenance.

Although an increasing proportion of application software will be purchased as standard software, there will always be a large number of systems developed or adapted specifically for the particular application environment. The large diversity regarding types of products, types of production, industrial equipment, organizations, etc. must be reflected in the application systems. There is also a growing trend of decentralization and adapting systems to each particular group of users.

Within production management, application of computer systems has mainly been at operational level. Now there is an increasing use of computer systems also for management control and planning and analysis.

As a result of all this, there will be an increasing need for systems development, which is at present a bottleneck. The only way to overcome this bottleneck is to increase productivity of systems development drastically. Thus there will be an increasing need for high-productivity systems development tools like fourth generation languages.

Although today's fourth generation languages form a major step of improvement, there are still several drawbacks. The most important one is probably the lack of flexibility. For many languages it has turned out to be very difficult to program some special logics. In addition, data dictionaries are used only for a few languages, and data are stored in internal files not accessible to external applications. Graphic features are, in general, primitive, and most languages are implemented on one or very few computer types only.

Fourth generation languages are under continuous development. Much effort is made to increase flexibility, especially by combining procedural and non-procedural code and extending high-level specifications. More languages are also utilizing data dictionaries and general databases also accessible from external applications. Several languages are implemented on various computer types, including mini- and microcomputers, and advanced colour graphics will also be available to more languages. Interpreters and compilers as well as run-time versions are expected to become more generally available, and the languages will be more integrated into complex systems development tools. Probably some fourth generation languages will also include features for knowledge-based systems.

6 References

1. Martin, James, Application Development without Programmers. Prentice-Hall, New Jersey, 1982
2. Martin, James, Fourth-Generation Languages. Prentice-Hall, New Jersey, 1985

Chapter 15
Design of a Generalized Job Shop Control System and PM Packages

Harinder Jagdev

Dr. Harinder Jagdev received his Batchelor of Technology, in Mechanical Engineering (1974), with First Class Honours from the Indian Institute of Technology, Madras, India. He joined Mercedes Benz where he worked for over two years as a Production Engineer. In 1976 he joined UMIST, Manchester, from where he received his M.Sc. and Ph.D. in Manufacturing Technology in 1977 and 1980, respectively. Having worked as a research associate for some time, he is currently a Projects Officer in the Control Systems Centre of UMIST. Since 1980 he has been consultant in all aspects of Production Management to many organizations in Europe.

1 Introduction

Batch manufacture occurs where the demand for individual items is insufficient to justify the installation of dedicated resources and other mass manufacturing techniques for the production of each item. The batch manufacturing environment, or job shop, consists of general-purpose machine tools capable of being fitted with various tools and fixtures to perform a wide variety of jobs. Individual items are grouped in batches, and their production proceeds in a series of discontinuous steps.

Because of the flexible nature of job shops, their operating concepts have penetrated into all types of manufacturing industries. These industries range from the much-cited examples of aircraft and machine tool manufacturing, to ones engaged in simpler products, and into the process industries. Industries which on first consideration seem to be engaged in mass manufacture run many of their operations with job shop concepts. Typical examples of this are press shop operations in many automobile plants; forge and foundry shops; plants for the processing, storage, and packing of most chemical, food or other related products. As a result of such wide applications, the definition of a job shop and its operating concepts tend to be modified to suit individual requirements.

Because of this wide variety of environments in which a job shop can exist, a generalized solution can only be developed for a hypothetical job shop. The operating environment of this hypothetical job shop will consist of a union set of operating conditions expected to be found in the targeted set of job shops. The generalized job shop control solution described here pertains to such hypothetical job shops.

2 Hierarchical Nature of Job Shops

It is convenient to divide the control or management of the system into a hierarchy comprising a number of different levels. One such hierarchy is shown in Fig. 1, where each level is characterized by the length of the planning horizon and the kind of data required for the decision-making process. Higher levels of the heirarchy usually have longer horizons, use highly aggregated data, and encompass and hence influence wider areas of manufacturing operations, whilst the lower levels have shorter horizons and use more detailes information. The nature of the uncertainties at each level of control also varies. The longest-term decisions involve capital expenditure or redeployment, whilst the shortest involve the times to load individual parts or even to control robot arm trajectories. Even though these decisions are made separately, each long-term decision presents a target for the next short-term decision.

The term Job Shop Control usually refers to the control of activities in the Mid-Range and Short-Range horizons of Fig. 1. The Short-Range horizon encompasses all day-to-day control of job shop operations, whilst the Mid-Range horizon concerns itself with the planning of activities. The projected lengths of these horizons depend on such external factors as market environment and internal factors like product cycle time, number of machines and operations involved,

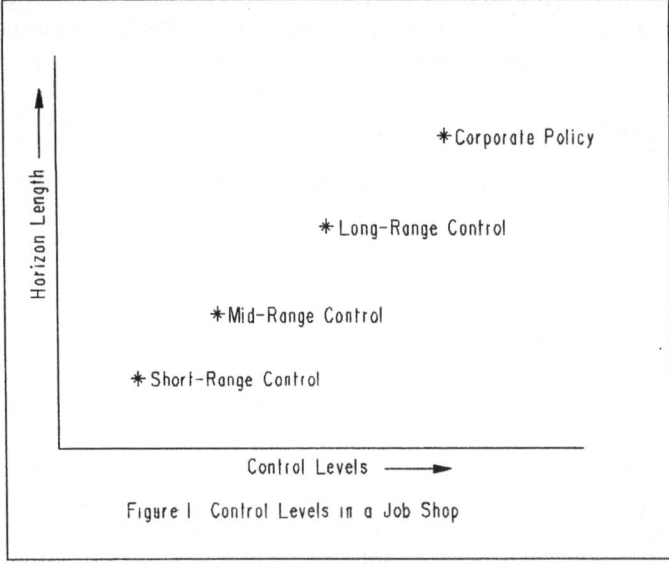

Figure 1 Control Levels in a Job Shop

number of live batches, accuracy of processing times, etc. Typically, the horizon lengths for Short-Range and Mid-Range are one of four weeks, and two to six months, respectively.

3 The Structure of a Job Shop Control System

The structure of major activities in the Mid-Range and Short-Range is shown in a SADT model describing only the Activity Diagrams. Figure 2 gives the Node Index of the SADT diagrams depicted in Figs. 3.1 through 3.8. The first diagram, A0, puts the activity of job shop control, "Manage Products and Production, A3", in perspective with other major activities in the factory manage-

Node	Node Details
A0	Manage Job-Shop Affairs
A1	Manage Finance
A2	Design Products
A3	Manage Products/Production
A31	Plan Products
A32	Plan and Control Production
A321	Review Mid-Term Strategy
A3211	Maintain Calendars
A3212	Maintain Predictive Techniques
A3213	Colligate Planning Conflicts
A32131	Understand Planned Load
A32132	Re-adjust the Load
A32133	Develop Action Reports
A3214	Colligate Production Conflicts
A32141	Understand Current Production Status
A32142	Understand Predicted Status
A32143	Understand Planned Status
A32144	Assign Global Priority
A32145	Develop Action Reports
A322	Plan Production
A3221	Maintain Schedules
A3222	Plan Requirements
A3223	Develop Action Reports
A323	Control Production
A3231	Review Short-Term Strategy
A3232	Monitor Production
A3233	Monitor Resources
A3234	Monitor Ancillaries
A33	Manage Procurement of Materials
A34	Manage Stocks
A4	Market Products

Total nodes in the reduced model are: 31

Fig. 2. The node index

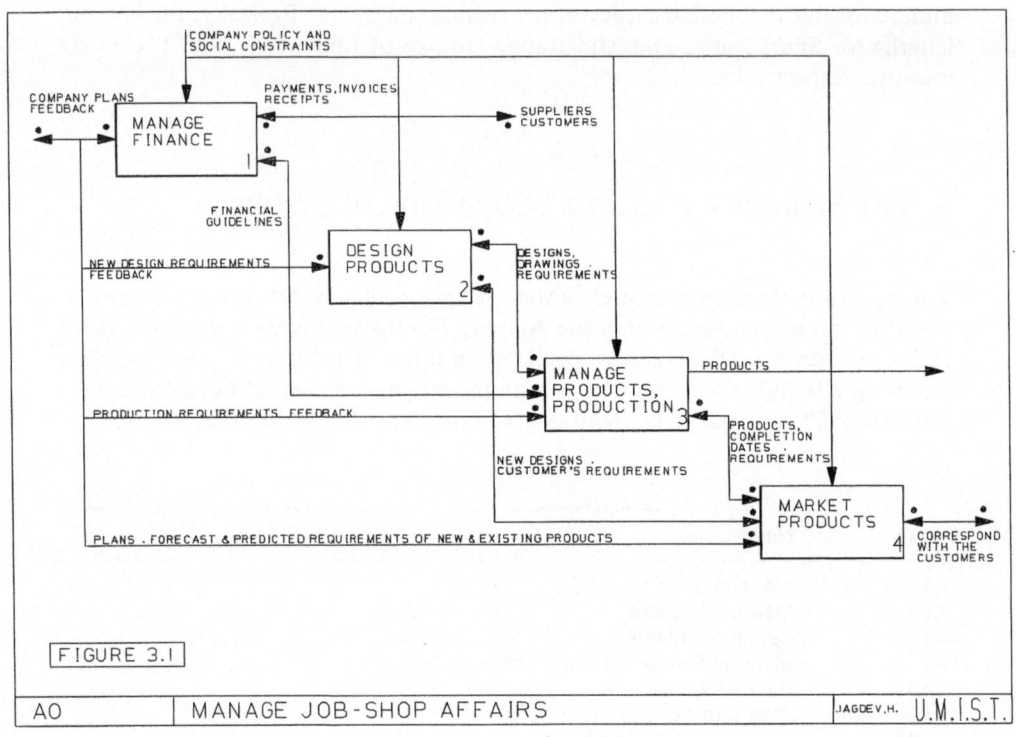

FIGURE 3.1

A0 — MANAGE JOB-SHOP AFFAIRS

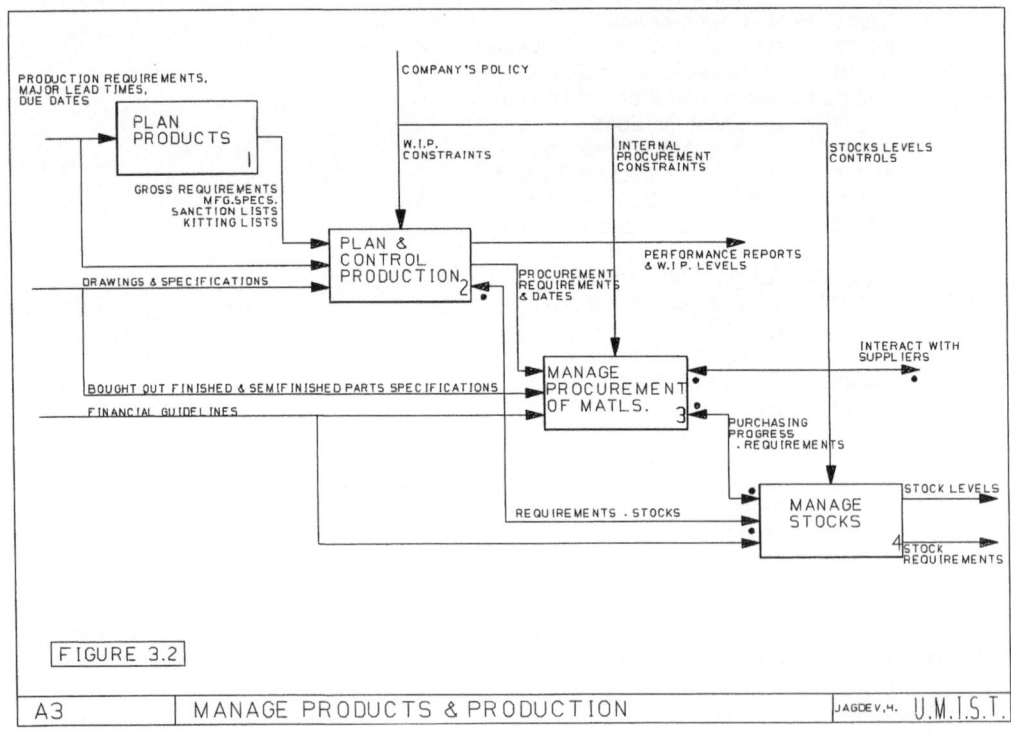

FIGURE 3.2

A3 — MANAGE PRODUCTS & PRODUCTION

15. Design of a Generalized Job Shop Control System and PM Packages

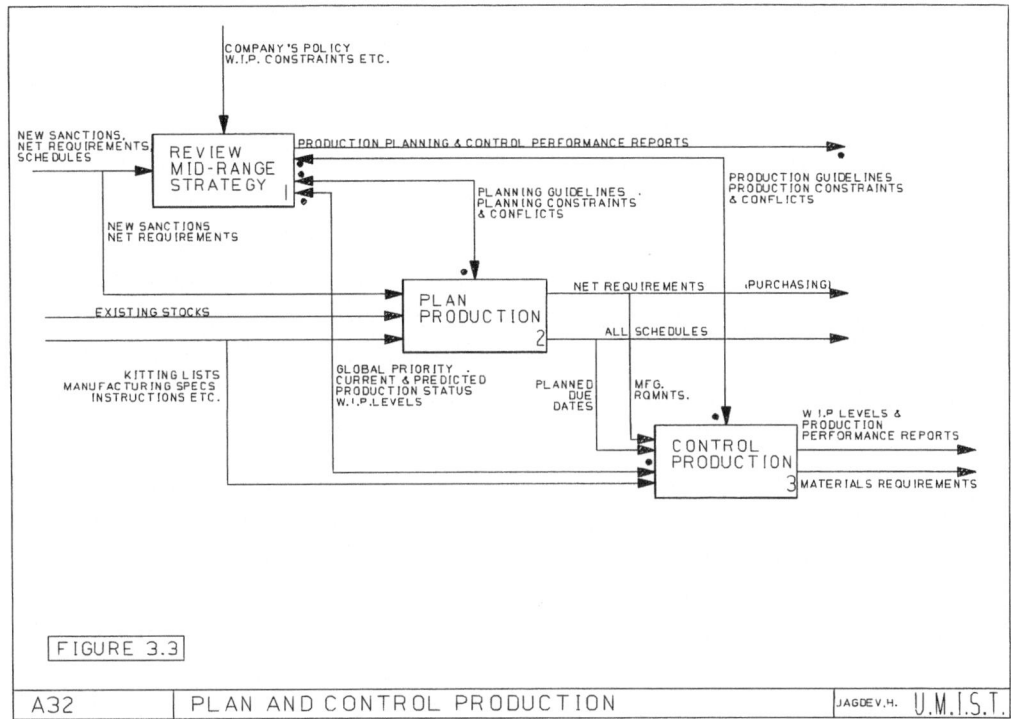

FIGURE 3.3

A32 — PLAN AND CONTROL PRODUCTION — JAGDEV,H. U.M.I.S.T.

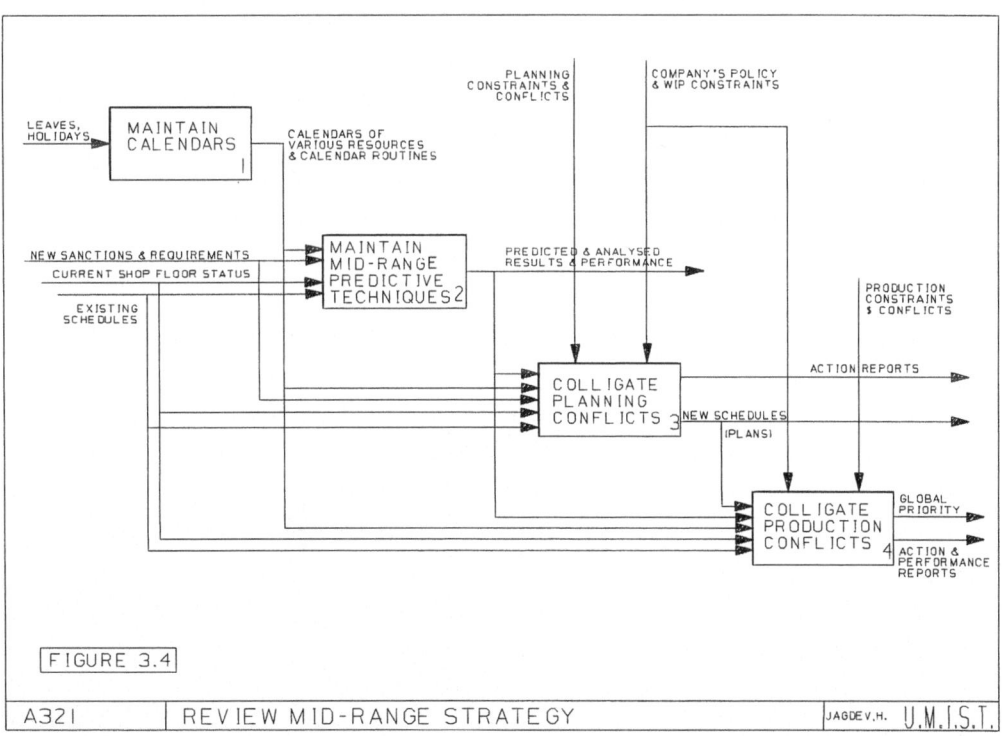

FIGURE 3.4

A321 — REVIEW MID-RANGE STRATEGY — JAGDEV,H. U.M.I.S.T.

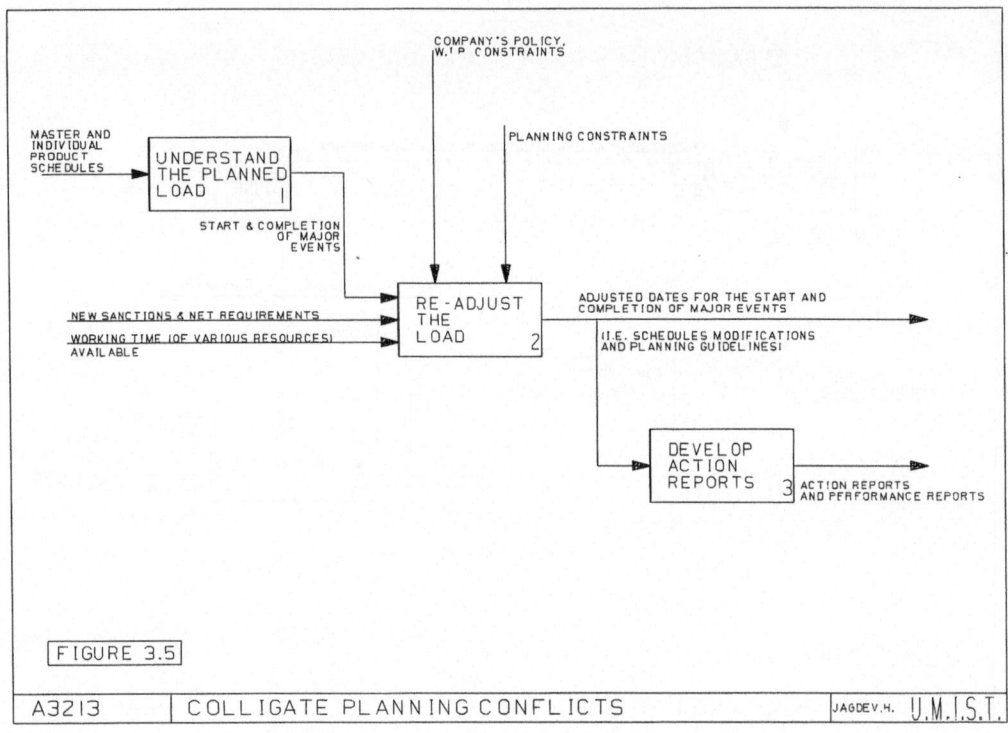

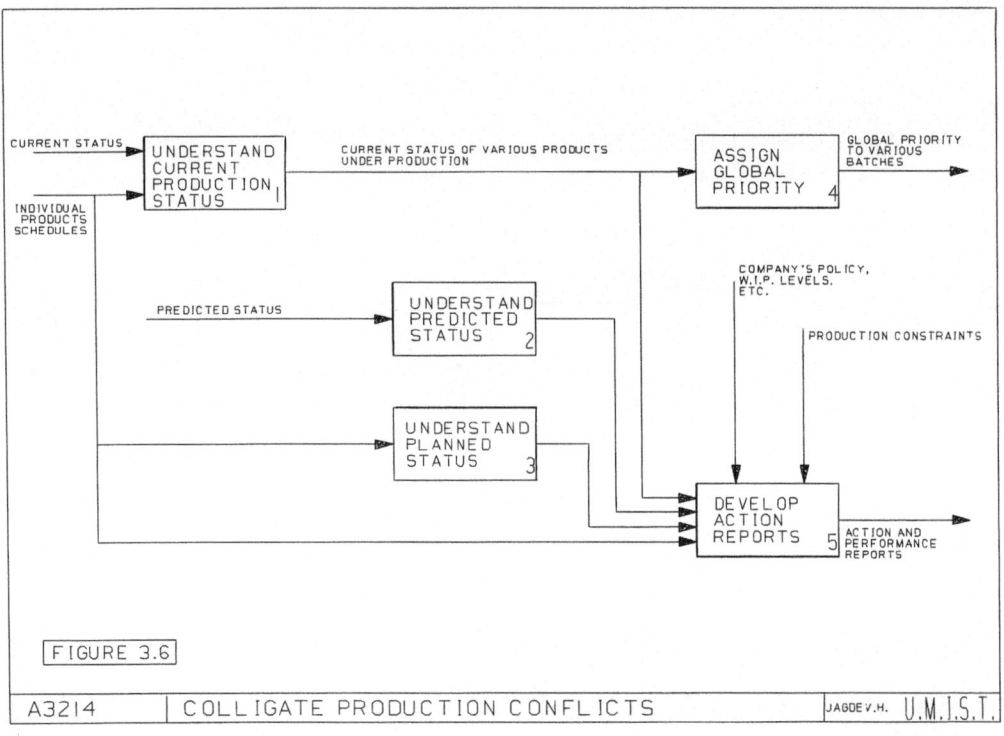

15. Design of a Generalized Job Shop Control System and PM Packages

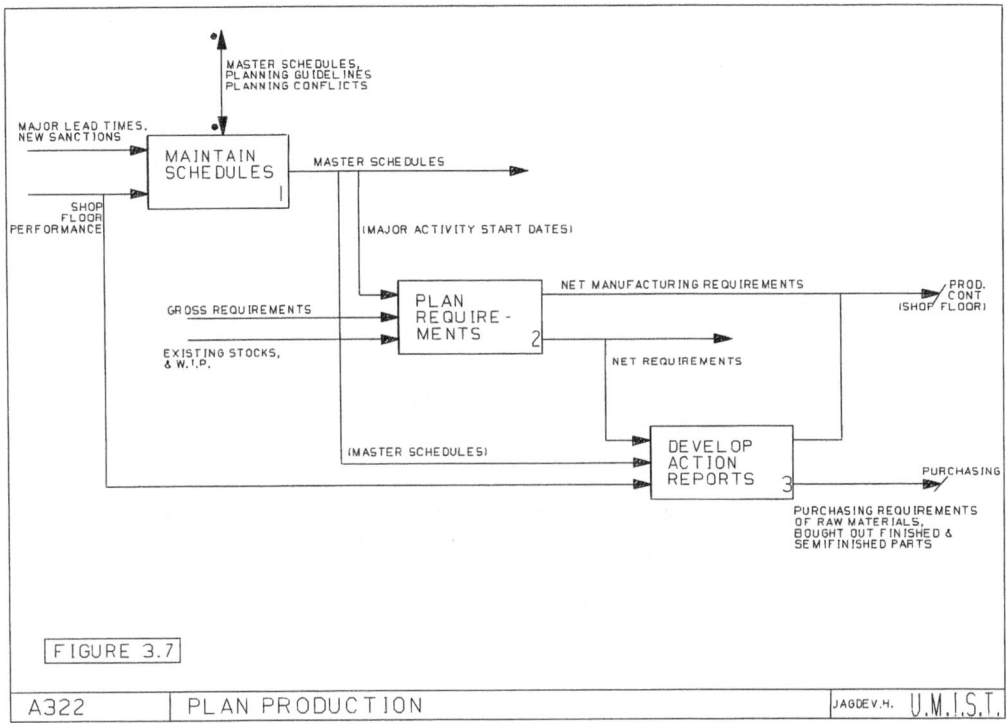

FIGURE 3.7 — A322 PLAN PRODUCTION

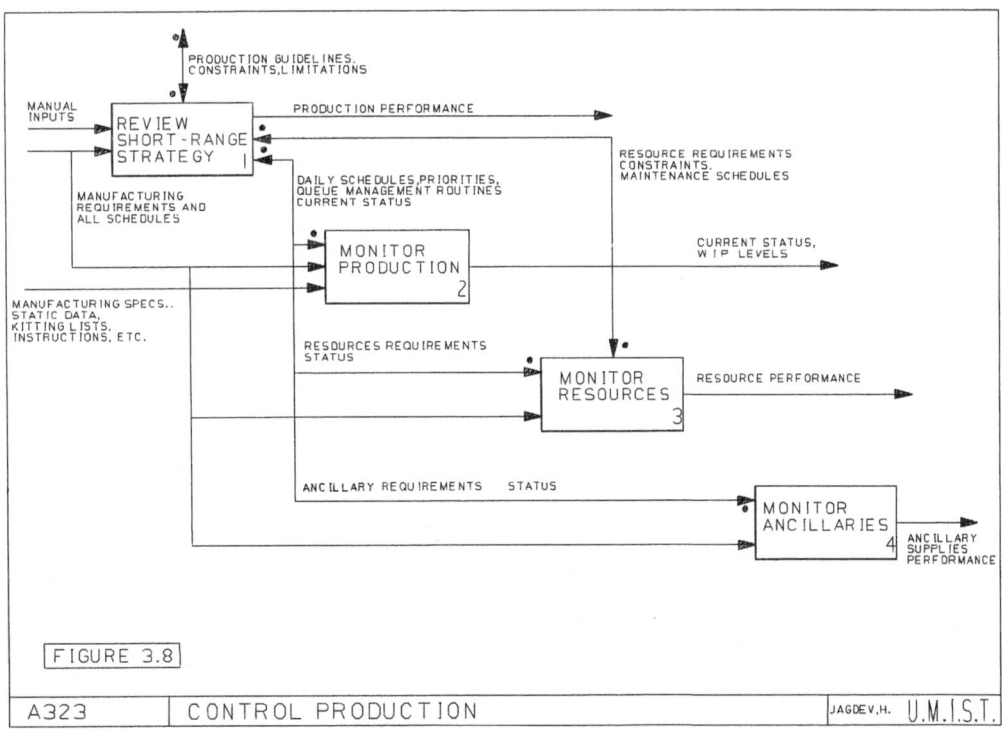

FIGURE 3.8 — A323 CONTROL PRODUCTION

ment. The remaining diagrams are exploded in standard top-down decomposition for only those activities which are relevant to the planning and control of production. Moreover, details are restricted to only the top few levels to highlight the philosophy behind the modularization. This decomposition takes care that different levels in the model fit appropriately in the organizational structure of most job shops, and it also ensures that the linking mechanisms are feasible and consistent at all levels of disaggregation. The details of all related activities (e.g. Stock Control, A34) have been omitted. Moreover, only the specific inputs and outputs are retained to maintain clarity.

The salient feature of this model is that it recognizes that most decisions in job shops are based on their internal (state of the system) and external environments, and, above all, these conditions will often change, requiring new operating rules. These varying operating conditions are considered by two separate modules, one for each horizon, called Strategy Control. The existing algorithms in these modules can be "tuned" or specifically written to suit individual requirements, without altering the rest of the package. This feature not only enhances the portability of the package, but the package can be implemented in a wider variety of job shops. Another advantage is that unlike most of the commercial production control packages, it does not always force the job shop to conform to its standards but "modules" itself around specific job shop requirements.

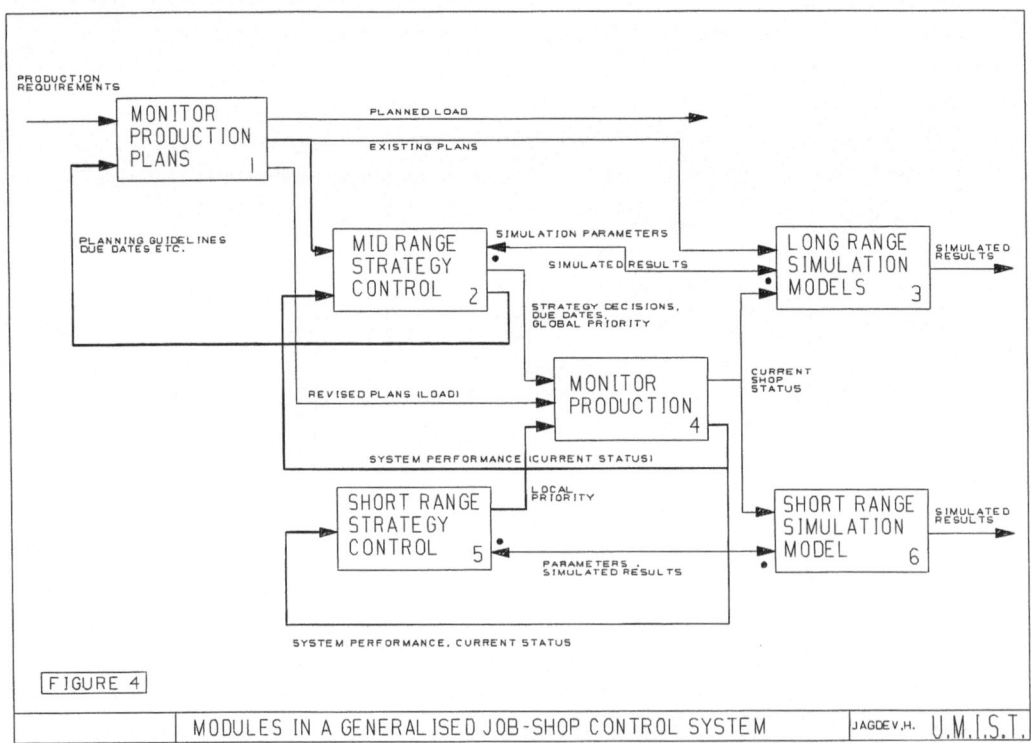

FIGURE 4
MODULES IN A GENERALISED JOB-SHOP CONTROL SYSTEM

As the time rolls into the future, the cascading of activities from Mid-Range (Plan Production, A322) into the Short-Range (Control Production, A323) can be clearly observed in the Actagram A32. The flow of information in this direction is controlled by the Mid-Range Strategy Control A321, while the feedback from the Short-Range is controlled by the Short-Range Strategy Control A3231. Besides Strategy Control there are two more modules, Monitoring and Predictive Techniques, for each horizon. Although these three modules perform similar functions in their respective horizons, the nature of events performed and the structure of the software logic, including the data structures, is distinctly different. The interrelationship within this two-level hierarchy, especially the control of feedback mechanisms within each level and between the two levels is highlighted in Fig. 4. The main functions of these modules are discussed below.

3.1 Strategy Control

Each of these modules consists of a suite of algorithms which act as decision support modules to the activities of the respective horizons. The Strategy Control modules are built within the Intelligent Knowledge-Based Systems framework. Generally, in this framework, the IKBS comprises a Knowledge Base and an Inference Engine which operates on the Knowledge Base. The Knowledge Base considers knowledge representation with different levels of granularity and it consists of a combination of IF-THEN rules and certain causal and mathematical models. The Inference Engine is capable of operating on both rules and models.

Mid-Range Strategy Control performs such activities as:

- Lot Size determination
- Monitoring and maintenance of Lead Times: lead times will be different for various items belonging to the same final product
- Maintaenance of various master schedules and interaction with them with respect to:
    - monitoring, ordering, and receipt of raw materials, bought out finished and semifinished parts
    - order scheduling of major events like the release of sanctions, machining of key components, and assembly operations
    - Modifications of the standard design and its planning to the customer's specifications
- Capacity requirements planning and related activities of smoothing and balancing of the load; it may also include decisions on subcontracting to accommodate capacity limitations of the system
- Controlling of Scrap and Rework on faulty components
- Monitoring of WIP levels according to the set guidelines
- Production Cost Control
- Tracking of anticipated shortages, especially the ones on the "critical path" of the products
- Maintaining the statistical performance of all the internal as well as external resources; this information will be used to maintain lead times
- Generation of the aggregate models of the system.

Short-Range Strategy Control performs such activities as:
- Generation of Decision Arrays. Decision Arrays form a part of the sophisticated database of the IKBS. These arrays contain a series of instructions for the Monitoring Module to perform non-routine decisions on reaching a specified target. Although these arrays are generated automatically through simulation, experience indicates that mannual modifications are sometimes necessary. Typical information contained in these arrays is when to take a machine off-line for preventive maintenance, decisions on the routings of batches with several feasible routes, Pre-emption of batches, changing of priority pointers for rush jobs, changing of queue-serving rules, etc. Routine decisions are made by a separate suite of algorithms maintained by the Strategy Control and coupled to the Monitoring Module
- Identification and close monitoring of the performance of "current" bottleneck resources in the system; also, monitoring resources upstream and downstream of the bottlenecks
- Generation of work-to-lists and "chasing" such necessary actions required by them as:
  - materials release from the stores
  - ancillaries requirements
  - transportation requirements
- Location, identification and notification all forms of shortages
- Monitoring the current capacity of the sytem by considering worker absence, machine down, and immediate calendars
- Performing Short-Range capacity planning by considering the current and anticipated states and capacity of the system
- Scheduling of components requiring re-work.

3.2 Predictive Techniques

These modules consist of various predictive algorithms and other programs to predict the performance of activities at each level. In Mid-Range they may include such modules as MRP, MRP II, MRP III, Forward and Backward Capacity Planning, and Stochastic Simulators. At this level the capacity of the system is considered in aggregate form which will, in turn, require sophisticated aggregation techniques. As a result, only the major event schedules are generated and possible capacity requirements conflicts resolved.

In semi-continuous process type plants, where the lead times and throughput times are generally small compared with the traditional job shops, it is not necessary to have predictive techniques at the Mid-Range level. Strategy Control algorithms which will generate the aggregate models and determine the lot sizes furnish sufficient information to the Short-Range level.

At Short-Range level, predictive techniques mainly consist of a deterministic discrete event simulator. This simulator should initialize from the monitoring module to reflect the current state of the system and then progress the work through the queues and resources according to the established rules. Any known, machine downtimes, like preventive maintenance, should be modeled in this simulator. The outputs of this simulator will give the precise times of each event occurring on the shop floor. From these details, relevant sets of information can be filtered.

Deterministic simulators may have two flaws. Firstly, any variance between actual times, which occur on the shop floor, and the stored times, which are used by the simulator, will result in the wrong prediction of event times. Secondly, if the rules used for loading the machines on the shop floor are not the same as in the simulator, wrong prediction of the sequence of events will result. Due to these two flaws, the correlation between the simulated results and reality will diminish with time. The first flaw can be corrected either by correcting the processing times, or if it is not possible and the variance is small, recognizing the factors (for each resource) causing the variance, while analysing the results. There is only one solution for the second flaw: either adhere to the work-to-lists generated by the simulator or, if they are not acceptable, establish the way decisions are made and the dynamics of the shop floor and incorporate them into the knowledge base of the simulator. According to the author's experience, simulators built on these lines give around 85 percent correlation with reality for up to 25 Batch-Operations per machine. This is, on the average, ten shifts each of eight hours.

In structuring Short-Range simulators for semi-continuous process type plants we have to consider not only all the resources and their capabilities but also the coupling constraints among them. Experience in developing such systems indicates that it is more efficient to incorporate most of the Strategy Control algorithms in the knowledge base of the simulator.

3.3 Monitoring

As the name implies, these modules monitor all the events that occur at each level. In essence, this is a database management task which for Mid-Range will include the maintenance of schedules and related activities, and for Short-Range the monitoring of the "current" state of each resource and batch on the shop floor.

Monitoring concerns itself only with the current state of the system. Hence, it operates on only those arrays which define the current state. In the Short-Range the arrays pertaining to the various queues reflect the exact state as it exists on the shop floor. Other information such as the expected arrivals of batches into the queues (Forward Load Lists) will be generated and used by the Strategy Control algorithms alone, which may, if necessary, update Decision Arrays. The monitoring software makes either routine decisions which are independent of the state of the systems or it is driven by the Decision Arrays.

Unlike the Mid-Range horizon, the Short-Range horizon is at the front end of the system where the capacity and Operation Routings cannot be considered in aggregate form. The effectiveness of Short-Range scheduling very much depends on how its software and data structures relate to the actual material and information flow on the shop floor.

One possible modular structure is to discretize the system into several sub-systems on the lines of shop floor layout and its inherent materilas flow network.

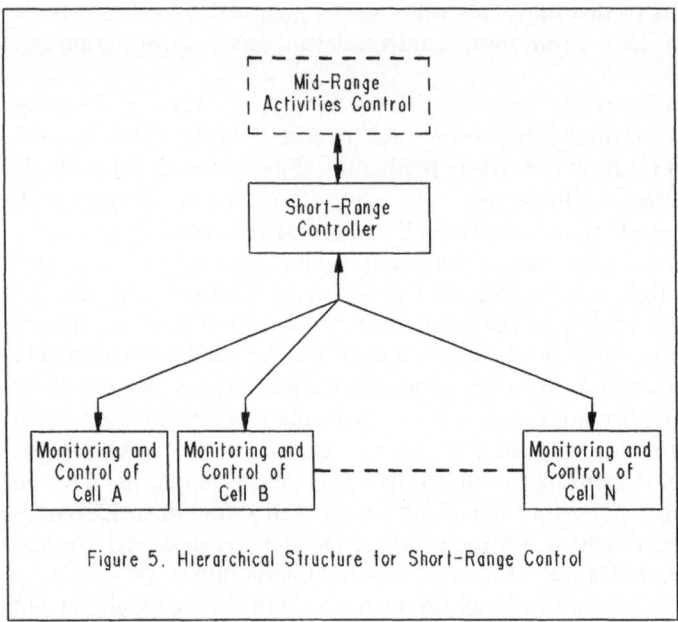

Figure 5. Hierarchical Structure for Short-Range Control

Such a hierarchical system, which has been successfully tested by the author, is shown in Fig. 5.

The monitoring, control, and scheduling of individual cells on the shop floor are performed by separate algorithms specializing in the particular structure of the cell. Separate optimizing (mathematical or Intelligent Knowledge Based) algorithms can be incorporated to enhance the performance of a cell. All cells have their own dynamic databases. The maintenance of this local database as well as all the intra-cell communications are performed by the cell controller. The communications between the cell and the outside world are controlled by the controller at the upper level. In fact, such a hierarchy can be extended further into a cell with complex structure. The most useful feature of such a structure is that it can be easily converted into a decentralized control system.

Typical cells common among the job shops, which a generalized system will have to incorporate, are:

- Functional machines cell, where all the machines are similar but need not be identical
- FMS Cell, which is a miniature integrated manufacturing system
- GT Cell, which a family of similar components visit, all with similar routing direction
- Flow or Transfer Line Cell
- Heat Treatment type Cell, where possibly a whole batch can be processed simultaneously, or batches can be mixed
- Assembly Cell, where various components and subassemblies arrive according to established schedules to be assembled

15. Design of a Generalized Job Shop Control System and PM Packages

Process Groups	Machines Contained
P1	M1, M2
P2	M1, M2, M3 (P1, M3)

Fig. 6a

Machine	Process Groups it belongs to
M1	P1, P2
M2	P1, P2
M3	P2

Fig. 6b

Figure 6. Definition of Functional Process Groups

- Control of transportation, of recources common to the whole shop. Materials movement facilities dedicated to a cell, as in FMS, form an integral part of the cell, to be controlled only by the respective cell controller.

Differing capabilities of similar machines in a Cell can be taken into account by using the concept of Process Groups. The main features of Process Groups are that a Process Group can contain more than one machine, a machine can belong to more than one Process Group, Routings of jobs indicate the Process Group number and not the machine number where each operation can be performed, and, finally, the jobs are queued under Process Groups and not under individual machines.

For example, consider a system of three machining centres M1, M2, and M3. If M1 and M2 are identical and of larger capacity than M3, then all jobs capable of being performed on M3 can also be performed on M1 or M2, but not vice versa. In this case there will be two Process Groups P1 and P2 with the structure given by Fig. 6.

The machine loading under this setup will be done in such a way that when machines M1 or M2 become available, the first priority will be given to batches in queue P1. In the case there are no jobs waiting or none of the waiting jobs are urgent, queue P2 will be considered. When M3 becomes available, there is one queue (P2) needing to be searched. The Process Groups in Fig. 6(b) can be arranged in a given order of priority.

If there is some form of priority system used to serve the queues, the type of rule used will depend on the state of the queue. Since this state will vary from one Process Group to another, and also for a given Process group from time to time as it moves in and out of the bottleneck state, the Strategy Control should be able to assign thr priority selection criteria separately for each Process Group. This feature, which is very important, ahs been ignored by most of the commercial production control package developers.

3.4 Remarks on the Model

The philosophy described here has very sound foundations. The software developed for traditional job shops within this framework has been tested and validated in real shops, with very encouraging results. The software developed for process plants is under rigorous field trials. The international company which originally funded the project intends to implement it in over one hundred of its plants around the world.

4 Available Packages

Production Control packages available on the market today offer a very wide range of facilities covering almost all aspects of manufacturing. The trend in computerization of manufacturing activities has been from relatively simple and straightforward modules covering the peripheral activities like Finanacial Control, to manufacturing operations-related activities like Stock Control, Order Processing, Bill of Materials Processing, Requirements Planning, etc., and to the very core of manufacturing operations − Shop Floor Control.

Table 1 shows a list of some of the packages available (with the supplier's name in parentheses) on the British market. Even though some of the packages are meant for specific types of shops − such as the Foundry control package number 31, PLANTRAC (96) for project management, SASM-PCB (113) and SCICARD(R) (114) for the manufacture of PCBs − most of the packages claim to operate under traditional job shop environments. This list considers only the shop floor control facilities offered by the packages, for two reasons. Firstly, the packages offering the shop floor control module invariably offer all other supporting modules for complete production control.

Secondly, because of the very nature of shop floor activities the quality of this module varies most among the available packages. Each package reflects the philosophies of the designers, and hence their understanding of the job shop operations and how they should be controlled. Among the available packages these philosophies vary very considerably indeed. The facilities and flexibility offered by the shop floor control module reflects, in general, the sophistication of the complete range of modules offered by the supplier.

This classification under shop floor control merely reflexts the requirements of traditional manufacturing job shops. Certain environments may not require shop floor control in its entirety. Process, semi-continuous production plants, and project management type of operations may not need the constant real-time monitoring of all the resources. A job shop mainly engaged in manual operations such as overhaul and repair, where the work content and the processing times can never be accurately predetermined, will not benefit from the real-time facility, and a good MRP and MRP II could replace the shop floor module. On the other hand, a job shop consisting of numerically controlled machine tools and clearly defined work-in-progress could greatly benefit by implementing a real-time con-

15. Design of a Generalized Job Shop Control System and PM Packages

Table 1. Production control packages

No.	Package and Supplier	#	$	%	*
1	ACSPIC-I (ACS Commercial Systems)	Y	Y	N	N
2	ACSPIC-II (ACS Commercial Systems)	Y	Y	Y	Y
3	ACT (Applied Computer Techniques)	N	N	N	N
4	ACTION FILE (Action file Software)	–	–	–	–
5	ALPHA (Alpha Omega Business Sys.)	Y	–	–	Y
6	ALPHA PROCON (Alpha-Numeric Mgmt. Services)	Y	Y	N	Y
7	AMACS (ADP Network Services)	N	N	N	N
8	ANDROMEDA (Marshall Spencely Ltd.)	N	N	N	N
9	AOMEGA (Samleco/TC Systems)	Y	Y	Y	Y
10	APICS (LMR Computer Services)	N	Y	Y	N
11	ASPECT (Ospery Computer Services)	Y	Y	Y	Y
12	ATLAS (Scan Computers)	Y	Y	Y	Y
13	BESPOKE P.C. (Geest Computer Services)	N	N	N	N
14	BR-PAC (Business Research)	Y	Y	Y	Y
15	CAC/P.C. (Claisse-Allen Computing)	Y	Y	N	N
16	CAPE (National Engg. Lab.)	N	N	N	N
17	CAPITOL (Computer Tech.)	Y	Y	N	N
18	CCS P.C. (Computing & Comm. Services)	Y	Y	Y	N
19	CEEQUEL (Scott-Grant Ltd.)	Y	–	–	–
20	CLASS (Compower Ltd.)	Y	N	N	Y
21	CONCEPT P.C.S. (Concept Computer Systems)	Y	Y	Y	Y
22	COPICS (IBM)	Y	Y	Y	N
23	DATALINE (Systemshare)	Y	Y	N	N
24	DELTA (Compaddress Ltd.)	–	–	–	–
25	EBS 11 (Computer Encapsulated B.S.)	Y	Y	Y	Y
26	ECCSYS (Compex Computer Services)	N	N	N	N
27	EMMA (Tripont Associated)	Y	Y	N	N
28	EPIC (Safe Computing)	Y	Y	Y	N
29	EXPAC/STOKPAC (Data Services Int.)	Y	N	N	N
30	FEROS (Nixdorf Computers Ltd.)	Y	Y	Y	Y
31	Foundary P.C. Package (BNF Tech. Centre)	N	N	N	N
32	GP 90 (British Olivetti)	Y	N	Y	N
33	HDMS (Honeywell)	Y	Y	N	N
34	HI-PROD (Memec Systems)	N	Y	N	N
35	HOCUS (P-E Information Systems)	Y	Y	Y	Y
36	IMCS (NCR Ltd.)	Y	Y	Y	N
37	IMPCON (Stavely Computing)	Y	Y	Y	N
38	IMP/BOMP (CPU Computers Ltd.)	N	N	N	N
39	IMS (Honeywell Information Systems)	N	N	N	N
40	IMS/PCS (Honeywell Information Systems)	Y	Y	N	N
41	IMS-TD (Honeywell Information Systems)	Y	Y	Y	Y
42	INCA (Software Services)	Y	Y	Y	N
43	INFOFLO (Interactive Inc.)	Y	Y	Y	Y
44	Integrated M.C. Systems (Allen Computers)	Y	Y	Y	N
45	Integrated P.C. (Security Computing)	Y	Y	N	N
46	Interactive Prod. Scheduler (Protech Data)	Y	Y	Y	N
47	INTERNET 80S (Computation Res. & Dev. Ltd.)	–	–	–	–
48	IPICS (IBM)	Y	Y	Y	N
49	MAAPICS (IBM)	Y	Y	Y	N
50	MAC-PAC Cobol/RPG (Arthur Anderson)	Y	N	Y	N
51	MAC-SYS (Disc Computer Services)	–	–	–	–
52	MANMAN (Scicon Consultancy Ltd.)	Y	Y	Y	N
53	MAN-TRAC (Computerline P.M.S. Ltd.)	Y	–	Y	Y

Table 1 (continued)

No.	Package and Supplier	#	$	%	*
54	M.S.C. (Osprey Consultancy)	Y	Y	N	N
55	M.C.S. (Computer Machinery Co.)	Y	Y	Y	N
56	MMC (Adserve Computing)	Y	Y	N	N
57	Mfg. Package (Compex Computer Services)	N	N	N	N
58	MRPS (Cincom Systems)	Y	Y	Y	Y
59	Mfg. & Stock Systems (A & P Appledore)	Y	Y	Y	N
60	MANUMARK (Hallmark Associates)	Y	Y	Y	Y
61	MAP/3000 (Cara Consultancy)	Y	Y	Y	N
62	MAPS (Datasab Ltd.)	Y	Y	N	N
63	MARC (Deritend Computer Bureau)	Y	Y	Y	N
64	MAS (Hoskyns Group)	Y	Y	Y	N
65	Material Control System (System by Design)	N	N	N	N
66	Material Mgmt. 3000 (Hewlett-Packard)	Y	Y	Y	Y
67	MICRO-SaFeS (Safe Computing Ltd.)	Y	–	–	–
68	MCS (Hill Price Davidson)	–	–	–	–
69	MICRO-SCOPE (Systemshare)	Y	Y	N	N
70	MICROSS (Selven Systems)	Y	Y	Y	Y
71	MIDAS (Cara consulting Ltd.)	N	N	N	N
72	MIMER II (Ericsson Info. Systems)	Y	–	–	–
73	MINIBOMP (Incomputer)	N	Y	N	N
74	MIMS (Geisco)	Y	Y	Y	N
75	MINIPICS (Globe Computing Ltd.)	N	N	N	N
76	MIPS (Business Computers Ltd.)	Y	Y	N	N
77	MISSION (NCR)	Y	Y	Y	N
78	MMAS (IBM)	Y	Y	Y	N
79	MMPS (Cincom Systems)	N	N	N	N
80	MMS Net Change M.C.S. (Package Programmes)	Y	N	Y	N
81	MSA (Mgmt. Serv. America)	–	–	–	–
82	MSS P.C. (MSS Computing & Business Consl.)	–	–	–	–
83	MULTIBUS-MANUMARK (Allied Business Systems)	Y	Y	Y	Y
84	NIPS (Nixdorf Computer)	Y	Y	Y	Y
85	OMAC (ICL)	Y	Y	Y	N
86	ODESSY (OD Engineering Systems)	Y	–	–	–
87	OPAIC (Enterprise Systems)	Y	–	–	–
88	PACS (Pactel)	Y	Y	Y	N
89	PARTSPLAN (Baric Computing)	N	N	N	N
90	PASCS (Hourds Computing)	Y	Y	Y	N
91	PC II (Mgmt. Control Systems)	Y	N	N	Y
92	PCS II/III (Burroughs)	Y	Y	Y	Y
93	PECS (Loscher Associates)	–	–	–	–
94	PIPA (Safe Computing Ltd.)	Y	Y	Y	N
95	PLANIT (Sheffield Micro)	Y	–	–	–
96	PLANTRAC (Computerline)	Y	N	Y	Y
97	PMC Eurosoft (Eurosoft-Gothenberg)	Y	N	N	Y
98	PMCS (Security Computing Services)	Y	Y	N	N
99	PMS (Phillips Data Systems)	Y	Y	Y	N
100	PRESS (BMG Micro Systems)	Y	N	Y	Y
101	PRIORITY MANAGER (Ariadne Systems)	Y	–	–	Y
102	PRO II (Sys. Designers Industrial)	Y	Y	–	Y
103	P.C.S. (Concept Computer Systems Ltd.)	Y	Y	Y	N
104	P.C.S. (Gama Software Products Ltd.)	N	N	N	N
105	P.C.S. (Sys-Com Computers Ltd.)	Y	Y	Y	N
106	Production Scheduling System (Hourds)	Y	Y	N	N

15. Design of a Generalized Job Shop Control System and PM Packages

Table 1 (continued)

No.	Package and Supplier	#	$	%	*
107	PROTOS 2000 (Powell Duffryn)	Y	Y	Y	N
108	PPS (TDS Business Systems)	Y	Y	N	N
109	PSL 6 (Factory Management Systems)	Y	Y	Y	N
110	RCSL (Roeder Computer Systems)	Y	Y	Y	Y
111	REPORT (Nexus Data Services)	Y	Y	Y	N
112	SAFES (Safe Computing)	Y	Y	Y	Y
113	SAM-PCB (Dionysys Ltd.)	Y	–	–	–
114	SCICARD(R) (Scientific Calculations)	–	–	–	–
115	SCOPE (Systemshare)	Y	Y	Y	Y
116	SCP (Loscher Associates)	N	Y	Y	N
117	SIMACC P.C. System (Lincoln House)	Y	Y	N	N
118	Small Business System (Tube Plastics)	Y	Y	Y	N
119	SMART 1 (Whessoe Tech.)	Y	Y	N	N
120	SMM (Stibbe Computer Services)	Y	N	Y	Y
121	Standard P.C. Systems (SPIS)	Y	Y	Y	N
122	STAPL (Universal Computers)	Y	N	N	N
123	Stores Recording (GMS Computing)	N	N	N	N
124	Stores Level 1 (Dayton James)	N	N	N	N
125	SWORD (Compeda Ltd.)	Y	Y	Y	Y
126	SYSIMP (Systime)	Y	Y	Y	N
127	SYSTEM 400 (PERA Assoc.)	Y	Y	Y	Y
128	SYSTIME Mfg. System (Systime)	Y	Y	Y	N
129	TBOMP (Time Utilising Business Systems)	Y	Y	N	N
130	TELEPLAN (Teleprocessing Computers)	Y	Y	N	N
131	Time Based Net Rqmnts. Plg. (Economic Data)	Y	Y	Y	N
132	TOPICS (MSS)	Y	Y	Y	Y
133	TRIFIED (Trified Software)	Y	Y	Y	Y
134	UNIFORM (Sperry Univac)	Y	Y	Y	N
135	UNIS 1100 (Sperry Univac)	Y	Y	Y	Y
136	UNIS 80/80E (Sperry Univac)	Y	Y	N	Y
137	VISOR (Comshare)	Y	Y	Y	Y
138	WASP (AERE-Harwell)	Y	N	Y	Y
139	WIP/POWER (Dataskill)	Y	Y	N	Y
140	WISDOM (Minerva Computer Systems)	–	–	–	–

Key: # ⇒ Shop Control; $ ⇒ Real Time; % ⇒ Simulation; * ⇒ Finite Capacity

trol system. The absence of a shop floor control module in some packages does not reflect on the quality of other facilities offered. Some do offer good capacity and materials planning facilities without offering any shop floor control.

Any shop floor control module will have two main components: database management and decision making. These two components can reside in separate modules or they can be mixed together in respective software programs. The advantage in the former architecture is that the resultant package can be designed to be more generalized, and it can be further "tuned" to suit individual requirements. Because of the complexity of the decision-making process, these requirements will vary more for the shop floor control than for any other module. Although almost all packages claim to be very modular, it is difficult to ascertain whether this modularity is restricted to entirely different functions such as order processing, BOM, stock control, etc., or whether individual modules are also composed of modular components which can be selectively connected to suit individual requirements.

Since the quality of the database management component is very high among most top contenders, one very useful way to check the sophistication and, above all, the suitability of a shop floor control module is to see the flexibility offered by the decision-making component, and how it relates to individual requirements.

Table 1 also includes the following three features necessary in a package purporting to perform shop floor control:
- Capability for Real-Time Operation
- Simulation facility: simulation seems to be the most efficient (and popular) way of scheduling a real-life-size job shop and answering other "what-if" type of questions
- Finite Capacity: at planning stage, capacity may be considered in aggregate form, but for-day-to-day shop floor control, only finite capacity can give the realistic load profiles, work-to-lists, etc.

For each of these facilities, the answer can be Yes that facility is offered by the package (Y), No, this facility is not offered (N), or ($-$) implying not enough information is available from the brochures.

It is not a very easy task to analyse the full capabilities of any package from its brochure. The information contained can sometimes be confusing and even contradictory.

The term Simulation has been very widely interpreted. Academics understand this term as processing the work through the system with capacity and capabilities of the resources fully grasped. Simulation will give results in the form of events anticipated on the shop floor and the performance of resources. This should not be confused with Capacity Planning which predicts the capacity required by the given work-load over a time period. It does not determine how the available capacity will be utilized to process the work. Simulation and Capacity Planning will complement but not replace each other. Yet, 40 packages offer Simulation without ever considering the Finite Capacity of the system. In fact two packages, APICS (10) and SCP (116), offer simulation but not shop control. Of the 21 packages which offer both Finite Capacity and Simulation, it is not clear which of them actually consider the capacity in their simultation programs. When considering the capacity, most of the packages consider only the machines. Only two packages, EBS11 (25) and SYSTEM 400 (127), are known to consider the finite capacity of both men and machines.

Five packages (10, 26, 34, 73, and 116) offer Real-Time Operation but no shop floor control.

In all packages the scheduling of work-load is done by simple rules like SPT, FCFS, NOROP, DD, Critical Ratio, etc., or some combination of these rules. Most packages offer a fairly wide variety of the rules, but once the priority selection criteria are made (manually), this selected rule applies to the whole shop. The author is not aware of any package, in which different priority rules can be used, where the different areas of the shop or system select the priority rule automatically depending on the state of the queues.

No package offers any form of optimization in schedule generation for the whole shop or parts of it. A possible reason could be that standard optimization techniques are at best very time consuming and at worst useless for the real size and complexity of problems. The recent advances in hierarchical techniques and Perturbation analysis show good promise in solving job shop type problems, but it may take a few years before they become commercially available. It is also not clear how modular and flexible most packages are and to what extent they can adapt to standard or specific optimization algorithms.

A detailed comparison of the packages claiming to offer identical facilities can be done only after thorough study of the packages and their architectures. This study may not always be possible, and, if done, will always be subjective.

Undeniably, some of the packages are excellent. It would not be fair for the author to point out some names while others, not analysed in detail, could be equally good. It should also be noted that all packages may be under a continuous state of development. Existing modules are being enhanced and extra modules added on a regular basis. It is up to the prospective user to determine his current and future requirements and only then decide on the package.

5 References

1. Axsater, S., Schneeweiss, Ch., and Silver, E. (eds.), Multi-Stage Production Planning and Inventory Control, Lecture Notes in Economics and Mathematical Systems, No. 266. Springer, Berlin, Heidelberg, New York, 1986
2. French, S., Sequencing and Scheduling. Ellis Horwood, Chichester, 1982
3. Buchanan, Bruce G., and Shortliffe, Edward H. (eds.), Rule-Based Expert System. Addison-Wesley, Reading, MA, 1984
4. Singh, M. G., and Titli, A., Systems: Decomposition, Optimisation and Control. Pergamon Press, Oxford, 1978

Chapter 16
Validation of Job Shop Control Software – A Case Study

Harinder Jagdev*

1 Introduction

Verification and Validation form a crucial component of systems development. While verification concerns itself with the process of checking whether the conceptual model has been acceptably translated into an operational program, Validation refers to the process of confirming that the conceptual model reflects the reality to an acceptable level. This chapter describes the experience gained during the initial validation process, performed in a company, of the Generalized Job Shop Control System which has been developed at UMIST by the author, hereafter referred to as the Model.

2 Company Background

The company where the Model was validated is a major manufacturer of large diesel engines for use in industrial power generation and marine propulsion. The engines are built to standard design and range from 3-cylinder 1220 bhp to 18-cylinder 10800 bhp. The machine shop consists of 67 machines and 121 personnel. Total number of employees is around 2000.

There were two reasons for selecting this company for the validation. Firstly, this company was not considered while developing the Model and its verification, thus providing a "true" test for its generality.

Secondly, this company has been one of the pioneers in CAM in the North-West of England. Since the early 1970s they have been using an in-house developed software, hereafter called the System, to control various aspects of their manufacturing operations. As a result, all the data required were readily available in the form of computer printouts, and, more importantly, detailed discussions could be held on the technical requirements of a successful job shop control system.

---

* Dr. Harinder Jagdev is also author of Chapter 15. His biography and photo appear on p. 233.

For production control, the System consists of four modules performing such functions as Product Explosion (BOM), Stock Control, Purchase Order Progressing, and Machine Shop Scheduling.

In the scheduling module, the focus of our attention — the status change of the resources and batches — is reported to the computer through a "Collect-Data system" which links six terminals, sited at convenient locations on the machine shop floor, to the central computer. A days's transactions are recorded in a sequential event file called, not to be confused with its usual meaning, "work-in-progress" file. At the end of each working day, the information collected in this event file is used to update the System's database. As a result, the database should reflect precisely the state of the shop floor. Finally, the whole System is re-scheduled using the priority rules which particularly suit this company.

3 Selection of an Area

To save the enormous time required to build the database, it was decided to perform the validation on a part of the shop floow which fulfilled the following objectives:

- There should be no material and information flow from the selected resources to the rest of the shop floor and vice versa.
- The selected area should contain a variety of machine types to reflect the whole shop floor.
- There should be sufficient production rate and "live parts" to give an adequate test.

Using the above criteria, twelve machines were selected. Six of these are used for the machining of Major Components such as cylinder blocks, crank cases, and bed plates. The remaining machines process Small Components like connecting rods, valve cages, etc. The batch sizes of Major Components are always one, and these of Small Components are normally equal to the number required in one assembly, though this cannot be generalized.

The performance of the shop floor and that of the Model were closely monitored in periods of one week at a time. At the start of each comparison run, the dynamic information on the state of the queues and batches was extracted from the System printouts. During the run all the events were physically observed on the shop floor. If there was any discrepancy between the observed event completion times and those reported by the System outputs, the latter were selected to understand the extent of the influence of delayed reporting of event completion on the System.

Observations made can be classified into three categories: Influence of Processing Times, Operations Sequencing, and Miscellaneous influences. To support the observations, whereever necessary, the results of shop floor performance and those of the Model are shown in the form of bar charts (Figs. 1 to 7). In each diagram, the top bar chart represents the progress of batches through a machine

16. Validation of Job Shop Control Software

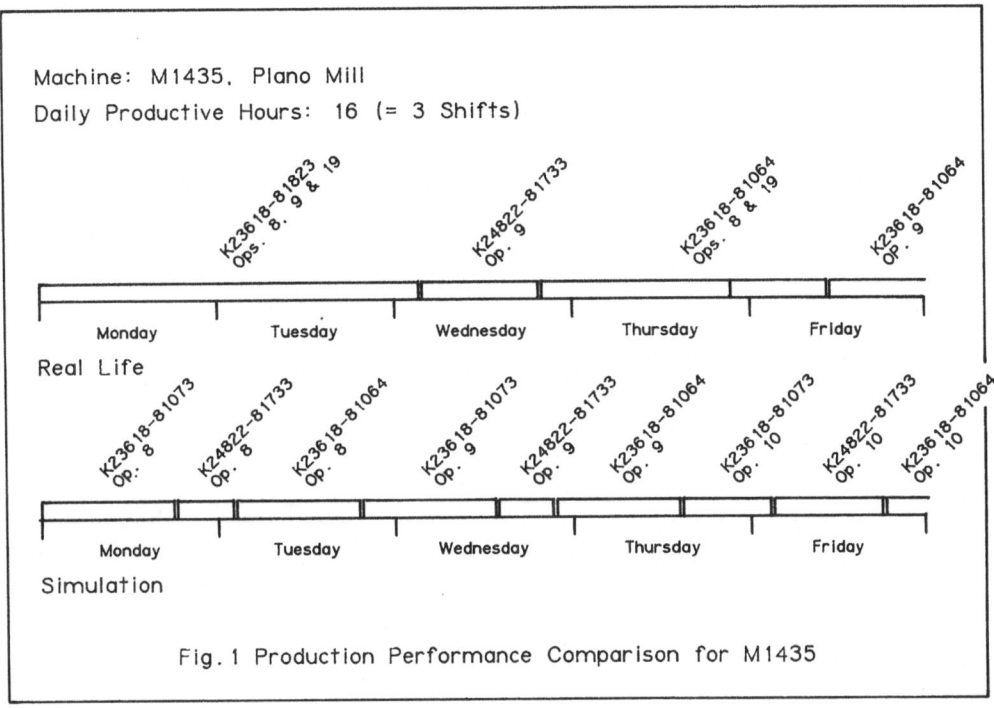

Fig. 1 Production Performance Comparison for M1435

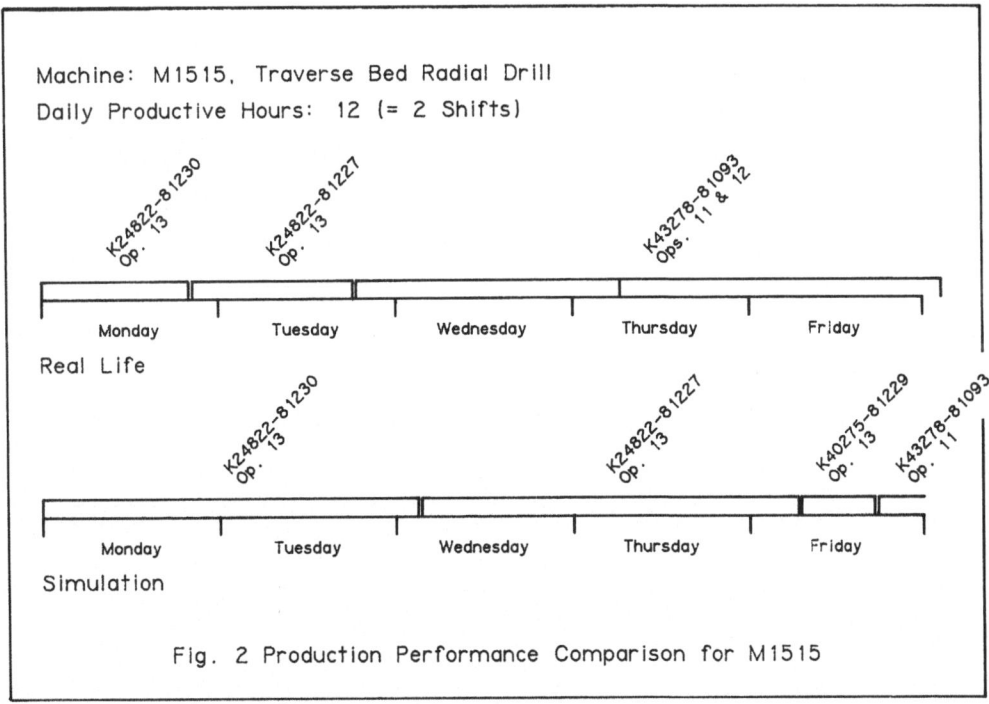

Fig. 2 Production Performance Comparison for M1515

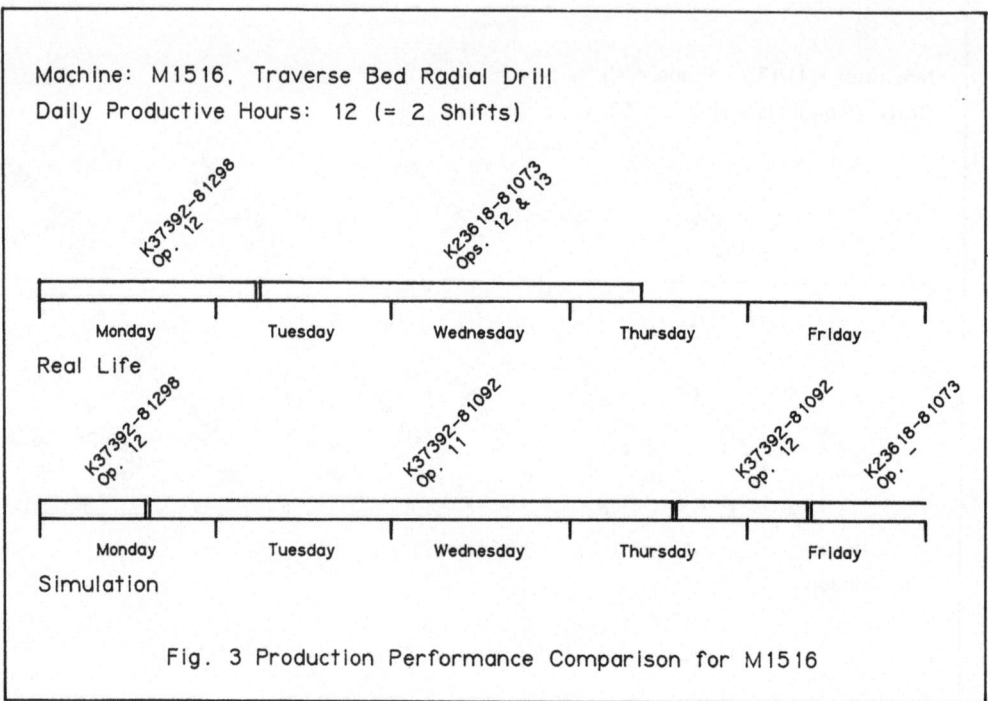

Fig. 3 Production Performance Comparison for M1516

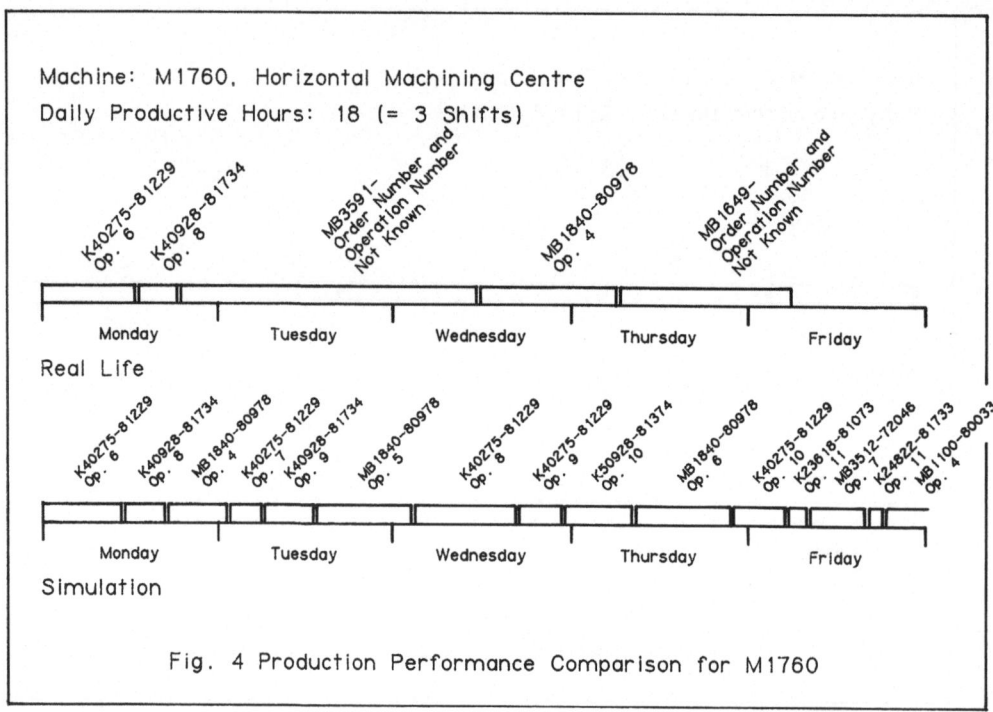

Fig. 4 Production Performance Comparison for M1760

16. Validation of Job Shop Control Software

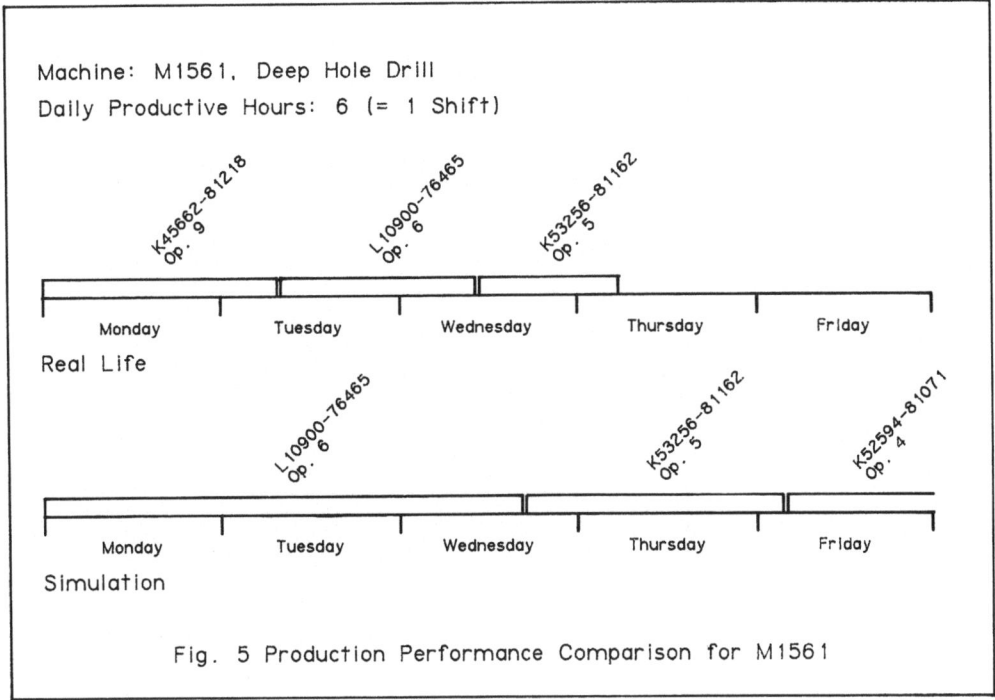

Fig. 5 Production Performance Comparison for M1561

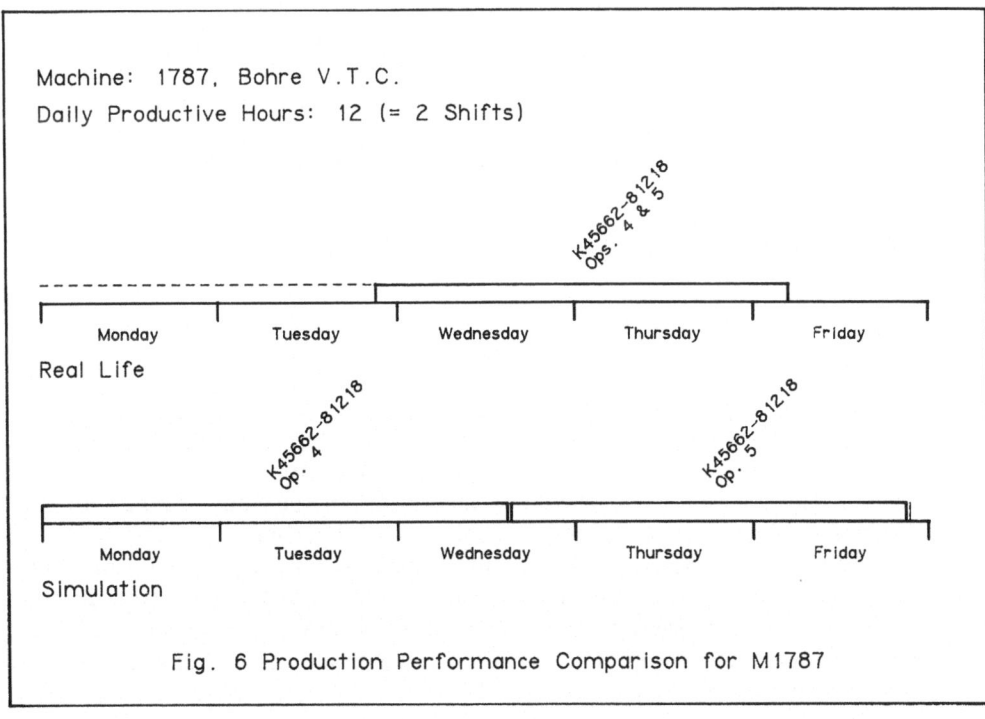

Fig. 6 Production Performance Comparison for M1787

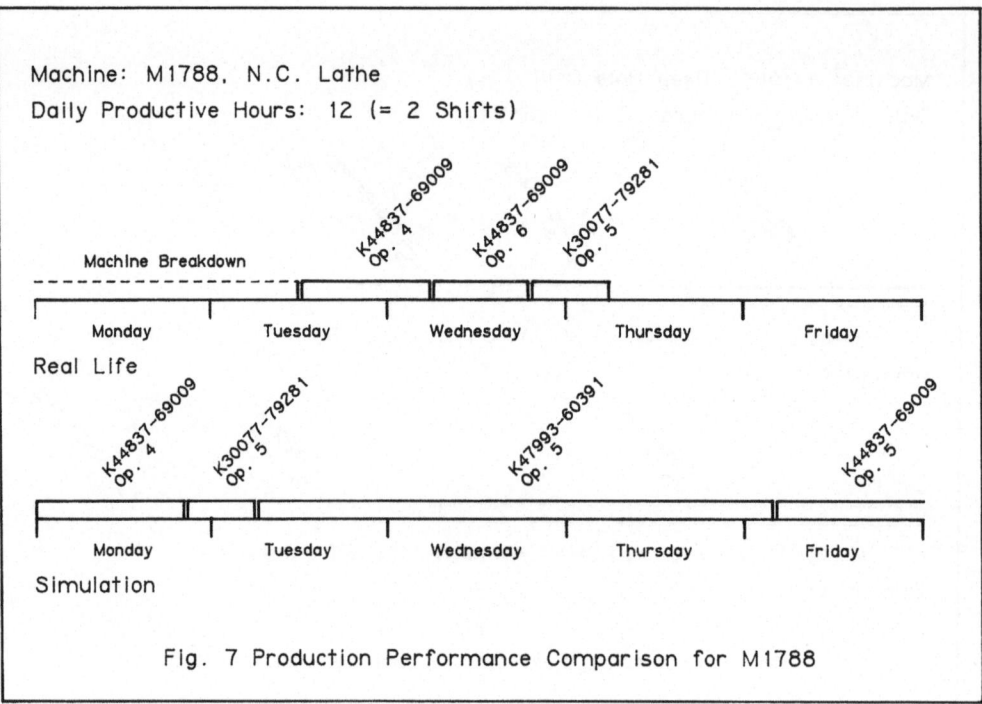

Fig. 7 Production Performance Comparison for M1788

in real life, and the bottom bar chart depicts the simulated run for the corresponding queue. The details of route cards and the state of the queues at the start and finish of test periods are not given. The list of batches pertinent to the following discussions can be derived from the bar charts of each machine. Also note that the processing times given by these charts are not strictly to scale. However, the interrelationship between real life and simulation depicted by these charts still holds true.

4 Influence of Processing Times

The accuracy of processing times is one of the most critical requirements in the scheduling of job shops. Unlike in the Mid-Range horizon, where the times can be approximated and aggregated, the performance of day-to-day schedules generated by deterministic scheduling algorithms in the Short-Range horizon depends very much on the accuracy of the processing times.

Assuming that unexpected events, like breakdowns of machines (or men) or the unavailability of the necessary ancillaries such as tools and fixtures, do not influence the processing times, the times observed have been found to deviate from their expected values. This deviation could arise from any one or more of the following reasons.

16. Validation of Job Shop Control Software

A direct cause arises from the very definition. There are two types of processing times: Actual and Allowed. Actual times are calculated through work study techniques, and they reflect the time a batch will take for a particular operation. Allowed times are those which an operator is assumed to take for the same operation. Allowed times, used to calculate the bonuses for the machine operators, are normally from 1.5 to 3.5 times the Actual times. Although the Actual times do not change, the Allowed times change frequently because operators expect a certain level of bonus each week. If this level is slipping behind in general or relative to that of other operators, pressure is put on the Rate Fixers to alter the Allowed time. In many organizations the route cards depict these negotiated Allowed times, and without the knowledge of the ratio between the two times – which can vary among the machines – it is difficult to ascertain the Actual times. There is no doubt that such systems go adrift over a period of time.

Although the company under consideration did claim to record only the Actual times, there were instances where the Author detected that the times actually taken for the operation were as low as 40% of those calculated from the Route Cards. It was not possible to determine the possible causes and, understandably, it was difficult to convince certain company personnel of these huge variances.

For many operations the processing times depicted by the route cards (and also the System's schedule files) were zero. This default value implied that the times were not finalized by the process planners. This observation becomes more significant with the knowledge that the company has been operating the computer system for many years, and the parts with unfinalized route cards were mostly standard. To resolve such cases, the Author was directed to the machine shop foremen who could specify "approximate" times. Although such cases may be acceptable in "urgent" circumstances, they should never be considered as a norm – which seems to be the case here.

A source of variance in the processing times could also arise from delayed reporting of operation completion to the System. In fact this cause was very prevalent in the company under consideration.

Another factor that could deviate the processing times from their standard values is what can be called the Just-In-Case philosophy – from the operator's point of view. Under normal circumstances when there are plenty of jobs waiting for machining, the operator works at what can be called the normal speed. When this queue of jobs starts diminishing due to lack of new arrivals the operators sense this and correspondingly reduce the speed of current jobs – just in case they may be forced to sit idle. Idle time does not fetch any bonus. Although it was not possible to confirm this practice at first-hand due to its clandestine nature, this practice is prevalent not only in this company but in several other companies the Author has visited.

Some examples where the times on the shop floor are greater than the expected values are as follows:
- Two operations, 8 and 9, for batch K23618-81073 were completed on machine M1435 but were not reported to the system until operation 19's completion on the same machine (Refer to Fig. 1). Furthermore, it is noticeable that operation 8 of the same batch has taken almost a day longer than predicted by the Model simulator. Part of this delay can be attributed to the tools pro-

curement, but the rest is due to erroneous data. Reference to Fig. 1 also shows that operation 8 on batch K23618-81064 was actually completed on Thursday night on machine M1435 but remained unreported until the following morning. This delay of approximately two hours can be considered to be acceptable.
- Figure 3 indicates that batch K37392-81298 incurred twice the expected time of 6.62 hours for operation 12 on machine M1516.
- Similarly, in Fig. 4, it can be seen that batch MB1840-80978 took an extra 4 hours for operation 4 on machine M1760.
- It was observed that for at least 50% of the cases, there was a serious delay in reporting the operation completion to the System. For example, as many as five batches processed by machine M1760 (Fig. 4) during the week, were never reported to the System. In another case, the breakdown on machine M1788 (Fig. 7) for 1.5 days at the start of the week was never reported to the system.

Examples where the times on the shop floor are less than the expected values include the following:

- Batch K24822-81230 actually takes only 11.5 hours for operation 13 on machine M1515, whereas the database gives this time to be 27.62 hours (see Fig. 2).
- Figure 5 shows that two batches, L10900-76465 and K53256-81162, for operations 6 and 5, respectively, on machine M1561 take less than half the expected time in real life. In fact, the times given on the route cards for this machine were found to be always greater than the times incurred in reality.
- Two operations 4 and 5 are performed on batch K45662-81218 on machine M1786 (Fig. 6) − in real life as well as in simulation. In the former, it takes less than half the predicted times for these two operations.

These observations justify the need to constantly monitor the processing times by a computer. The computer should generate exception reports where the variance is more than a certain percentage. These reports should result, if necessary, in the updating of the operation times on the "Route Cards" of the database.

5 Operations Sequencing

Two distinct, but related, activities fall in this category. First is the Route Sequencing which considers the sequencing of operations of a single part/batch from its raw material stage to the final inspection. The second type is the Batch Sequencing which entails the sequencing of batches awaiting one or more operations in the machines queue. The desired effect of the Route Sequencing is to smooth the work-load among the available resources. Batch Sequencing, on the other hand, is the result of scheduling algorithms which may take into account

16. Validation of Job Shop Control Software

not only the shop floor state but also such functions as the Due Dates (and hence the tardiness and slack times), number of remaining operations, Critical Ratio, Cost Functions or other objective functions which are being optimized.

Route Sequencing may include either the selection of alternative machines for a given operation, or in cases where there are no precedence constraints among some operations, the sequence of operations given on the route card could be modified to suit the current shop floor state. The selection of alternative machines poses no great problem for manual or computerized scheduling systems. It is well known that process planners indicate their favourite machines, normally the latest ones, on route cards. Frequently, some batches are transferred, at the discretion of the foreman, to older and under-utilized machines to relieve load on bottleneck machines. A computer can easily consider the "second choice" machines and generate acceptable schedules. The alternative machine selection may necessitate redefinition of the operation. For example, if one operation can be performed on an NC machine, the transfer of this operation onto a manual machine may result in several distinct operations, resulting in increased machining times. Moreover, the route cards depict the process routes based on a certain form of raw material. If the raw material form changes, say from a casting to a billet, it will require the modification of the route card, which may result in the inclusion of more operations and hence increase the lead times.

The modification of the order of operations can pose some problems. A complex part, say with 50 discrete operations, could, in theory, have hundreds of feasible routes from raw material to its required finished state. Consideration of all the feasible routes of a part will be beyond the scope of a day-to-day scheduler. Moreover, order of operations is altered on account of some very subtle reasons which require the use of a foreman's knowledge and experience – and they are known to make occasional mistakes. It may not always be possible to incorporate all this knowledge into the knowledge base of the scheduling system. Even if there is success in doing so, the resultant system will be difficult to generalize. This highlights the necessity of a company-specific submodule in the scheduler, developed solely for a given environment, which will help in smoothing the schedules.

Validating the effect of such decisions can also be very difficult. As the following examples show, the routes are altered very frequently in the company in question. The System schedules serve only as guidelines, and almost all loading decisions are made by the foremen. The Author could not find any clear set of rules which will always emulate the foreman's decision-making process for this company. Such environments necessitate the need for interactive submodules in the scheduler where the foremen can modify the computer-generated schedules to their satisfaction. Possibly, by recording such complex decisions and the subtle reasoning behind them, a truly knowledge-based company-specific scheduler can be evolved.

Some examples on Route Sequencing comparison are as follows:

– In Fig. 1, operation 19 follows operation 9 on machine M1435 for batch K23618-81073 even though there is operation 10 on the same machine and operations 12 and 13 on machine M1516.

- For batch K23618-81064, machine M1435 (Fig. 1), operation 19 follows immediately after operation 8, without performing operations 9 and 10 on this machine and further intervening operations on other machines.
- For batch K23618-81733, operation 9 is performed prior to operation 8, even though both of these operations are performed on the same machine M1435 (see Fig. 1).
- Reference to Figs. 1 and 3 shows that batch K23618-81073 has suddenly become urgent during operation 19 on machine M1435. Through the foreman's intervention it is moved quickly from machine M1435 to M1516 for operations 12 and 13, even though there are several batches already waiting in the queue. In the Model simulation, this batch does not arrive from machine M1435 until Friday.
- Machine M1561 (Fig. 5) processed operation 9 of batch K45662-81218. However, in reality, this batch was in the queue of machine M1787 (Fig. 6) awaiting the completion of operations 4 and 5. It was transferred to machine M1561 for operation 9, without any instructions from the System, and then transferred back to M1787, without informing the System. During this duration of two days, machine M1561 was kept idle to await the arrival of batch K45662-81218 from M1787.
- Figure 7 shows that for batch K44837-69009, operation 6 follows operation 4 without performing operation 5. All three operations have to be performed on the same machine — M1788.

Batch Sequencing, strictly speaking, does not fall within the realm of the validation process. The idea of using sophisticated algorithms is to generate improved schedules with respect to some objective function. These improved schedules will have different sequencing of batches through the machines. Having generated the schedules by a new algorithm, it is necessary to check their global validity. A realistic environment and data in testing the schedules will shed light on their performance and highlight the weaknesses in the algorithm used.

Even though the schedules generated by the Model were endorsed by the company management during validation, some weaknesses were discovered. The main drawback of the Model schedules was related to the interrelationship between two main components of the setting time. A typical setting time is composed of the time required to set the components, directly or through fixtures, onto the machine and the time required to prepare the machine by changing to the necessary tools. The interplay between these two times can become significant, as is explained by the following examples.

Consider a batch which requires more than one successive operation on one machine, and that there are several batches, belonging to a similar type of component, waiting in the queue for the machine. Such a scenario is depicted in Fig. 1, where each of three batches — K23618-81073, K24822-81733, and K23618-81064 — are waiting for operations 8, 9 and 10 on machine M1435. Of these three batches, two are of an identical part, K23618, and the third, K24822, is similar in form, though with different processing times. If the component setting times are smaller than the tools changeover times and the batches are of similar priority, as is the case here, it makes sense to perform operation 8 on all three batches followed by operations 9 and 10. In doing so the schedule has minimized the downtime of machine M 1435, and the resulting Batch sequence is depicted in the lower bar chart of Fig. 1. This type of Batch Sequencing has analogous influence to pre-emption.

The logic described above will not be globally successful. The adverse effects can be observed for machine M 1788 in Fig. 7. For this machine, the tools changeover times are generally very small compared with the component setting times. Batch K 44837-69009 is waiting for operations 4 and 5 along with other batches. A good schedule should perform these two operations in one setting to eliminate the extra changeover. However, the sequence generated by the Model shows that the two operations are separated by two different batches, giving a very unsatisfactory result.

6 Miscellaneous Influences

The timing of the communications with the computer can be important. The necessity to inform the computer of event completion has already been shown. Two interesting examples observed during validation show the importance of accuracy and completeness of information fed to the computer:

- There are two batches MB 3591-* and MB 1649-* waiting in the queue for machine M 1760 (see Fig. 4). The system database does not even know of the existence of these batches. As a result, they cannot be scheduled by the computer. These batches have been present on the shop floor for weeks without established or default order numbers, and their progress known to very few people on the shop floor.
- In Fig. 7, machine M 1788 breaks down for a day and a half at the start of the test period. Since this was not informed to the System, and hence the Model, full capacity was assumed. This resulted in theoretical over-utilization of the machine. With such practices no machine utilization or its maintenance records can be maintained.

Working practices can also influence the operations of computer-controlled production. To highlight the importance of this, it is worth mentioning the case of another company which manufactures aircraft engine control systems. Besides the manufacture of such components as fuel pumps, they also manufacture a considerable quantity of spare parts of their products. The agreed policy with the unions states that 33% of manufacturing capacity will be devoted to the production of spares. This policy has to be adhered to irrespective of the actual demand proportions in a given calendar month. As a result, very frequently, the production schedules of components required for assemblies have to be curtailed to revert to the production of spare parts – whether they are urgently required or not. Occasionally the reverse is the case. In fact, it is extremely rare that the demand pattern is such that established proportions of work can be maintained and the schedules are adhered to. This company uses WASP for its Short-Term scheduling and control. The schedules are manually altered to maintain the proportion of work.

7 Concluding Remarks

This validation study and discussions with other computer-aided production control users has brought into focus several points. The important requirement for a computer system is that it should be capable of accepting and possibly exploiting the particular features of a manufacturing environment. In order to perform its function effectively, the scheduling module should be capable of being "plugged" into manual decisions and/or specialized algorithms developed solely for a given topology.

Once the computerized production control system is installed, the factors that will determine its success are the commitment to and, above all, the discipline required to maintain it. Without them no computer system can give benefits to its fullest extent – however sophisticated and "user friendly" it may be.

Part V

Some Important Aspects of Production Management Functions

Chapter 17
Production Scheduling

John R. King

John R. King is at present Reader in Industrial Engineering at the Management Science department, Imperial College, London, UK. He has his education from Durham and London universities. His industrial experience covers shipbuilding and light engineering industry.

1 Introduction

Scheduling is an activity that is important in many fields of endeavour ranging from the construction of school timetables to the processing of information in computers. There is, however, no area in which the problems of scheduling are dawn more sharply into focus than in the field of production. Indeed, much of the terminology employed in scheduling theory derives from the production context, simply because in the past the bulk of research activity has traditionally been directed at problems identified in manufacturing terms.

In manufacturing it is necessary to determine a master production schedule which – taking account of existing inventories – will specify what products to make, when to make them and in what quantities, in order to meet actual or forecast customer demand in the next planning period. What makes this a complex problem in practice is the requirement that the schedule must, if it is to be feasible, take account of the constraints imposed by the capacity and availability limitations on the resources that will need to be employed. Engineering products are built up from component parts which are themselves formed into subassemblies and assemblies. Some of these items may be bought out complete and ready to be used, others may require further processing, the remainder will need to be manufactured "in-house" from purchased raw materials. Provision thus has to be made for these raw materials and other bought-out items to be available in sufficient quantities in time for them to be transformed into the required finished products and by their customer due dates: this too is a scheduling problem.

At the shop floor level the scheduling problem is concerned with organizing the flow and sequencing of raw materials and parts for processing through the various manufacturing and assembly work-centres.

The problems described above are clearly interrelated and hence are integral parts of the overall production scheduling problem facing any manufacturing organization. Ideally what is needed is a computer algorithm that would, within all the myriad of constraints on the amount and availability of human, physical, and monetary resources, establish an optimal overall schedule for the specified planning horizon. In this context we imply, by optimal schedule, the solution that will generate the maximum profit, which, in all cases where sales are not production constrained, may be equivalently expressed as the solution that will incur the minimum total cost. This is a formidably complex global optimization problem which current research suggests may for ever be beyond our capability to solve. It is the recognition of this complexity that has forced researchers in this field to pursue a piecemeal approach in which the constituent parts of the problem have been considered separately under the generic titles of aggregate scheduling (or production smoothing), and machine scheduling/sequencing.

2 Aggregate Production Scheduling

The aggregate production scheduling problem may be described as that of determining a minimum total cost production schedule that will meet the forecast demands in each of the periods in the specific planing horizon ahead. The solution depends upon achieving a balance between the various and conflicting costs involved.

One classic formation of the problem is due to Wagner and Whitin [1] in which they employ dynamic programming to provide an optimal solution which resolves the conflict each month as to whether demand should be met by production (thereby incurring setup costs), or from inventory carried over from an earlier period (thereby incurring inventory holding costs).

If there are options available to meet demand each month in the planning horizon, such as regular production, overtime, inventory from an earlier period, or subcontracting, and if the associated costs are linear, then the aggregate scheduling problem can be formulated as a linear program and an optimal solution obtained within the available capacity and resource constraints.

It is the efficiency of linear programming computer codes and their ability to handle large-scale problems that makes linear programming particularly important in practice for generating, on a routine basis, optimum aggregate production schedules.

In what has become a classic study, Holt, Modigliani, Muth and Simon [2] devised a method of jointly smoothing aggregate production and work-force requirements over any specific future planning horizon to minimze total costs. This is often also referred to in the literature as the production and employment scheduling problem. This method originated by Holt et al. and further developed by others requires detailed and lengthy analysis of costs, which allows in turn the determination of the coefficients for two "linear" decision rules. The first rule determines the aggregate production and the second the size of work-force required in the next period. The only data required on each occasion the calculation is performed are the size of work-force employed in the previous period, the aggregate inventory remaining at the end of the previous period, and the aggregate demand forecasts for each period in the planning horizon. Thus, once set up, the decision rules can be routinely employed on an ongoing period-by-period basis; modifications to the simple decision rule formulae are necessary only when changes in costs arise which necessitate the recomputation of the coefficients to be employed. The assumption of quadratic forms for the various cost components gives rise, as a consequence, to linear form to the decision rules (a result of the processes of differentiation in determining an optimal solution).

In a broadly parallel approach, except that they assumed linear instead of quadratic forms for the cost functions, Hanssmann and Hess [3] employed a linear programming formulation to determine the optimum aggregate production and work-force decisions to minimize total costs based on a specified planning horizon.

The linear decision rules developed by Holt et al. [2] were employed in a paint factory where the study was carried out. The subsequent literature in this area has concentrated on the theoretical aspects of the methodology, and there is a dearth of information as to why the use of linear decision rules was discontinued at the paint factory, and there is little evidence that they have found practical application elsewhere. There have also been critical comments which call into question the basic assumptions under which the methodology was developed. Perhaps more fundamentally, the assumption of Holt et al. that work-force levels can be changed each period, if necessary by hiring and firing people, may be questioned as being totally unrealistic, thereby making the results impractical.

3 Scheduling, MRP, MRP II, and OPT

It is true to say that optimum aggregate scheduling techniques of the kind outlined have had relatively narrow application in practice and hence little impact on the industrial world at large.

It was in the 1960s that more pragmatic data-processing approaches were developed employing network structures for the job/machine system and network techniques combined with heuristic methods for setting priorities and generating schedules. These methods were used to create schedules that endeavoured to organize production to comply, insofar as that was feasible, with due-date commitments of the finished products. These early computer methods were stand-alone packages which employed their own systems for file organization and database generation.

It was also in the 1960s that computerized materials requirements planning (MRP) systems were developed that were to lead on to more extensive systems including production scheduling and capacity loading. The MRP system converts the master production schedule requirement via the bill of materials processor into the material requirements plan, which specifies the net-of-inventory materials and parts requirements necessary to fulfill the needs of the master production schedule. The materials requirements plan thereby determines what needs to bought from suppliers and when, and what needs to be made internally and when.

Materials requirements planning (MRP) is now an important and integral part of the much broader closed-loop systems of manufacturing resources planning (MRP II) that are now being widely employed in industry. MRP II employs backward scheduling of items from their due dates, assuming that each job has to queue the same average time in front of the work-centre and ignoring capacity constraints on the work-centres' resources (infinite capacity scheduling). This may of course produce schedules that could result in some or all of the work-centres becoming overloaded. MRP II also embodies a capacity requirements planning module which translates the "in-house" parts requirements schedule into the capacity demands, in standard hours, that will be loaded on to each of the work-centres within the plant to carry out the necessary manufacturing operations. Capacity requirements planning will identify work-centre overload and underload in each time period; it will not, however, determine what changes need to be made to create a feasible production schedule that will not violate due dates: this has to be done by adjustments to the master production schedule. MRP II normally employs "rough-cut" capacity planning, as a preliminary part of the master production scheduling process, in order to establish that the available capacities of the resources are broadly in line with requirements, before more detailed computations are undertaken. This "rough-cut" evaluation is usually a manual process and carried out as an adjunct to the running of the program.

MRP II systems generally have a finite capacity scheduling capability by which the computer endeavours to reschedule work by loading and reloading the job networks in a structured and iterative manner, taking into account the work-centre capacities available, to reschedule work so that preference is given to ensuring that jobs with the highest priority will be completed on or before their due dates. Often in practice, however, only scheduling to infinite capacity, together with rough-cut capacity planning, are employed.

MRP II allows replanning but restricts this to the regular periodic running of the system, it also regulates the flow of work by generating a priority dispatch list but the detailed machine scheduling within the work-centre is not generally part of the MRP II.

OPT (Optimized Production Technology) is a bottleneck "optimizing" method for planning and scheduling production, developed by Creative Output Inc. The basis of the OPT system is a set of elementary rules (or "thoughtware") that have to be applied. The proprietory Creative Output software comprises three modules: SPLIT, OPT BRAIN, and SERVE. The first module identifies and sorts out the bottleneck from the non-bottleneck resources. The second module employs a secret finite-capacity forward-scheduling algorithm to produce what, it is claimed, is an optimized schedule for the bottleneck resources which simulta-

17. Production Scheduling

neously takes account of capacity and priority. The last module, which has been described as a "smart" MRP system, is used to backward schedule the non-bottleneck resources from the key end-dates, established by the optimal schedule for the bottleneck resources, so that time buffers are provided. OPT schedules jobs to arrive in advance of when they are needed to act as buffer inventory in front of critical operations so that delays will not disrupt the throughput of the plant. The schedule shows which due dates will be missed and by how much. The system requires as inputs the usual planning information such as: bills of materials, job routings, resource capacities, etc. These data are converted into a product – process network model representation of manufacturing on which OPT operates.

OPT, it is claimed, "optimizes" throughput whilst minimizing inventory and operating expense. OPT has an impressive and growing list of major industrial users worldwide, many of whom have reported substantial improvements in throughput and work-in-process inventory. It is the publicity generated by the reported successes of certain users that has, to some extent, overcome the natural scepticism from some academic and industrial quarters. There still remains, however, an understandable concern about placing the control of the manufacturing operations of a company in the hands of a black-box module employing a secret algorithm. The software is expensive, which has generally restricted its adoption to large companies.

In many ways the Toyota Kanban system and other forms of just-in-time production have a strong conceptual affinity which bottleneck-optimizing methods. All have a common aim of achieving productivity improvement. In the Kanban system, for example, inventory reduction is constantly pursued in order to show up bottlenecks in the system which then come under scrutiny to see whether or not they can be reduced or eliminated to allow the required level of work throughput but at a reduced level of inventory.

In the Kanban system the master schedule for the month ahead is fixed and "levelled" so that the total planned production of each model type is spread uniformly over the period in a continually repeating final-assembly model-sequence cycle. This differs from MRP II, where the entire process batch would normally be run off for each model. Under OPT too, process batches of the required models whould be produced but these could be subdivided into smaller sizes if required. The Kanban system is highly sensitive to fluctuations in demand, and while it is appropriate to repetitive production it is not suited to job shop production. The beauty of the Kanban system is its simplicity and its "transparency". People can actually understand and relate to the system, they can see how the work is flowing and where there appears to be "unnecessary" inventory and production bottlenecks. There is an immediacy with Kanban that is often lacking in computer-based systems because of data-processing requirements.

4 Machine Scheduling/Sequencing

Machine scheduling is concerned with setting the rules or procedures that will govern the processing priorities at, and transfers between, the stages of production, which will thereby determine the manufacturing timetable. In its generalized form, where different work may require different routes through the machines, this is the job shop scheduling problem. The term job, in this connection, may be applied to describe a single item or, as is more usual, a batch of items that require processing on the machines. Important special cases of the general job shop scheduling problem are (1) flow shop scheduling, where every job follows the same route through the machines, and (2) permutation scheduling, where the same job sequence obtains on every machine.

A fundamental feature of machine scheduling is that no item can be processed by two machines at one and the same time. In addition, the routings of items through the machines are predetermined by the operations that are required to be performed. This means that some non-productive time will almost inevitably be induced into every schedule because of "interference" since some machines will have to wait for the transfer of items from the previous stage of manufacture. The effect of these "induced delays" is most pronounced, of course, when transfer takes place only when processing out a machine stage has been completed on an entire process batch of items. In practice, job splitting, where a batch is divided for processing between similar machines, and job lapping, where a completed part of a process batch of items — the transfer batch — is moved to the next stage of processing before the whole batch is completed, are employed to reduce induced delays.

The machine scheduling problem, although very simple to define, seems, paradoxically, even in its most basic forms, to defy algorithmic solution in all but a few simple special-case situations.

Important among the early successes were Johnson's algorithms to generate minimum makespan-time schedules for two machines and certain special cases where three machines are employed. Setup times can be handled providing that these are sequence independent, because they can then be treated as simple additions to the corresponding machine processing times.

Particularly important results deriving from the study of the single-machine-scheduling problem were the shortest processing time (SPT) rule and the earliest due date (EDD) rule. The SPT rule requires that the jobs are scheduled on the machine in increasing processing time order, beginning with the job with the shortest time. Operating this rule will minimize average completion time (i.e. over all the jobs) and equivalently will minimize average waiting time, average flow time, and average lateness.

The EDD rule in which jobs are scheduled in increasing due date order, beginning with the job with the earliest due date, minimizes maximum tardiness, where tardiness is defined as positive lateness (i.e. completion time minus due date when this value is positive).

The single-machine problem, in which the setup times are sequence dependant and where it is necessary to determine a work sequence that will minimize the total time lost due to setting-up operations is particularly difficult to solve. This is not surprising, since it is simply another form of the notorious travelling salesman problem in which ports of call are replaced by job-batches and distances by setup times.

The inherent difficulty of the machine-scheduling problem was widely appreciated but did not stop some researchers in the field from hoping that a breakthrough would ultimately be achieved and a universally applicable algorithm developed. If only the small and simple problems on which research was centred could be solved then it was believed by many that this would provide the necessary insights that would ultimately lead to the cracking of the more complex industrial forms of the problem. This goal was never actually achieved, for reasons that are discussed below, but nevertheless important progress was made in the development of the theory of scheduling, leading to the publication of a collection of what was considered to be the most important papers in a book entitled *Industrial Scheduling,* edited by Muth and Thompson [4]. It was some years later however, before an attempt was made by Conway, Maxwell, and Miller [5] to present, for the first time, the subject as a coherent theory with the publication of their book *The Theory of Scheduling.*

Complete enumeration of all alternative schedules in order to select an optimal one is not feasible, except for very small problems. For large industrial scheduling problems, the computational demands of complete enumeration are impossible to fulfill even with the largest and fastest computers.

An infinite number of schedules can be generated with different degrees of idle time between the job time-blocks. A schedule of this kind might be relevant for reasons other than time: because, for example, specific labour resources to attend the machines might be available only intermittently and for specified times. It is always possible to produce a unique "semi-active" schedule in which, without changing the sequence of jobs on the machines, the waiting times between jobs are minimized by "left-shifting" the start times of the operations by the maximum possible amount.

If "global" shifting of the start times for jobs is permitted so that a change of sequence is allowed providing that no job completion time increases as a result of the change then a subset of the semi-active schedules – the "active schedules" – will be produced. The importance of the active schedules is that they constitute the smallest set guaranteed to contain an optimum solution for all regular measures of performance. It was Giffler and Thompson who provided a systematic partitioning procedure to enable the entire set of active feasible schedules to be generated, by resolving all job conflicts in all possible ways. A job conflict arises whenever two or more jobs overlap in contention for processing on a specific machine, so that giving priority to one job will mean other job(s) will have to wait. Resolving the conflict entails specifying the priorities to be given to the conflicting jobs.

Giffler and Thompson also showed how to construct "non-delay" schedules in which no machine is kept idle when it could begin processing on a job. These schedules constitute a subset of the active schedules but, unlike the latter, cannot be guaranteed to contain an optimal solution.

A major breakthrough in the quest for optimal solutions to combinatorial problems in general, and scheduling problems in particular, came with the development of branch and bound tree-search methods.

The formation of schedules can be represented by a tree, where the nodes in the tree constitute the jobs, and any branch connecting a series of nodes and running from the root of the tree upwards will constitute a unique schedule sequence, so that the complete tree will designate all the possible schedules. The branch and bound technique involves the stage by stage construction of schedules, in a manner which allows, at each stage, the determination of a bound on the value of the complete schedule, even though the sequence for the remaining jobs required to complete the schedule has not been determined. This procedure provides a mechanism for "looking ahead" to determine a solution value that cannot be bettered; it does this without the need to actually generate the solution and hence the subsequent nodes and branches of the solution tree. This reduces the amount of enumeration that needs to be carried out. An optimal solution is identified when a complete schedule sequence corresponding to a tree branch running from the root of the tree to the topmost node is determined that has a value that is better or equal to the bound values at the "pendant" nodes representing the last selections in the partial schedules already constructed.

The active schedules can be represented in terms of a tree depicting the various conflicts and the alternative ways in which they may be resolved. If the branch and bound technique is combined with the Giffler and Thompson method, then the amount of the tree that needs to be developed can be greatly reduced, with consequent saving in computer time. Even so, with increasing problem size the computational time requirements quickly become prohibitively large.

5 Efficiency of Algorithms

The computer time required to execute an algorithm program will depend on the size of the particular instance of the problem input, say n, which translates into the number of binary operations that have to be performed by the computer. There are two types of algorithm:

1. "Efficient" or polynomial-time algorithms where the computation time is polynomial in the size of the particular instance of the problem input. We are actually concerned merely with establishing the order of magnitude of the worst computational time that will required. For this purpose, the dominant term, involving the highest power and ignoring any constant coefficient, will suffice as the measure of the computational efficiency of an algorithm and may be expressed as O (n^k), meaning – of the order of n^k operations. In practice the complexity of polynomial time scheduling algorithms is generally, at worst, O (n^3).
2. "Inefficient" exponential-time algorithms where the computation time is an exponential in the size of the particular instance of the problem input.

It is clear from the forms of the order of complexity in the two categories above that although for cases where the particular instance of the problem input is small in size the polynomial-time function may be greater than the exponential-time function, there will always be a size at which the exponential function will overtake the polynomial whatever its form. Thus it is apparent that it is only the polynomial-time type of algorithm that is universally applicable to all problems, large as well as small.

Modern computational complexity theory has shown that all problems — not just scheduling problems — that can be solved by exponential-time algorithms are members of a class NP. Problems that have polynomial-time algorithms are members of a class P. P is a subclass of NP since enumerative-search exponential-time algorithms can be constructed for any problem that has a polynomial-time algorithm. Of course, as already indicated, where both types of algorithm exist it is the polynomial-time one that is preferred. It is possible to identify a subclass of problems within the class NP — termed NP-complete — which comprises the most difficult combinatorial problems for which no polynomial-time algorithm has been found. It has also be shown through a property called "reducibility" that if a polynomial-time algorithm could be found for any one of the NP-complete problems then polynomial-time algorithms could be constructed for all of them. As described by French [6], the reducibility property can further be employed to show that: if any polynomial time algorithm can be found for any NP-complete problem then all problems in NP can be solved in polynomial time, in which case P would be equal to NP. It is now generally postulated that P is not equal to NP since all the indications are against it: this is, in effect, equivalent to the assertion that no polynomial-time algorithm can be determined for NP-complete problems. The significance of the identification of a problem as being NP-complete is that it constitutes a formal recognition of the fact that an optimal solution can be determined only through the employment of an exponential-time algorithm employing some form of enumerative search.

The significance here is that except for a few simple special-case situations involving one or two machines the majority of scheduling problems, including the general job shop scheduling problem, are NP-complete. It is not surprising therefore that all attempts at trying to find simple optimizing solutions have failed when it is now clear that enumerative-search exponential-time algorithms, such as branch and bound, constitute the best methods available.

6 "Satisficing" Solutions and Heuristic Techniques

The general approach of the analytical methods already described is to consider the machine-scheduling problem in its total system form. The relative lack of success of this approach in providing a general optimization method of universal applicability has led to a switch of focus of attention to a more simplified, decomposed subsystem view of the problem as comprising a set of interrelated single-machine-scheduling problems. With such a viewpoint, the problem reduces to that of determining what form the priority dispatching rule should take for establishing the queue discipline for the jobs awaiting processing at the individual machines.

There is some theoretical support for this decomposed treatment of the problem which derives from queueing theory, and is expressed by the following decomposition principle which states that if the following assumptions apply, namely:
1. inter-arrival times for each job arriving from outside the system are negative-exponentially distributed,
2. processing times at each machine are negative-exponentially distributed,
3. jobs are routed to a machine by a fixed probability transition matrix which determines the probability of a job going from a particular machine to any other machine or out of the system because of completion,
4. the priority rule at each machine is first come first served (FCFS),

then the total system of jobs and machines can be decomposed and treated as a network of independent single-machine job-queueing problems.

This result is based on the fact that under FCFS priority, the output inter-arrival times of a single queue are negative-exponentially distributed. In the dynamic job shop scheduling situation the output from one machine job-queue becomes the input to another machine job-queue.

The assumptions on which the decomposition principle is founded do not, in general, correspond with what happens in practice. Nevertheless, the use of priority dispatching rules applied at the individual machines, as though the latter were independent, is very common practice in industry because it is such a very simple procedure to operate.

The success of analytical methods directed at the single-machine problem, although limited, encouraged the belief that perhaps the optimal priority dispatching rules so determined might be applicable, either directly or in some modified form, to the general machine-scheduling problem in its decomposed form. To a large extent this has turned out to be true, with the SPT rule and the various slack rules that incorporate due date, figuring very prominently among the best priority dispatching rules, as determined by computer simulation studies.

Priority dispatching rules are said to be static if the priority does not change with the elapse of time, and dynamic if it does. SPT therefore is a static rule and SPT-with-truncation a dynamic rule. In the latter case truncation refers to the cut-off time that is set so that when any job has waited this length of time it is given top priority and moved to the front of the job queue. This is equivalent to momentarily switching to an FCFS queue desciepline, which unfortunately gives rise to high mean-waiting-time and hence long mean flow-time but on the positive side produces low flow-time variance. The SPT rule produces low mean flow-time but suffers from high flow-time variance. It was Conway and Maxwell who showed that, by a judicious choice of the cut-off time, the SPT-with-truncation rule could provide an acceptable trade-off in reducing flow-time variance compared with the pure SPT rule, without a very significant increase in the mean flow-time.

Another composite priority dispatching rule is the COVERT rule (from c over t or cost over time) developed by Carroll and discribed by Buffa [7]. This endeavours to retain the high performance of the SPT rule, in terms of its low mean-waiting-time characteristic, whilst keeping to a minimum the extreme delays created by a few jobs. COVERT for any job is defined as the ratio of delay cost and the processing time for the job on the particular machine. Jobs are thus arranged for processing in COVERT priority order, starting with the one with the highest COVERT value – this job sequence minimizes minimum total delay costs. The COVERT rule is claimed to be superior to SPT-with-truncation in terms of mean tardiness (i.e. mean positive lateness – where lateness is defined as job completion time minus due date).

Conway [8] compared the SPT dispatching rule with over thirty other rules and found that although it did not provide a minimum value for any measures of work-in-process inventory being studied, nevertheless its performance was consistently good throughout.

Aggarwal, Wyman, and McCarl [9] developed a cost priority dispatching rule which incorporated inventory, setup, lateness, and processing costs and reported that the SPT rule dominated their own rule. They hypothesized that the SPT rule might provide the benefits of cost control without actually considering cost *per se*, just as it has been observed to confer benefits in terms of reducing lateness and work-in-process inventory without specifically considering these criteria.

Le Grande [10] in a simulation study at the Hughes Aircraft Company found that minimum dynamic slack-time per operation was the best rule when measured against a weighted set of ten performance criteria, dynamic slack-time for a job being defined as due date minus current date and remaining processing time. What Le Grande also found was that the SPT rule still ranked second to none of the other rules tested.

The SPT priority dispatching rule exhibits, as indicated by these and other studies, a tremendous robustness with accounts for its long and widespread adoption in industry. It has particular intuitive appeal to production controllers because it appears to place emphasis on production efficiency. Due-date-related priority dispatching rules, on the other hand, have intuitive appeal to management because they place emphasis on satisfying customer delivery promises.

A state-of-the-art survey of dispatching rules was carried out by Blackstone et al. [11], which identified the dispatching rules discussed above as being among the seven best overall (where the FCFS rule was used as the control rule). In an earlier survey, Panwalkar and Iskander [12], reviewed over one hundred dispatching rules and concluded that "the consensus among many researchers appears to be that a combination of simple priority rules or a combination of heuristics within a simple priority rule, works better than the individual priority rules".

7 Future Developments

Computer programs to carry out large-scale machine scheduling have existed for many years, initially in free-standing form and later as modules in integrated production control systems. These programs, employing usually a combination of network and heuristic procedures, involve an enormous amount of data processing and are designed to be run on a regular periodic batch-processing basis. They treat the machine-scheduling problem as an overall global problem and provide solutions that determine the detailed future timetable of work to be performed on the machines. Such schedules are too rigid to be entirely satisfactory in practice. If machines break down, if materials are not delivered as planned, if personnel are unexpectedly absent or any of a series of unscheduled events happen then rescheduling cannot be undertaken until the next batch run of the program: by which time it is too late anyway. It is because these machine-scheduling systems lack flexibility and the ability to respond quickly enough to change that schedules provided by them are frequently at variance with what is actually possible on the shop floor. It is not surprising, therefore, that most scheduling systems today, such as those employed in MRP II, provide a dispatch list of jobs to be undertaken and the priorities to be assigned to them but leave the detailed machine scheduling to be decided at the local work-centre level.

Uncoupling the detailed machine-scheduling problem in this way allows both flexibility and rapid response in dealing with any local unforseen circumstances that may arise. Traditionally, work-centre supervisors have employed heuristic dispatching rules in this scheduling task. With the advent of real-time systems and online facilities for data input and output on the shop floor, it is now possible for supervisors to interrogate their local computer terminal or PC and, hence, via a decision support system, provide a schedule for their work-centre that will take account of job priorities, as well as current commitments and resource availabilities. A mimic of the work-centre may be employed to display on the VDU the schedule and the progress made in meeting the schedule. "What-if" types of experimentation may be possible with some systems, to assess the consequences of alternative schedules before a final commitment to one particular schedule is made. Optimum solutions might theoretically be possible to achieve in some cases because of the smaller scale of the localized problem but, in practice, such approaches are likely to be ruled out because of the data-processing requirements and the time that would be involved. The need for fast response in the generation of feasible schedules is likely, at least in the forseeable future, to mean the continuing employment of heuristic methods for this purpose.

There are three categories of machine tools: manually operated, semi-automatic, and automatic. Before the advent of computer-aided machining (CAM), setting up on machines was a very time-consuming and skilled process necessarily carried out at the machine by specialist setters or setter operators. Automatic machines, such as single spindle and multi-spindle cam-operated automatic lathes, existed but required particularly long setting-up time and could be justified only on very long production runs. Automation in general was "hard" thereby making it very inflexible and hence applicable only to dedicated tasks. The bulk of production centred on manually operated and semi-automatic machine tools since this was the only way to handle small-volume and large-component variety. The concept of the "economic" batch was developed to determine an "optimal" manufacturing quantity that would achieve a cost balance between frequency of setting-up and amount of component inventory. The organization of machine tools for batch production was functional and the problem of machine scheduling was of the traditional jobshop kind — being concerned with "economic" batches of components that had to be routed, in accordance with their particular and usually different processing needs, through a variety of machine tools in a spaghetti-like work flow.

The advent of NC and CNC machine tools changed the nature of batch manufacture. The development of pre-set tooling and the redesign of jigs or fixtures and procedures has considerably reduced the time required for setting-up and product changeovers, so that even in the traditional job shop form of organization small processing batches have become "economically" justified and a move towards just-in-time production made.

The big development has been in the flexibility of production that is now possible with CAM machine tools individually, but more particularly when arranged together as a manufacturing system. A CAM machine which, under the same computer control, has workpiece probing, tool management and exchange, and is supported by storage and conveying mechanisms that will provide a supply of workpieces, load them for processing and remove and store or pass them on when completed, is the most basic form of flexible manufacturing module. This is the building block of modern "soft" automation that can be expanded to create the flexible manufacturing cell (FMC) or more complex flexible manufacturing system (FMS). The machining centre is a particular form of this building block of automation which, because of its ability to perform a range of machining operations that would normally require a variety of different machines, means that machine variety as well as the number of machines and the space required is thereby reduced.

With FMS, a single-unit processing batch becomes "economic", since the system treats each palletized workpiece as an individual entity. Workpieces may thus be fed into the FMS in any order, but such decisions clearly have consequences for the effectiveness and efficiency of the system overall and its ability to complete required numbers of individual products within a specific time. It is possible to leave this decision-making to a human being and this is what happens currently in some systems. FMS, unlike the old style job shop, is, including the computer control system, entirely machine based. This means that apart from the occasional malfunction or breakdown it is a completely deterministic system. When a workpiece is entered into the FMS its route through the machines is predetermined and controlled by the system: its progress depends, however, on the state of the FMS, as represented by the other workpieces in the system and any consequent "interference" that may be generated by workpieces unable to move on because one or more workpieces are waiting for processing or are simply blocking the path. What governs this interference and hence the overall performance of the FMS are the schedule sequencing decisions regarding the entry of workpieces into the FMS.

If the entire workload for a production period were to be fixed in advance and the released to the FMS without change or addition then the deterministic nature of the system could be exploited to provide a means of determining the most appropriate work sequence. In practice, however, the release of work is continuous and not discrete so that decisions on the work sequence need to be constantly re-evaluated in the light of the changing circumstances and in relation to the current state of the FMS. This means that real-time online control of the workpiece input to the FMS is what is ideally needed, and this becomes imperative if computer-integrated manufacturing (CIM) is to be achieved.

It is clear that new and innovative computer-based decision support techniques for schedule sequencing are needed, and the most promising approach towards the achievement of this goal is likely to be in the development and employment of expert systems.

8 References

1. Wagner, H. M. and Whitin, T. M., Dynamic version of the economic lot size model. Management Science, Vol. 5, pp. 89–96
2. Holt, C. C., Modigliani, F., Muth, J. F., and Simon, H., Planning Production, Inventories and Work Force. Prentice-Hall, New York, 1960
3. Hanssmann, F. and Hess, S. W., A linear programming approach to production and employmentd scheduling. Management Technology, Vol. 1, January 1960, pp. 46–52
4. Muth, J. F. and Thompson G. L. (eds), Industrial Scheduling. Prentice-Hall, Englewood Cliffs, New Jersey, 1963
5. Conway, R. W., Maxwell, W. L., and Miller L. W., Theory of Scheduling. Addison-Wesley, Reading, Mass., 1967
6. French, S., Sequencing and Scheduling: an Introduction to the Mathematics of the Job-shop. Ellis Horwood, Chichester, 1982
7. Buffa, E. S., Scheduling and control for intermittent system. In Basic Production Management, 2nd edn. John Wiley, New York, 1975
8. Conway, R. W. Priority dispatching and work-in-process inventory in a job shop. Journal of Industrial Engineering, Vol. 16, 1965, pp. 123–130
9. Aggarwal, W. C., Wyman, P. F., and McCarl, B. A., An investigation of a cost-based rule for job-shop scheduling. International Journal of Production Research, Vol. 11, 1973, pp. 247–261
10. Le Grande, E., The development of a factory using actual operating data. Management Technology, Vol. 3, No. 1, 1963, pp. 1–19
11. Blackstone, J. H., Jr., Phillips, D. T., Hogg, G. L., A state-of-the-art survey of dispatching rules. International Journal of Production Research, Vol. 20, No. 1, 1982, pp. 27–45

Chapter 18
Production Planning and Scheduling in Flexible Manufacturing Systems

Kathryn E. Stecke

Kathryn E. Stecke received the B.S. degree in Mathematics from Boston State College in 1972. She received an M.S. in Applied Mathematics in 1974 and an M.S. in 1977 and a Ph. D. in 1981 in Industrial Engineering from Purdue University. She is now an Associate Professor of Operations Management at the Graduate School of Business Administration at The University of Michigan. She is the Editor of the *International Journal of Flexible Manufacturing Systems*. She has authored numerous papers on various aspects of planning and scheduling of flexible manufacturing systems in numerous leading journals and several proceedings and book contributions. She was Chairperson of the First and Second ORSA/TIMS Conference on Flexible Manufacturing Systems: Operations Research Models and Applications, held in Ann Arbor, Michigan in August 1984 and 1986. She spent Fall 1985 at General Motors Research Laboratories, spent Fall 1984 at the Centre d'Etudes et de Recherches de Toulouse, France and spent Fall 1987 and Winter 1988 at COMAU in Torino, Italy.

1 Introduction

A flexible manufacturing system (FMS) is typically defined as a set of computer numerically controlled (CNC) machine tools and auxiliary equipment that are integrated with some type of automated materials handling equipment [1]. This is because the early applications were in metal-cutting industries, such as Caterpillar Tractor (Peoria, Illinois), Ingersoll Rand (Roanoke, Virginia), SCAMP (Colchester, UK), and Messerschmitt-Bölkow-Blohm GmbH (Augsburg, West Germany), for example.

However, subsequently, different types of applications of flexible manufacturing have been implemented, for example, in electronics assembly, integrated circuit board manufacturing, and mechanical assembly (see, e.g., Hall and Stecke [2]). In these latter FMS types, the flow of work is often largely unidirectional and high volume. Hence they are often called flexible assembly systems (FASs) or flexible flow systems (FFSs). In these two basically different types of flexible systems, the hardware is different and the software to operate them should be. Browne et al. [3] and Stecke and Browne [4] classify such flexible systems according to various design and operating characteristics.

The focus in the remainder of this chapter is on discussing the production planning and scheduling problems in the metal-cutting type of FMS. For additional information on flexible assembly systems, see Hall and Stecke [2] and Ranky [5]. The FMS planning and scheduling functions are those involved with system operation. For a particular system, the FMS design problems had been addressed during the lengthy design and implementation phase [6]. Once the system has been purchased and is running, concern should focus on developing appropriate means to efficiently operate the system, while meeting due dates and required quantities if these are applicable to the particular setting.

The usefulness of flexibility in the manufacture of component parts has been noted in part because of the possibilities of quickly creating new designs and the potential improvements in process plans that can continuously be made. Increased and better use of automation can allow a rapid response to a changing market or a quick introduction of a new or an improved product. Flexible manufacturing allows the capability to simultaneously machine various subsets of families of somewhat similar part types in very low batch sizes, if desirable. Because the parts are palletized and fixtured, there are certain restrictions as to size and maximum weight and volume.

A flexible manufacturing system is not very large relative to conventional systems (usually about 4 – 20 CNC machine tools). However, it is a difficult and lengthy process to design such a highly automated system to plan on meeting projected production requirements in an efficient manner. Because of the diversity of the part types being machined, it is also difficult to attain capacity utilization of the equipment.

Sufficient thought and prior analyses of the many operating problems and design issues involved in flexible manufacturing implementation will help to provide an effective system. Here, short descriptions are provided of the various FMS management decisions that have to be made during operation. In particular, the problems and decisions that have to be addressed during the planning, scheduling, and control of an FMS are overviewed. Many of these problems have been outlined previously [5]. Additional details are provided here.

Some of these problems have been examined to some extent from varying points of view using different modelling aids and various solution techniques. Different levels of decision-making for some of these problems are outlined in Suri and Whitney [7].

2 FMS Planning Problems

The production management problems involved with operating a flexible manufacturing system can be disaggregated in terms of the two distinct but related functions of planning and scheduling. In an FMS, a part cannot be routed to a particular machine tool of the correct type unless all of the cutting tools that are required for that part's next operation are already loaded into the machine tool's limited-capacity tool magazine. This indicates that some decisions have to be made before production begins.

The *FMS planning problems* are those decisions that have to be made before the system can begin to produce parts. When the FMS has been "set up", and all of the cutting tools that are required are loaded into some tool magazine(s), production can begin. The *FMS scheduling problems* are those concerned with actually running the system, of scheduling the flow of parts throughout the FMS over time.

At the FMS planning stage, the system has already been designed, implemented, and is in production. From the part numbers which can be processed on the system, there is a subset for which production orders exist — perhaps requirements from another department in the factory or from a sister plant, or customer orders, or maybe forecasted demand.

Stecke [8] has suggested five production planning problems, with the aim of aiding an FMS manager in setting up the system, in advance of production, so as to work well with the FMS's subsequent scheduling procedures. The *first FMS planning problem* is to select, from those part types for which production orders of various sizes are specified, a subset of part types for immediate and simultaneous manufacture. This *part type selection* decision can be made in various ways. Due dates may be considered. Or part types may be chosen that are compatible in the sense that each type mainly utilizes a different machine type, and so, when machined simultaneously, they help attain a good, overall system utilization.

When the part types have been selected, some aggregate information can be calculated concerning the total processing requirements and total number of tool slots required from all machine types. Relating this information to the capacity available per machine of each type, both processing time and tool magazine capacity, can help determine an appropriate amount of pooling. In particular, the *second FMS planning problem* is to partition the machines of each type into machine groups. Machines in a particular group are said to be pooled when they are identically tooled, and are each capable of performing the same operations during real time. Pooling machines can improve most system performance measures. However, pooling does not have to be performed. Duplicating operation assignments can achieve a portion of the benefits from pooling.

The *third FMS production planning problem* is to determine the *production ratios* at which the selected set of part types should be maintained over time on the system. Related to this, the limited numbers of pallets and the fixtures of each fixture type have to be allocated among the selected part types. Minimum inventory requirements to maintain these production requirements should be determined.

A minimum-inventory philosophy is increasing in importance and concept. This can possibly be achieved with the aid of greatly reduced setup times on individual parts, increased tool commonality between different parts, and very small batch sizes — for example, batches which tend to size one. The movement towards just-in-time production can increase the complexity of the planning problem.

The *final FMS planning problem* is to allocate all operations and associated cutting tools of the selected part types among the possibly grouped machines. Different loading objectives help to guide this decision, and each could be appropriate under different circumstances.

When the FMS planning problems have been solved and all cutting tools have been loaded into the specified tool magazines, production can begin. These FMS planning problems can be solved sequentially, or iteratively, or several simultaneously. They can be re-solved as often as every couple of days or weeks. They may require re-solving if one of the machine tools is down for a long time or a part type needs to be expedited. When the production requirements for one of the selected part types are complete, the planning problems are solved again. For example, the cutting tools for the finished part type are no longer required. Either the reduced set of part types can be machined, possibly allowing a regrouping of machines to enable more pooling, or one or more part types can be added to the set of selected part types, signalling a repartitioning of the machine tools into different groups.

3 FMS Scheduling Problems

FMS scheduling problems are different from those of job shop scheduling. Considerations are different. These differences are discussed in Rachamadugu and Stecke [9]. FMS scheduling problems are concerned with the operation of the FMS in real time after it has been set up during the planning stage in advance of actual production. There are many possible approaches that can be taken to schedule the operations of parts through the system. Different approaches might be applicable in different situations. The problems include the following.

An appropriate input sequence at which the individual parts of the selected part types are to be input into the system should be determined. Sometimes part types have to be produced in certain relative ratios, say for subsequent assembly purposes. A periodic input sequence might also be appropriate for some dedicated types of FMSs. Producing to maintain certain production ratios of part types on the system may be appropriate in some situations. Also, a fixed, predetermined input sequence may be appropriate. In other situations, a flexible, real-time decision concerning which part to input next may be best.

Appropriate scheduling methods and algorithms have to be determined. Such scheduling aids can range from simple dispatching rules to complex algorithms or procedures incorporating look-ahead features. Most of the classical scheduling literature has been concerned with generating *off-line* schedules for a manager to use. In particular, often a scheduling algorithm is applied to some input data to result in a fixed schedule specifying which operations would be performed on which machine tools and *when*. More appropriate for an FMS might be a *real-time, on-line* scheduling policy, with scheduling decisions based on the actual state of the system. If the FMS were carefully set up during the planning stage, then a real-time scheduling function might be easier to apply. Perhaps more important during the planning stage, due date criterion also applies during the scheduling.

It can be desirable to solve some of these FMS scheduling problems in a real-time, off-line mode. The FMS manager may be able to use a decision support system in conjunction with appropriate scheduling algorithms to generate a schedule based on real-time data on the state of the system. This could provide up-to-date due date and production quantity requirements as well as quality and yield information.

Other types of scheduling problems include the following. If there are several parts waiting to be processed by the same machine tool, the priorities among these parts can be determined. Random processing of parts in batches of size one is often referred to regarding FMS scheduling. However, some intelligent ordering of parts can greatly improve FMS performance. It may be appropriate to determine an optimal (dynamic) sequence at each machine tool. Perhaps this sequence could be periodic. Perhaps a simple (or complex) dispatching rule suffices. Also, it may be useful to assign machine priorities according to the current bottleneck.

Many of the typical performance measures are important, such as maximize production or machine utilization (the systems are very expensive) or minimize inventory or flow-time, subject to meeting due dates, in systems in which due data criteria are relevant. In a just-in-time situation, minimizing earliness might also be appropriate.

4 FMS Control Problems

FMS control problems are those associated with continuously monitoring the system and tracking production targets to be confident that requirements and due dates are being met as scheduled, expected, and promised.

In particular, one control issue is to determine policies to handle machine tool and other breakdowns. These policies should be determined during the design phase. Implementation of these policies occurs during the control of an FMS. Things go wrong that could never have been anticipated during the design phase. There is a significant maintenance learning curve during system implementation.

It may be that one or more machine tools are down for a long period of time. There are several approaches that one might take to handle such random occurrences. Perhaps the schedule could be revised. Or perhaps some procedures could be developed that will help to enable production to return to the original, planned schedule as soon as possible. Or perhaps a new schedule could be developed. If a machine fails and other machine tools have not been previously tooled to produce those parts, then either production is halted for some part types, unless the cutting tools can be quickly moved and loaded into another machine's tool magazine. But this may displace other cutting tools. Shifting tools takes time and planning and scheduling, and the production of the other part types will most often be reduced.

Pooling machines into functionally similar machine groups during the production planning stage, in conjunction with real-time scheduling, allows machine tool breakdowns to be *automatically* handled.

Another related problem is the following. If several breakdowns need attention, what should the maintenance person address his or her attention to next? This is the classical machine interference problem [10].

Scheduled and/or periodic preventative maintenance policies have to be determined. Preventative maintenance can be scheduled along with the production requirements as part of the weekly schedule, or it can be planned to occur on an off-shift or a weekend.

In-process and/or finished goods inspection policies have to be determined. The points of quality inspection of part dimensions for tolerances as well as the frequency of each inspection have to be determined. Inspection equipment has to be selected and implemented. Further procedures for dealing with unacceptable parts through rework, recycling, and/or scrapping must be worked out and built into the control software. There are various yield issues that have to be dealt with.

Procedures for tool life and process monitoring and the consequential data collection, as well as for updating the estimates of tool life, have to be specified. This includes determining how to track tool wear, as well as policies for replacing worn and broken cutting tools.

5 Modelling and Solution Aids

Many models are available that have been applied to help answer some of the preceding questions. Various models have been useful to identify key factors that will affect system performance. Each model structures the problems differently. Each model ignores or aggregates some features of the system to focus on particular aspects. The models have provided either operational or qualitative insights into system behaviour and into possible procedures to solve some of the FMS decision problems.

Some of the models that either can, have, or will be applied to these FMS operational problems include: simulation, group technology, computer-aided process planning, queueing networks, mathematical programming (linear, nonlinear, integer, dynamic), perturbation analysis, Petri nets, and artificial intelligence. Buzacott and Yao [11] review some early FMS research that uses analytical models – in particular, open and closed queueing networks. Suri [12] provides an overview of some of the models that can be used to evaluate potential solutions. Stecke [6] also describes the use of many of these models, that both suggest and/or evaluate potential solution methods. There is an extensive literature review, outlining much of the research done on these FMS planning, scheduling, and control problems.

Different collections of these models have been developed and used to address many of the FMS problems overviewed here. The review papers listed in the bibliography provide many additional references to these investigations and to the potential uses of these models.

There is a wide variety of FMSs. Additional research is required to address all of the problems presented here for the many different FMS types. The solution procedures need to be integrated into a framework to effectively handle the FMS planning and scheduling functions.

6 References

1. Stecke, K. E. and Solberg, J. J., Scheduling of Operations in a Computerized Manufacturing System, Report No. 10, NSF Grant No. APR74 15256, Purdue University, W. Lafayette, Ind., December 1977
2. Hall, D. N. and Stecke, K. E., Design problems of Flexible Assembly Systems. In Proceedings of the Second ORSA/TIMS Conference on Flexible Manufacturing Systems: Operations Research Models and Applications. Ann Arbor, Mich., 12–15 August 1986, Elsevier Science Publishers, Amsterdam, pp. 145–156
3. Browne, J., Dubois D., Rathmill K., Sethi S. P., and Stecke, K. E., Classification of Flexible Manufacturing Systems. The FMS Magazine, Vol. 2, No. 2, April 1984, pp. 114–117
4. Stecke, K. E. and Browne J., Variations in Flexible Manufacturing Systems according to the relevant types of automated materials handling. Material Flow, Vol. 2, Nos. 2+3, July 1985, pp. 179–185
5. Ranky, P., A Program prospectus for the simulation, design and implementation of flexible assembly and inspection cells. In Proceedings of the Second ORSA/TIMS Conference on Flexible Manufacturing Systems: Operations Research Models and Applications, Ann Arbor, Mich., 12–15 August 1986. Elsevier Science Publishers, Amsterdam, pp. 157–168
6. Stecke, K. E., Design, planning, scheduling, and control problems of Flexible Manufacturing Systems. Annals of Operations Research, Vol. 3, 1985, pp. 3–12
7. Suri, R. and Whitney, C. K., Decision support requirements in flexible manufacturing. Journal of Manufacturing Systems, Vol. 3, No. 1, January 1984, pp. 61–69
8. Stecke, K. E., Formulation and solution of nonlinear integer production planning problems for flexible manufacturing systems. Management Science, Vol. 29, No. 3, March 1983, pp. 273–288
9. Rachamadugu, R. M. V. and Stecke, K. E., Classification and Review of FMS Scheduling Procedures. Working Paper No. 481, Division of Research, GSBA, The University of Michigan, Ann Arbor, Mich., November 1986
10. Stecke, K. E., Machine interference: the assigment of machines to operators. In the Handbook of Industrial Engineering, Gavriel Salvendy (ed.). John Wiley & Sons, New York, 1982
11. Buzacott, J. A. and Yao, D. D. W., Flexible Manufacturing Systems: a review of models, Working Paper No. 7, University of Toronto, Ontario, Canada, March 1982
12. Suri, R., An overview of evaluative models for Flexible Manufacturing Systems. Proceedings of the First ORSA/TIMS Special Interest Conference on Flexible Manufacturing Systems: Operations Research Models and Applications, Ann Arbor, Mich., 15–17 August 1984, pp. 8–15
13. Browne, J., Chan, W. W., and Rathmill, Keith, R., An integrated FMS design procedure. Annals of Operations Research, Vol. 3, 1985, pp. 207–237
14. Buzacott, J. A., The fundamental principles of flexibililty in manufacturing systems. In Proceedings of the 1st International Conference on Flexible Manufacturing Systems, Brighton, UK, 20–22 October 1982
15. Morgan, T. M., Planning the introduction of FMS. In Proceedings of the 2nd International Conference on FMS, London, UK, October 1983
16. Stecke, K. E., Useful models to address FMS operating problems. In Proceedings of the IFIP Conference, Advances in Production Management Systems, Budapest, Hungary, 27–30 August 1985. Republished by Elsevier Science Publishers (North-Holland), Amsterdam 1986, pp. 271–283

Chapter 19
Forecasting and Stock Control

Birger Rapp

Dr. Birger Rapp is Professor in Economic Information Systems at Linköping Institute of Technology, Sweden. He has international experience as a consultant and a lecturer for many years and is the author of several articles and books. He is a member of the following professional societies: Swedish Operations Research Association (SORA), The Institute of Management Sciences (TIMS), American Production and Inventory Society (APICS), International Society for Inventory Research (ISIR). He is Past President of EURO (Association of European Operational Research Societies within IFORS) and SORA and an editorial associate in *European Journal of Operational Research* (EJOR) and *Belgian Journal of Operations Research, Statistics and Computer Science* (JORBEL).

1 Introduction

This chapter contains a brief survey of the usefulness of forecasting and stock control systems, with special emphasis on different computer-based procedures. The subject is so extensive that it cannot be dealt with exhaustively in this chapter.

The topic of this chapter is introduced in Sect. 2, along with assumptions and part of a reference system which is aimed at computer-based manufacturing planning and control (MPC). The module for inventory control of this system is also discussed. The functions and procedures of the module are then used for further analysis. Sect. 3 deals with different ways of carrying out valuation of inventories and classification of items. Sect. 4 treats various ways of estimating demand. The quality of different forecasting methods and systems is also discussed. Order quantities and order points are important decisions in regard to inventories; different ways of calculating these decisions are presented in Sect. 5. Other functions and procedures that belong to the inventory module are introduced in Sect. 6. The chapter is summarized in Sect. 7.

2 The Inventory System

The components of this and the following sections are based on Fig. 1. It shows that the stocking point smooths the differences in the two flows, one inflow and one outflow. Somewhat simplified, decisions at the stocking point in regard to order quantities and order points may be said to depend primarily on the characteristics of these two flows. However, it should be kept in mind that the different parts are dependent. Thus the minimum costs of the total inventory system must be calculated as the sum of the costs of the different parts.

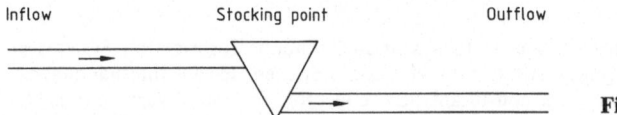

Fig. 1. Inventory system

The outflow system is characterized by demand, the value of promised or desired delivery lead time, t_D, and the size of the orders for different items. Since items in the outflow system have differing characteristics, they have to be dealt with in different ways.

The inflow system is characterized by the production lead time or the replenishment lead time. This lead time depends on different factors in the inflow system. Such factors are, e.g. utilization and lot sizes. The production lead time varies with different items and groups of items.

Delivery lead time, t_D, can also vary with items. If the difference between production lead time and delivery lead time is positive, that is, $L = t_p - t_D > 0$, there has to be at least one stocking point. The time L is called lead time and can be positive or negative. If it is negative, then the number of stocking points is determined solely on economic grounds.

In a computer-based system there are often a number of procedures related to materials planning and inventory control. A reference system for manufacturing planning and control (MPC) is introduced in Refs. [1] and [2]. This system comprises modules as shown in Fig. 2. Each module is divided into functions and procedures. The inventory control module is a prerequisite for achieving good inventory control. In addition, functions and procedures from purchasing and sales order management modules are sometimes required, but these modules are not described in this chapter.

The prime concerns of inventory control are to ensure materials supply through order quantity and order point decisions and to maintain accurate inventory records. This applies to both purchasing orders and production orders. Order quantities are created for purchasing and capacity planning. Inventory valuation such as ABC analysis (items distributed by value) may lead to differentiated ordering policies. Planned orders cannot correspond to actual requirements unless accurate inventory records are maintained. Inventory records should reflect current inventory status in terms of physical inventory availability. In practice, it is almost impossible to maintain complete data accuracy. Therefore, physical inventory counts are necessary, in the form of full physical

19. Forecasting and Stock Control

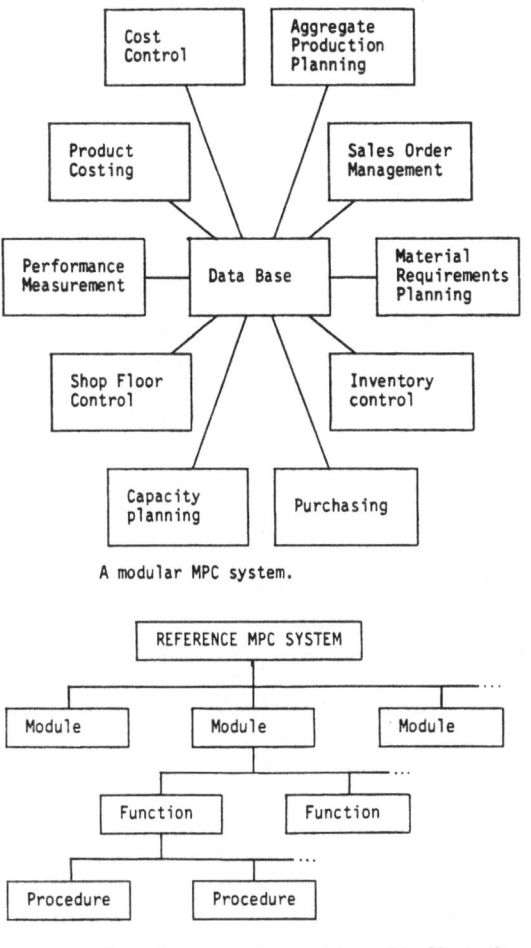

Fig. 2. Reference system; sources [1, 2]

A modular MPC system.

The reference system is hierarchically built up by modules, functions, and procedures.

inventory counts or cycle counting. Cycle counting should be performed with respect to different frequencies for different items, where classification may be based on the so-called ABC classification. The flow of information from material planning modules is shown in Fig. 3 [1].

Japanese manufacturing techniques such as JIT (Just-In-Time) try to fit the inflow of material and items to the outflow system. The better the fit, the less the amount of stocking required. However, this does not have any particular effect on the flow of information.

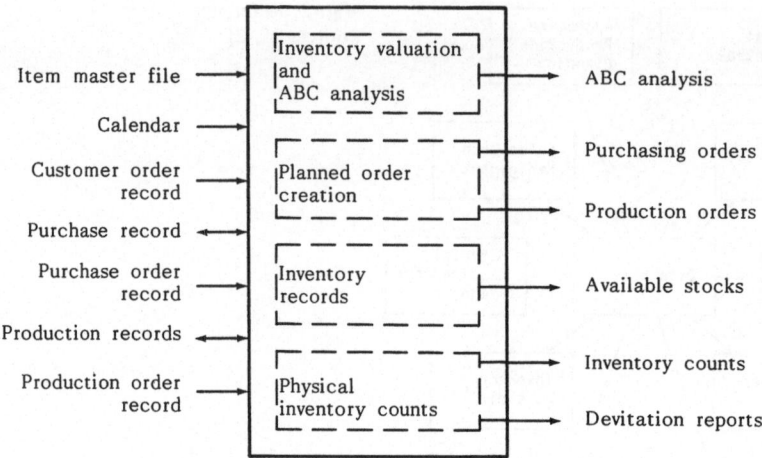

Fig. 3. Flow of information in the inventory control module

3 Inventory Valuation and ABC Analysis

Analysis of an inventory should provide an inventory valuation and link different items to their costing data such as standard, current, and actual costs, sales volume and tied-up capital curves, number of withdrawals, and lead time and its variation. In some cases there may be reason to distinguish between items on hand and items on order. Differentiation of items may also be needed: i.e. different items or groups of items may require different control strategies. In addition, comparison between sales value and stock value provides a good basis for estimating the amount of capital tied up in inventories, and turnover rates. However, there are numerous ways of classifying different items. Some companies may prefer an ABC classification as a basis for further analysis. Sometimes control strategies should also take the effects of different withdrawal rates and lead times into account.

Classification of items is not sufficient, however. Attempts should also be made to understand the demand patterns of different items due to different lead times and variations in withdrawals. In sum, stock and forecasting systems should be designed so that different items can be treated with flexibility. Such differentiation can also include different types of safety stocks. A computer-based MPC system can perform these types of analysis.

4 Forecasting and Demand Estimation

Accurate estimation of demand is a prerequisite for decisions about order quantities and order points. Total demand is the sum of dependent (derived) and independent demand.

Dependent demand is given by existing information within the company. For example, net Materials Requirements Planning (MRP) uses due-date plans for products which are then exploded through the product structure to give dependent demand for items. This is possible as long as the production plan is correct. This is seldom the case, however. In some instances the lead time might be negative. But in such cases – and sometimes when production is made-to-stock – demand can be derived with certainty.

In MRP [3] it is assumed that dependent demand exists and that the master plan is fixed during the lead time. Uncertainties in regard to the replenishment lead time and quantities are assumed to be negligible. In practice such cases are rare. Reference [4] has indicated that failure in demand specification can be a key factor behind an increasing number of shortages and increased safety stocks. Today, deterministic MRP planning is under reconsideration. Uncertainty and safety stock are discussed in the context of MRP [5–8].

It is often possible to separate the demand for an item into two parts: one which is known with certainty and one which is uncertain. If such uncertainty can be interpreted as independent demand, then a forecasting method can be used to estimate the independent demand. Forecasting methods are required, for example, in order to estimate the demand for spare parts. Some items can also be controlled by simple rules instead of MRP planning by assuming independent demand.

4.1 Forecasting

A complete forecasting system is rarely included in ordinary computer-based MPC systems. Instead such systems usually fit into specific requirements. Reference [9] contains an overview of different forecasting systems for computers.

During the last decade, forecasting research has undergone a shift in emphasis. Earlier attempts involved improving existing forecasting methods or finding new methods. Current efforts are aimed at explaining why simple methods such as simple exponential smoothing are useful procedures.

There are many reasons for producing a forecast. The type of forecasts used here are all intended as bases for decision-making. This means that the forecast and forecasting method should be related to decisions and their consequences rather than the outcome of the forecast.

In stock control, forecasts are used to reduce uncertainty about variations in demand during the lead time. The lead time is defined as the difference between replenishment time and delivery time: $L = t_p - t_D$.

The methods dealt with here are all time series. They are based on the assumption that the underlying structure will remain the same in the future. In addition, they all assign some weight to old observations in various ways.

A useful forecasting module can be obtained in three steps: (i) choice of an appropriate model; (ii) estimation of all the parameters; and (iii) determination of suitable starting values for the variables in the model. Forecasting errors have to be monitored automatically.

Table 1. Time series; source [2]

Let

X_t	=	actual observed demand in period t
P_t	=	forecast of demand in period t
a	=	smoothing constant
w_t	=	expression for weighting the actual demand in period t
N	=	number of periods
MAD	=	Mean Absolute Deviation
ME	=	Mean Error
MAPE	=	Mean Absolute Percentage Error
MSE	=	Mean Square Error
MAD	=	$\dfrac{1}{n}\sum_{t=1}^{n}\|X_t-P_t\|$
MAD_t	=	$\dfrac{1}{n}\sum_{t=1}^{n}\|X_t-P_t\|+(1-a)\,\text{MAD}_{t-1}$
MAPE	=	$\dfrac{100}{n}\sum_{t=1}^{n}\|(X_t-P_t)/X_t\|$
ME	=	$\dfrac{1}{n}\sum_{t=1}^{n}(X_t-P_t)$
ME_t	=	$a(X_t-P_t)+(1-a)\,\text{ME}_{t-1}$
MSE	=	$\dfrac{1}{n}\sum_{t=1}^{n}(X_t-P_t)^2$

All smoothing methods assign earlier actual demand a certain weight for the forecast one period ahead.

$$P_{t+1} = \sum_{i=t-N}^{t} \omega_i X_i$$

For the moving-average method it is true that

$$\omega_i = 1/N$$

For simple exponential smoothing

$$\omega_t = a,\ \omega_{t-1} = a(1-a),\ \omega_{t-i} = a(1-a)^i$$

The forecast for one period ahead can be written

$$P_{t+1} = P_t + a(X_t - P_t) = (1-a)P_t + aX_t$$

If $|X_t - P_t| \leq A\,\text{MAD}_t$ where $A \approx 4$
or $|\text{ME}| > B\,\text{MAD}_t$ where $4 \leq B \leq 6$
or $|\text{ME}_t| > C\,\text{MAD}_t$ where $0.2 \leq C \leq 0.5$
it is likely that the system is no longer under control.

Table 1 shows two forecasting methods. Several other methods are discussed in, e.g., Refs. [10–13].

A forecasting error measure is required in order to discuss the accuracy of different forecasting methods. Some measures of accuracy are shown in Table 1. As decisions and consequences of decisions vary, it is obvious that there is no "best" measure. Measures in use include Mean Absolute Deviation (MAD), Mean Absolute Percentage Error (MAPE), and Mean Square Error (MSE). MAD is the most frequently used measure. However, it is often argued that MSE should be used for stock control. One reason is that the error costs for decisions based on a forecast are more closely related to the square error than to the linear error of the forecast.

The smoothing parameter in simple exponential smoothing is $a = 2/(N+1)$ where N is the number of periods in the moving-average method. Simple exponential smoothing can also be regarded as an ARIMA (0,1,1)-process and is thus a special case of a general ARIMA process. Under special conditions, it can also be shown that the Kalman filter reduces to simple smoothing.

In the so-called "M-competition" [14], where 1001 different time series were used to compare different forecasting methods during different time periods and for different measures of accuracy, deseasonalized single exponential smoothing yielded very good results. In fact, it was the best of all the choices for one-period-ahead forecasting. It was still a good choice for some-periods-ahead forecasts, especially in terms of its simplicity. The investigation also showed that Holt's method was superior to Winter's method. Of course, the best forecasting method varied with the situation under study. The ranking of forecasting methods depends on the time period for the forecast, types of data used, and measures of the degree of accuracy.

Today, if they apply any method at all, large Swedish manufacturing companies use simple smoothing for stock control. References [15] and [12] give several reasons for using exponential smoothing in practice. In short, exponential smoothing methods are simple and robust.

An exponential smoothing of degree g can be regarded as an ARIMA (o, g, g)-process. Since, e.g., the Box-Jenkins method uses general ARIMA processes, it could be expected to be superior to exponential smoothing. However, in Ref. [14] it is shown that simple smoothing gives a more accurate one-period-ahead forecast than the Box-Jenkins method. This seems to depend on how forecast models and parameters are chosen. Furthermore, Ref. [16] emphasizes that *ex post* fit has little correlation with *ex ante* accuracy.

The smoothing parameter, a, in simple smoothing is often determined on the basis of past data where a is chosen in order to minimize forecast errors, e.g. MAD. In practice, a is chosen between 0.05 and 0.3. In Ref. [17], it is argued that another method should be used if a turns out to be larger than 0.3. However, Ref. [11] argues that the a-values should be chosen in a wider range, $0 < a < 2$.

The Box-Jenkins method starts by analysing the structure of the relevant data. Then a model is chosen and parameters are selected. It follows that simple smoothing and the Box-Jenkins method do not necessarily give the same forecasts. For instance, analysis of data might lead to a process other than an ARIMA (0, 1, 1)-process. The Box-Jenkins method might identify the current structure of the data correctly, but this does not help if the future structure of the data changes. The best method would then be the one which is robust and can take future unexpected changes into account.

In simple smoothing the starting value must be determined in the proper way. In practice, the starting value in simple smoothing is estimated from the mean of earlier observations as first suggested in Refs. [18]. Other possibilities are discussed in, e.g., Ref. [11].

When using a forecasting method, the forecasting error has to be monitored automatically to ensure that the system remains under control. This is important in stock control because signals about significant changes in demand require special stock-control action, such as additional or cancelled orders.

Table 1 lists some monitoring measures. There does not seem to be any overall "best" measure. Further discussions about different measures may be found in, e.g., Ref. [11], [12], and [15].

Many companies currently use simple forecasting systems. They are monitored continuously in order to react as soon as changes have occurred. Special action is then taken. Sometimes the parameters used in the forecasting model are also changed.

In order to improve the accuracy of forecasting, attempts have been made to find different ways of combining various methods. Simple smoothing has also been extended to multivariate time series and intermittent time series. These extensions will probably have an impact on stock control, and simplified methods may soon be derived. Today, considerable interest is also being devoted to *ex ante* analysing and classification of time series; see, e.g., Ref. [19].

5 Order Quantity and Order Point

Effective stock control cannot be achieved without administrative rules and procedures for determining order quantities and order points for every item. For inventories with many items, this is almost impossible to perform unless a computer-based system is used. A first step towards effective control is classification of items. Different means are mentioned in Sect. 3. A simple ABC classification is sufficient for some companies. This usually has to be extended through one or more of the following types of classification: number of withdrawals, lead time and its variations, type of items such as on-hand or order items, etc. A successful classification of items can then serve as a basis for control strategies where different items are treated in accordance with their characteristics. This may well lead to a decrease in total inventory costs.

One purpose of inventory systems is to keep down inventory costs. Another purpose is to eliminate shortages. Decisions in regard to order quantities and order points should be made simultaneously. So far, however, there are no simple analytical solutions to these integrated questions, although many approximate solutions have recently been introduced.

In practice, inventory decisions are performed in two steps. First, the size of the order quantity is determined under deterministic demand conditions. Second,

Table 2. Different lot-sizing techniques and their relationship with changes in demand and capacity utilization; source [22]

Lot-sizing technique	The average relationship between lot size and demand rate	Does the model include capacity utilization?
Fixed Order Quantity (FOQ)	–	No
EOQ	$q \sim \sqrt{d}$	No
EOQ-modified (EOQM)	$q \sim \sqrt{d}$	[1]
Silver-Meal (S-M)	$q \sim \sqrt{d}$	No
Wagner-Within (W-W)	$q \sim \sqrt{d}$	No
Period Order Quantity (POQ)	$q \sim \sqrt{d}$	No
Least Unit Cost (LUC)	$q \sim \sqrt{d}$	No
Least Total Cost (LTC)	$q \sim \sqrt{d}$	No
Part Period Balancing (PPB)	$q \sim \sqrt{d}$	No
Lot For Lot (L4L)	$d \sim d$	No
Fixed Period Req's (FPR)	$q \sim d$	No
O-R model (O-R)	$q \sim d$	Yes

[1] The model includes a finite replenishment rate m, and the ratio d/m can in some sense be interpreted as an expression for capacity utilization.

the order point is determined by considering uncertainty in demand over the lead time. Reference [20] shows that a stepwise procedure is acceptable for most B and C items. It is above all for A items that simultaneous determination of order quantity and order time is needed. However, such items often have to be handled and controlled individually. Additional factors also have to be taken into account with respect to A items. Thus many managers prefer simulations in order to understand the consequences of different strategies for A items.

There are several single-level lot-sizing rules assuming deterministic conditions. Table 2 lists ten different rules and shows how they are affected by changes in demand and available capacity [2]. For example, Refs. [21] and [22] emphasize that changes in these factors can have an impact on lead time and turnover rate. Further discussions about different lot-sizing rules may be found, in, e.g., Refs. [2], [3] and [23]. Some of these rules can be used for lot sizing in multistage systems if the parameters of the different rules are adjusted accordingly [24].

In principle, there are five different inventory systems for controlling items with independent demand. They are listed in Table 3. For a further description of these methods, see, e.g., Ref. [13] or [23].

There are several reasons for coordinated replenishment of items, e.g. in the sense that management expects savings in purchase costs, unit transportation costs, ordering costs or discounts. However, coordinated replenishment implies a rise in inventory-holding costs. Many approximate solutions to the coordinated replenishment problem have appeared during the last few years. Near optimal lower-bound solutions have been shown in Refs. [25] and [26]. Within companies, so-called "can systems" are popular; see Fig. 3 and Ref. [23]. However, such a solution is sometimes far from optimal.

In MRP planning, three different methods provide protection against the risk of shortages in inventories. These are safety stocks, safety time, and overestima-

Table 3. Different control systems

---

1. (s, Q)-system, order-point system
This system comprises continuous review of the stock level. As soon as it drops below a certain level a fixed order quantity, Q, is ordered. In its simplest form this is a two-bin system, i.e. the item is stocked in two bins. A new order is launched as soon as one of the bins is empty. In order to operate this system, the order quantity, Q, should be larger than s. If this is not the case, cardex cards or computer-based systems can be used. The Kanbans in Japanese JIT systems may also be regarded as a modified two-bin system. The system is easy to operate but it cannot take into account individual transactions of appreciable magnitude close to the order point, s.

2. (s, S)-system, order-point system with variable order quantities
As in the (s, Q)-system, a continuous review of the stock level is performed. When it drops below a certain level, s, a quantity, S, minus the net stock is ordered. The method is commonly used. However, the fact that order quantities vary can disturb other parts of the inventory system.

3. (R, S)-system, period review system
Inspection of stock and replenishment are carried out in each period, R. The order quantity is equal to S, minus the net stock. One advantage of this system is that it facilitates coordinates replenishment. However, the system gives rise to higher holding costs as compared with other systems.

4. (R, s, S)-system
This is a combination of an (s, S)-system and an (R, S)-system. The stock is reviewed in each R period, and replenishment is ordered if the stock level is below s. Under certain conditions this is the best system [20]. However, the system may be difficult to operate in a company.

5. (S, c, s)-system, "can-system", coordinated replenishment
When a level for one item drops below s, all items which belong to the same group and with a stock level below c are reordered. The size of replenishment for all items is S minus the net stock.

6. Classification of different systems

		Order quantity	
		Fixed	Variable
Order frequency	Fixed		(R, S)
	Variable	(s, Q)	(R, s, S) (S, c, s) MRP system

---

tion of demand [6]. Shortages are due to uncertainty regarding demand and replenishment. According to simulation results it is claimed [27] that quantity uncertainty (both supply and demand) is best handled by safety stocks, and uncertainty regarding time (both supply and demand) is best handled by safety times.

Overestimation of demand can be an undesirable method, as all items are treated in the same way although they seldom require the same level of protection; see also Ref. [8].

There are several ways of taking shortages into account in an inventory system [6, 13, 23]. Theoretical approaches often use shortage costs as a measure for the

Table 4. Different concepts of service

The concepts of service is related to:

- Probability of being able to deliver from stock over a cycle. The safety stock is determined as a factor multiplied by the standard deviation of the forecast error of demand over lead time.

- Number of orders that can be delivered from stock. The safety stock is based on an acceptable number of shortages and lot sizes.

- Part of the demand that can be delivered from stock. The probability for shortages over a cycle is directly related to the lot size. The safety stock is determined by a desired service level and given order quantity; see also Ref. [23]

---

cost of not being able to deliver on time. It may be difficult to estimate this cost. Therefore, the size of the safety stock is often determined according to the service concept. This means that the shortage cost can be determined implicitly. The concept of service level can be defined in many ways; see Table 4. One possibility is to relate the concept to the probability of delivering items on hand over a cycle. It is then assumed that the variation in demand over lead time is normally distributed. Uncertainty, expressed as the standard deviation, σ, is estimated as $1.25 \times \text{MAD}$.

Somewhat simplified, shortages in inventories can only occur in connection with replenishment. Shortages arise if demand over lead time is underestimated. This means that infrequent replenishments create few risks of shortages. For some C items with simplified handling, this implies a high service level as purchases are performed once or twice a year. If these occasions are well monitored, almost no shortages need arise.

One tendency in production today is to achieve short lead times along with flexible production. This can result in small lot sizes. Such measures reduce the need for long-run forecasts and large safety stocks. However, small lot sizes imply an increased number of replenishments and thus require careful control of the flow. Unexpected variations in inflow or outflow can easily create shortages. This is a complex area and so far only subsolutions have been easy to find; e.g. models that stress shortages can sometimes give rise to large lot sizes. However, this may have undesired effects, such as increased stock levels and longer lead times.

Managers have to use simulation in order to handle some of the consequences of such decisions. Several simulation languages for simulation in production are currently available.

To sum up, a computer-based system can suggest order quantities and order times for all items. These suggestions should be calculated differently for different items. The system should also allow simulations of different decisions in separate files.

6 Other Inventory Functions and Procedures

The inventory module also has functions for inventory records and physical inventory counts.

Inventory records should reflect current inventory status in terms of physical inventory, allocations, issues, receipts, and projected inventory availability. Therefore all transactions must be registered as they occur at the sources.

When an allocation is performed for an item, the available stock decreases. A complete overview of the future available stock for different items can then be obtained by projecting orders on hand and planned orders. Such an overview is useful for all types of companies, even those which use MRP planning. An overview also makes it easier to foresee any lack of items and thus points out that special action has to be taken. Planned orders are uncertain and it is therefore advantageous if the computer-based system can allow for simulations.

In practice it is almost impossible to maintain complete data accuracy. Therefore physical inventory counts are necessary in the form of full physical inventory counts or cycle counting based on ABC classification. The system should also be able to give deviation reports.

Earlier, it was emphasized that there are no *a priori* "best" decision rules for total inventory systems. This area is also difficult to treat analytically. It is therefore particularly important that computer-based systems allow for simulation in specific files. This will give the decision-maker a chance to understand the consequences of different decisions regarding both forecasting and inventory systems.

7 Conclusion

The module for inventory control includes some fundamental functions in an MPC system. However, the module uses relatively few dynamic data. Item-inventory data are in fact dynamic and require updating in order to reflect current inventory status. In addition, a calendar is needed in order to project inventory demand.

Today, most companies do not try to improve their stock control by introducing advanced forecasting and inventory systems. Instead they turn to simple systems with quick reaction to unexpected changes. This in turn requires an administrative system that makes it possible to control different items individually. Computerized systems can assist managers in this respect. However, such systems do not exclude simple control systems such as two-bin systems for C items.

Improved administrative systems require accurate data. But there is also a need to understand the demand pattern of different items; e.g. demand for batches later in the outflow system can impair estimation of the demand for an item. Demand for batches can depend on internal stock production. Thus the demand for an item should be able to separate into two parts: dependent and independent.

It is important that computer-based systems allow for such differentation in regard to all items and that companies can acquire different computer-based procedures.

For a specific company, simulations for impact analysis of various decision rules seem to be the best method for understanding how different items should be managed. Thus, simulation facilities in separate files are necessary.

In an operating materials planning system, decisions about order quantities and order points may be based on more actual and correct information. This should cause fewer shortages and a reduction in on-hand stock, which in turn means less capital tied up in inventory.

8 References

1. Olhager, J. and Rapp, B., A reference system for manufacturing planning and control. In Proceedings of First World Congress of Production and Inventory Control. APICS, Vienna, May 1985, pp. 1–5
2. Olhager, J. and Rapp, B., Effektiv MPS. Studentlitteratur, Lund, 1985
3. Orlicky, J., Material Requirement Planning. McGraw-Hill, New York, 1975
4. Wemmerlöv, U., A time-phased order-point system in environments with and without uncertainty: a comparative analysis of non-monetary performance variables. International Journal of Production Research, Vol. 24, No. 2, 1986, pp. 343–358
5. Carlson, R. and Yano, C., Safety stocks in MRP-systems with emergency setups for components. Management Science, Vol. 32, No. 4, 1986, pp. 403–412
6. New, C., Safety stock for requirement planning. Production and Inventory Management. Vol. 16, No. 2, 1975, pp. 1–18
7. Wemmerlöv, U. and Whybark, D., Lot-sizing under uncertainty in a rolling schedule environment. International Journal of Production Research, Vol. 22, 1984, pp. 467–484
8. Wijngaard, J. and Worfmann, J. C., MRP and inventories. EJOR, Vol. 20, No. 3, 1985, pp. 281–293
9. Beaumont, C., Mahmoud, E., and McGee, V., Microcomputer forecasting software: a survey. Journal of Forecasting, Vol. 4, 1985, pp. 305–311
10. Armstrong, J. S., Long-Range Forecasting. John Wiley & Sons, New York, 1978
11. Gardner, E., Exponential Smoothing: The State of the Art. Journal of Forecasting, Vol. 4, 1985, pp. 1–28
12. Makridakis, S., and Wheelwright, S. C. (eds.), Forecasting, TIMS Studies in the Management Sciences. North-Holland, Amsterdam, 1979
13. Tersine, R. J., Materials Management and Inventory Systems. North-Holland, Amsterdam, 1976
14. Makridakis, S., Andersen, A., Carbone, R., Fildes, R., Hibon, M., Lewandowski, R., Newton, J., Parzen, E. and Winkler, R., The Forecasting Accuracy of Major Time Series Methods. John Wiley & Sons, New York, 1984
15. Gardner, E., Cusum vs. smoothed-error forecast monitoring schemes: some simulation comparisons. Journal of Operational Research Society, 1985
16. Fildes, R., Quantitative forecasting – the state-of-the-art: extrapolative methods. Journal of the Operational Research Society, Vol. 30, 1979, pp. 691–710
17. Johnson, L. A. and Montgomery, D. C., Operations Research in Production Planning, Scheduling, and Inventory Control. John Wiley & Sons, New York, 1974
18. Brown, R., Decision Rules for Inventory Management. Holt, Rinehart and Winston, New York, 1967
19. Steece, B., Ex-ante measures for evalutating forecasting models. Engineering Costs and Production Economics, Vol. 10, No. 1, 1986, pp. 25–34

20. Soarf, H., The optimality of (S,s) policies in the dynamic inventory problem. Mathematical methods in the social Sciences. K. Arrow, S. Karlin, and P. Suppes (eds.). Stanford University Press, 1960
21. Olhager, J. and Rapp, B., Balancing capacity and lot sizes. European Journal of Operational Research, Vol. 19, No. 3, 1985, pp. 337–344
22. Olhager, J. and Rapp, B., Lot sizing and turnover rate. In Operational Research '84, J. P. Brans (ed.). North-Holland, Amsterdam, 1984
23. Silver, E. A. and Peterson, R., Decision Systems for Inventory Management and Production Planning. John Wiley & Sons, New York, 2nd Edn., 1985
24. Blackburn, J., and Millen R., Improved heuristics for multi-stage requirements planning systems. Management Science, Vol. 29, No. 1, 1982, pp. 44–56
25. Jackson, P., Maxwell W., and Muchstadt, J., The joint replenishment problem with a powers of two restrictions. IEE Transactions, Vol 17, No. 1, 1985, pp. 25–32
26. Roundy, R., 98%-effective integer-ratio lot-sizing for one-warehouse multi-retailer systems. Management Science, Vol. 31, No. 11, 1985, pp. 1416–1430
27. Whybark, D. and Williams, J., Material requirement planning under uncertainty. Decision Sciences, Vol. 7, No. 4, 1976, pp. 595–606
28. De Bodt, M. and Van Wasserhove, L., Lot sizes and safety stocks in MRP; a case study. Production and Inventory Management, Vol. 24, No. 1, 1983, pp. 1–16

Chapter 20
Integration of PM into CIM

Gideon Halevi*

1 Introduction

Twenty years ago, when the third generation of computers was being introduced, the concept of total Management Information Systems (MIS) was conceived. The theory of MIS was to collect all the information a business used and make it available to all who needed it. Following MIS came Production Information and Control Systems (PICS) or Communication-Oriented PICS (COPICS) and many other names and abbreviations. All these systems were information systems using the computer as a number-cranking machine, for storage and retrieval of information. The starting point was the production engineering field's accepting the basic data of Bill of Materials and routine as constraints without questioning how and why they have been established and how real and efficient they are.

As time passed and computers became less expensive, smaller in size, and faster in performance, industry recognized that computers can be used as "machine members". A new era emerged, Computer-Aided Manufacture (CAM). CAM brought a new message. A message that a computer is a working tool not merely an information storage and retrieval tool and not only a number cruncher.

A computer can control a machine's motions, and thus CNC machines appeared. A computer can read sensors and replace the switching circuits software and hardware, and thus industrial robots appeared. A computer can display characters on the screen of the monitor – therefore, why not different shapes of characters and, if so, why not random shapes, and thus computer graphics appeared and Computer-Aided Design became one of the useful and beneficial applications of computers in industry.

The trend kept on spreading, and rightfully so, and today one can count at least 13 different computer-aided fields as follows:

CAA	–	Assembly	CAM	–	Manufacturing
CAD	–	Design	CAP	–	Process Control
CAE	–	Engineering	CAR	–	Robotics
CAF	–	Fabrication	CAS	–	Sales
CAH	–	Handling	CAT	–	Test
CAI	–	Information	CAW	–	Warehousing
CAL	–	Labour Relations			

---

* Gideon Halevi is also author of Chapter 6. His biography and photo appear on p.77

and probably more will follow, as this is a good and sound technique. It is basic engineering for and by engineers.

Naturally the developments did not stop here. The potential is much too great and leads towards the Automated Factory. In a factory one has to take care of customer orders, to create a master production plan, carry out materials requirements planning (MRP), and so forth. All these production management (PM) modules must be tied together in order to run a factory as one system. It is logic to do so, and today we have evidence of the beginning of a second stage: the integration of the separate CAM systems and the PM data processing to one Computer-Integrated Manufacturing system (CIM).

2 CIM Definition

There is unanimous agreement to call Computer-Integrated Manufacturing CIM. Beyond that, several interpretations exist concerning the meaning and scope of integration. Some examples are as follows.

- The goal of integration is not tying computers together, but tying organizations and their customers together.
- The basic engineering and manufacturing functions must be all-embraced into one operative system.
- CIM integrates *all* data-processing functions within the company.
- An important role of computer integration is to move functions closer together, not so much in terms of physical distance but in the completeness and timeliness of information flow.
- The goal of CIM is to improve product flow.
- Effort is made to focus on executing rather than replanning the plan.
- CIM is a technology that combines all advanced manufacturing technologies into one manufacturing system that is capable of:
  1. rapid response to manufacturing and market demands
  2. batch processing with mass-production efficiency
  3. mass-production with flexibility of batch processing
  4. reduction of manufacturing costs.

3 CIM Architecture

CIM is a new and unsettled technology. It is in the "boiling" stage where different ideas, concepts, and technologies are being proposed, and it will probably take a while until an acceptable architecture is adopted by the majority of researchers and users.

The general trend can be seen as follows: Figure 1 shows a traditional MIS system. Note that the system is basically an information system and that the engineering phases are completely disregarded.

20. Integration of PM into CIM

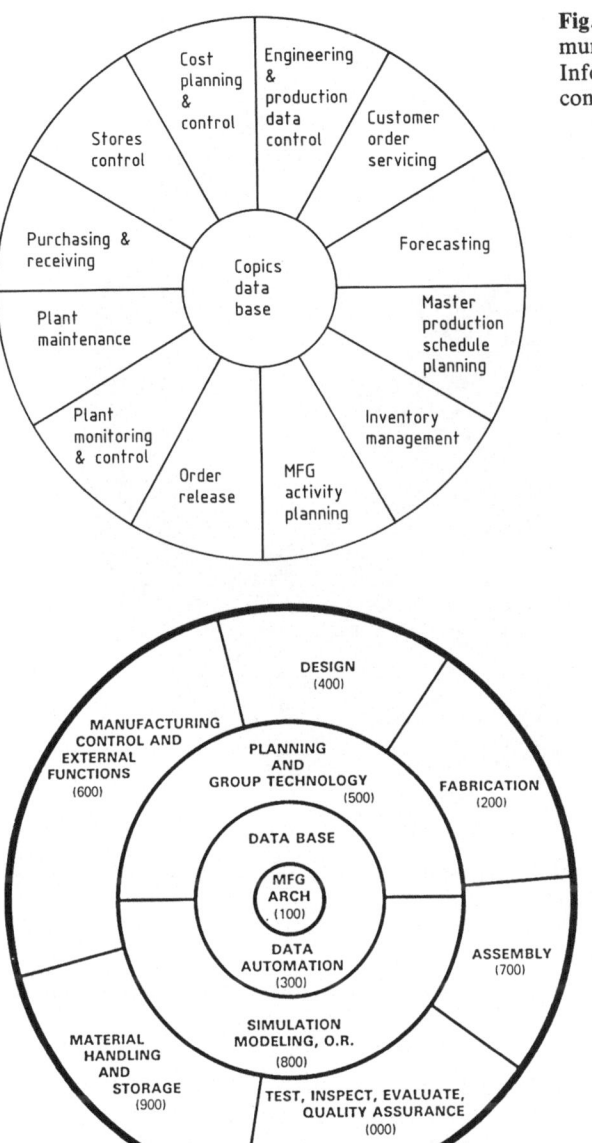

Fig. 1. IBM COPICS (Communications-Oriented Production Information and Control System) concept

Fig. 2. ICAM technology thrusts

Figure 2 shows an architecture for integration of Group Technology with PM. The engineering phases are being incorporated as databanks and serve PM and GT, but there is still a gap in communication between the product designer on one side and the PM on the other.

Figure 3 shows CASA/SME architecture in the form of a CASA "wheel". "The wheel expresses our belief that Computer Integrated Manufacturing encompasses the total manufacturing enterprise. Therefore, the outer ring of the

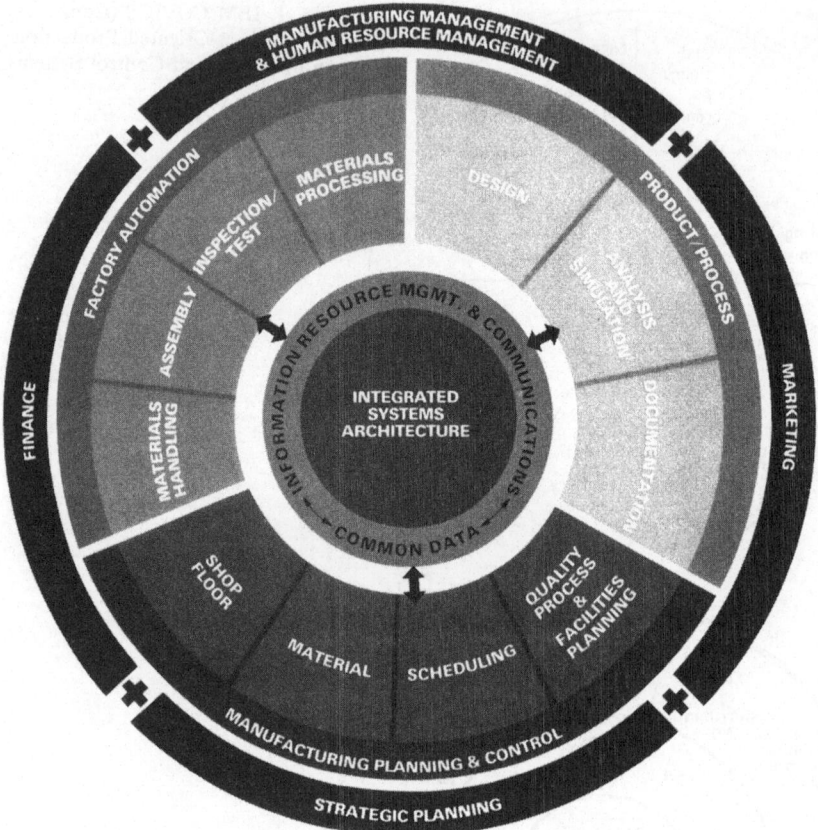

Fig. 3. CIM "wheel" developed by the Technical Council of the Computer and Automated Systems Association of SME (CASE/SME)

wheel includes marketing, finance, strategic planning and enterprise and human resource management."

Figure 4 shows CIM architecture of IPA – Institute of Manufacturing Engineering and Automation, Stuttgart, West Germany. "The plurality of goal conflicts which come up in the production field shows that the competitiveness of an enterprise can not be fully guaranteed if solutions are used which cover only part of the whole production system. All sections of an enterprise which are directly or indirectly involved in the production process have to be optimized all at the same time."

Figure 5 shows the "HAL Technology" concept, where all functions continuously communicate with one another, design, plan, and adjust the program according to the intermittent situation on the shop floor. "It does not contemplate the relationship between individual areas of activities, but rather dissolves them into one single system." It is a working system, not merely an information system.

There are many more wheels and charts with different concepts.

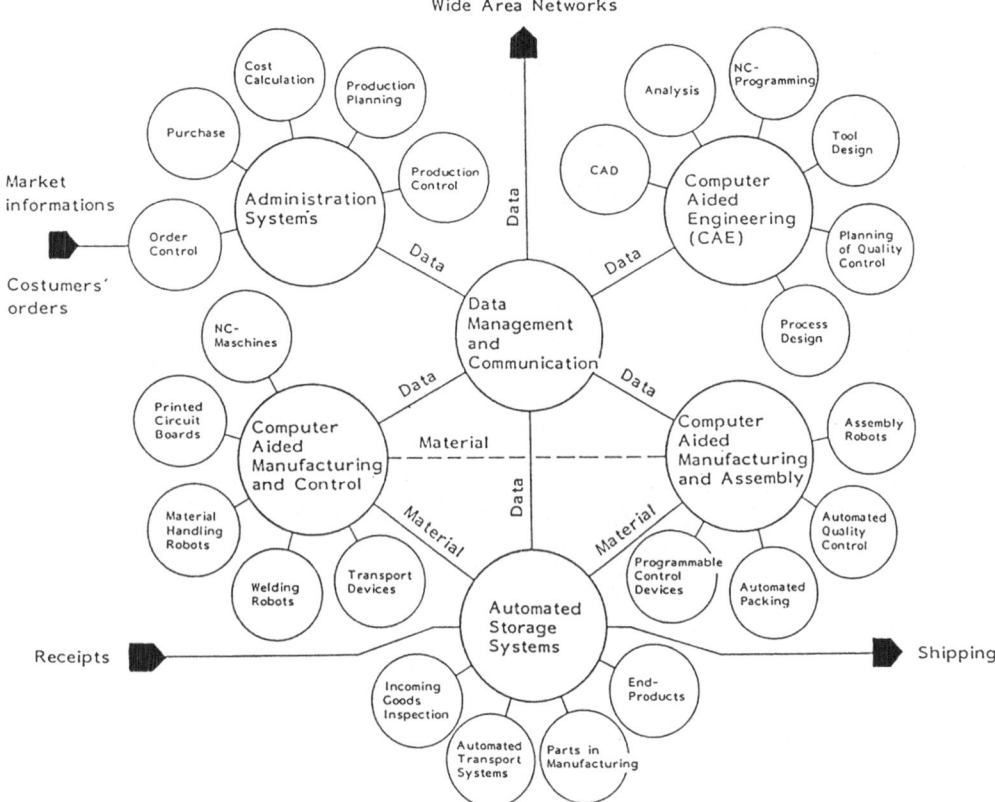

Fig. 4. CIM architecture

4 Implementation of CIM

Implementation of CIM requires knowledge, research and development in several disciplines:

1. Communication between computers, terminal and machines
2. Computer Science, to solve data storage and processing problems
3. Computer-operated facilities – such as CNC, industrial robots, AGV, etc.
4. Algorithms and methodology in the fields of basic engineering and production management.

4.1 Communication

"For the development and construction of a CIM system the transfer of data between computer systems with different specifications is an unalterable precondi-

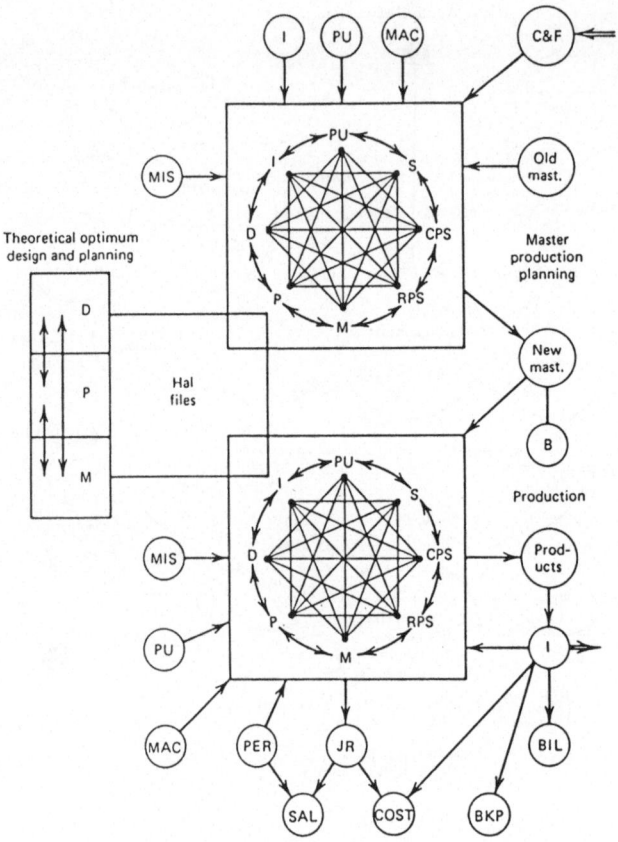

Fig. 5. Hal manufacturing cycle. Notation: D, product design; P, process planning; M, methods, time, and motion study; RPS, requirements planning; CPS, capacity planning; S, shop; I, inventory; PU, purchasing; MAC, machine file; C&F, customer orders and forecasting; MIS, miscellaneous; B, budget; PER, personnel; JR, job recording; SAL, salaries; COST, costing; BKP, bookkeeping; BIL, billing

tion. An essential characteristic of CIM is manufacturing with less paper. CIM applications refer not only to computer – computer communication but also to the coupling of dispositive and operative systems, i.e. shop floor control, cell computers, and machine control. This area of coupling within CAM (Computer Aided Manufacturing) is increasingly gaining in importance and is already currently realized in a variety of applications. Since especially in this area a multitude of different control and computer systems with very different interfaces are employed, the call for a standardization is here the loudest. A new magic word here is LAN technology (Local Area Network) which promises, in addition to the long overdue standardization, a substantial cost reduction and a supplementary expandability. More than forty vendors of such LAN systems are known on the world market. The systems offered are differentiated according to the network topology, the frequency range, and the method of access for the individual users. Characteristic for LAN technology is the procedure in which *one*

20. Integration of PM into CIM

Ebene	MAP-Definition	TOP-Definition
Funktions-bereich	Manufacturing Messaging Format Standard (MMFS) SASE	Textverarbeitung Graphik Verteilte Datenbanken
7. Application	ISO FTAM (DP) 8571 File Tranfer Protocol and CASE	
6. Presentation	Null	
5. Session	ISO Session (IS) 8327 Session Kernel, Full Duplex	
4. Transport	ISO Transport (IS) 8073 Class 4	
3. Network	ISO Internet (DIS) 8473 Connectionless	
2. Data Link	ISO Logical Link Control (DIS) 8802/2 (IEEE 802.2) Type 1, Class 1	
1. Physical	ISO Token Passing Bus (DIS) 8802/4 (IEEE 802.4) Token Passing Bus Medium Access Control	ISO CSMA/CD (DIS) 8802/3 (IEEE 802.3) CSMA/CD Medium Access Control 10 Base 5

Fig. 6. MAP standard for factory communication

data path is available for the exchange of data, that is to say, in which a logical connection in addition to the physical connection must be constructed in the preparation for data communication."[1]

Since 1980 there have been attempts to reach a standard among networks. The cooperation of the American companies DEC, INTEL, and XEROX resulted in a first preliminary standard – the ETHERNET.

Within the IEEE (Institute of Electrical and Electronic Engineers) the group 802 worked on questions concerning the standardization of local area networks. In the European sector, too, the ECMA (European Computer Manufacturers Association) focused its attempts on a standardization in this field.

In the production field there is a sudden increase of programmable control and computer systems. That is why General Motors, in 1980, founded a "Manufacturing Automation Task Force" which aimed at a common communication protocol for future industrial production systems. On the basis of the OSI-model this group worked out in the subsequent years a requirement profile and a multi-step concept for the realization as well as for the protocol specification of a system communication architecture concerning future automated manufacturing systems (MAP: Manufacturing Automation Protocol; see Fig. 6).

On the basis of a backbone network, single manufacturing cells (automation clusters) are combined. They consist of autonome solutions of various producers or other network structures (baseband, broadband, MAP-network). The use of networks in the manufacturing field places demands on response time, reliability, and flexibility of the communication system. MAP meets these demands since the token bus method is used. At the NCC, MAP, together with renowned producers, was first presented in a test installation. In February 1985, the version

MAP 2.0 was introduced, which, however, was extended by the compatible version MAP 2.1 as early as May. In November 1985 there was another presentation of MAP at AUTOFACT, which showed the stage of development and the possibilities of MAP. For 1987 or 1988 the specification of MAP 3.0 is announced, which would allow the connection of most of the computer and control systems to a MAP network.

In the meantime MAP is fostered by important producers of computer and control technology and is considered to be the future worldwide standard for factory communication. After the standardization of MAP is completed, we can therefore expect a common interface system, so that automation components of different producers are able to exchange data using a MAP network [2].

4.2 Computer Data Storage

The computer is the heart of the CIM system, while the communication to terminals, data collection units, machines, sensors, etc. are the veins. In a moderate CIM system one can expect to have hundreds of veins, and the heart should be able to supply instructions and receive data from all of them. As we know there is a limited number of terminals or work stations that a computer can serve before it clogs up. In the CIM environment things will be worse than in MIS, since in MIS humans are working interactively with computers. Humans are slow in reading messages, in absorbing their meaning, and in keying in information. Thereby the computer is I/O bound and can serve many users. In CIM the main computer communicates with other computers or sensors which are fast devices and we might get to a CPU-bound situation.

CIM must work only with a formal information system (as opposed to humans who work mostly with informal information). Information of any activity, any storage location, any machine command must be available to the computer in order to reach a smart decision. CIM must have a huge online storage capability. Some predict that the size of the manufacturing database will ultimately be anywhere from 20 to 50 times larger than most current data-processing databases. "With as many as 48 billion bytes devoted to manufacturing data compared with perhaps two billion bytes for data processing, it is quite conceivable that data processing will ultimately become a subset of the manufacturing operation once a company-wide CIM operation has been put into place" [3].

Others predict that the amount of data stored will be reduced because the computer will become intelligent. They will be taught to read engineering drawings and engineering technology. There will be no need to store engineering data as it will be more profitable to generate engineering decisions, design and process plan by the instantaneous state of the shop floor. "Don't store data, store concepts, rules, and generate data when and where it is needed" [4].

One thing is certain – that the factory will be managed by a hierarchy of computers, and CIM depends on the continuing development of high-speed intelligent computers and computer systems.

In parallel, Artifical Intelligence (AI) and smart terminals are being developed in order to relieve part of the load from the main computer.

"Evident or not, computers — or at least electronically programmable devices — are now built into just about every piece of electrical equipment. Soon, AI concepts also will have to be designed into just about all equipment intended for future factory operations. This will be true whether in robotic applications or in the automated factories of tomorrow — or even of 'later today'.

Goshorn points out, however, that all AI components will act as peripherals to the central unit. And all AI peripherals will be dramatically different from peripherals as we now conceive of them. They will not just be printers, plotters, and such. Instead, they will be human-like sensors and they will perform human-life activities: eyes for sight, fingers for touch, and more. But the design concept will remain as it is today. The central unit will need to be developed first — with provision for the artificial peripherals. Then those peripherals will be added." [5]

Working computer hierarchy and communication system pose a problem of reliability and a need for a backup system. On a smaller scale the problem of backup for DNC has not been fully solved. In CIM this problem is greater by a magnitude. There are not many papers on this topic but there is no doubt that a solution will be reached.

4.3 Facilities

The facilities in a CIM system must have the performance of standard machines, but in addition have the capability of communicating with their "foremen". The facilities must have a computer or a programmable controller that receives commands, and transmits data concerning its activities to the computer one step higher in the hierarchy.

Since the discovery that computers are machine members, a new breed of machines is rapidly being developed, with new capabilities and options. They can serve as stand-alone machines but they also fall in line with the CIM requirements.

4.4 Algorithms

A CIM system is built in order to achieve a better and more flexible manufacturing system and thereby reduce production cost, reduce WIP and manufacturing lead time, and reduce the time from an idea of a product to the delivery of that product. This is the main objective of CIM — not to build an aesthetically pleasing and effective communication network, or smart computer systems. They are merely tools to achieve the goals. However, as in many instances in real life, the tools get more attention than the goals. In the literature we can find papers about tools but very few about the algorithms of PM methodology and technology.

CIM is in essence and extension of production management in the direction of basic engineering. CIM is a combination of CAD/CAM/PM in one system.

CAD and CAM are very similar by nature but in spite of this it is very difficult to merge them into one system. Many articles have been published concerning the removal of the slash between the CAD and the CAM, but the slash is still

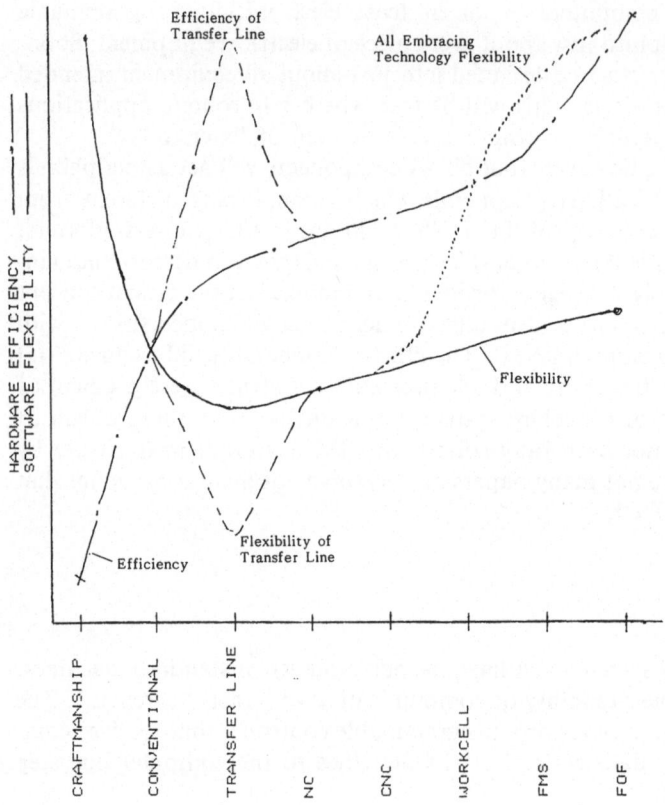

Fig. 7. Evolution of the manufacturing process

there. Now the ambition is to remove yet another slash, that of PM, and make a CIM of it.

The importance of PM in CIM was initially minimized. The idea was "let's leave the hardware and computers and it will adjust ifself".

However, it was soon realized that without appropriate PM, expensive hardware was idle. In spite of the hardware flexibility the total performance did not result in flexibility in production. On the other hand, it was realized that the existing PM has to be adapted to the new working environment.

Figure 7 shows the evolution of the manufacturing process, with the level of hardware efficiency and software (PM) flexibility. Initially, the craftsman was very flexible in his production management. He did not have a planned routine, since he did not even have a fixed product design. He had only the goals that the product should meet. Therefore, very seldom had he parts that did not pass inspection. If he missed his initial production aims, he just modified the design. The craftsman was the product designer, the process planner, the methods, the production planner (actually CAD/CAM/PM), and above all he was the machine power and operator. This method is very flexible (as the manufacturing process is by nature), but very inefficient and very costly.

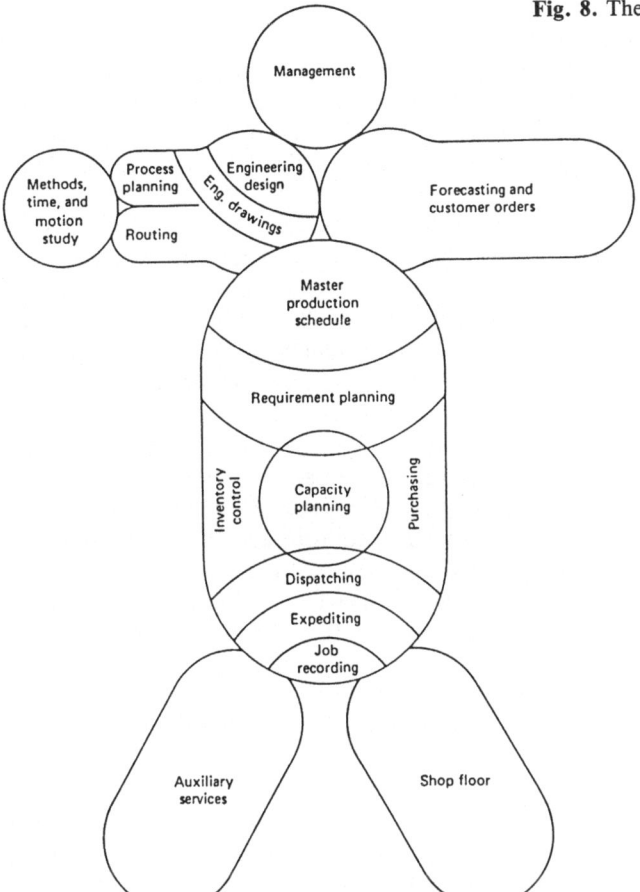

Fig. 8. The manufacturing cycle

The trend since the industrial revolution has been to seek increasing economies of scale through more capital-intensive manufacturing processes – trying to achieve lower cost per unit by building and implementing more and more specialized manufacturing equipment. On the graph we see an increase of efficiency for conventional machines, reaching a peak in transfer line manufacturing. It works, but we pay for this by decreasing PM flexibility. In scale production there are several functions to be performed, and it takes different skills and training to perform each task. No one can handle efficiently all the functions. Therefore a PM method of the chain of activities became the dominating one, and rightfully so. We first had the separation of technology (basic engineering) from PM. Then in PM we had stages of master production planning, requirements planning, capacity planning, etc. (see Fig. 8). Each stage receives input data, processes it by its optimized algorithms and generates output for the next stage. The input is regarded as constraint, as the present stage does not possess the information of how and why it was decided. The chain of activity methods suffer from reduction of flexibility, introducing artificial constraints into the system, and reduction in the overall optimization. However it was probably the only method to make factories run.

The numerically controlled machines which grew to FMC, FMS, and to the concept of the factory of the future (FOF) introduced a new capability of manufacturing small batches with the efficiency of mass production. We take a universal machine and convert it by software to a dedicated machine. The conversion does not take time, so why produce by scale when it is possible to produce what we need when we need it. This revolution was catching on much faster in the hardware than the software (PM). PM has not adapted itself as rapidly to the new era. What we actually need is the flexibility of the craftsman in the FOF environment and it can be done. This new type of factory imposes constraints on the PM which did not exist before. Capacity planning did not consider the accessibility of the part to the machine. We had shop floor personnel to take care of it. With FMS and FOF and CIM there are no personnel – programming has to be done by the system. Not the actual dispatching notes, but the capacity planning should take into account the location of the parts. Therefore it might be necessary to have two stages of planning: one, production planning, which tells us what parts to put in or near the transfer mechanism, and the second stage of scheduling which is the old capacity planning but with the limited number of parts. Or the PM could come with a completely new logic. The effect of the AGV and the AS/RS calls for more dynamic planning. It is appropriate to have 2–3 days for transfer between operations. Today, we relate high value to manufacturing lead time and want to cut it to a minimum. We have the means to control the transfer, but we miss the logic of how to apply it, and what the effect on the overall PM activities will be.

The chain of activities uses stand-alone optimization with today's computers, and we need to move to product-mix optimization. We want flexibility in manufacturing. We want alternatives and we can reach it; however, we need the PM algorithms. The CAD/CAM/PM–CIM era opens up a new degree of freedom to production management. There is no longer a fixed routine – actually no routine at all, no classification – inventory is run by shapes and not be catalogue number. It is a new world that has to be filled, and it is mostly PM.

5 Conclusion

CIM is a logical development and a natural evolution of the computer-aided era. One day it will be the dominating technology and control of factories. PM is not quite ready yet for CIM as it should be. There are missing links that cannot be overlooked. CIM must be a target and a research topic. Different approaches should be carefully studied.

Many modules and algorithms are needed, and it will take time to develop them. The more people work on it the shorter the time will be. However, there is no sense in being impatient and enforcing CIM today with the technological gap.

Some experts predict that CIM will not come of age for at least 5 or 10 years. Many others are even more pessimistic. They believe a minimum of 15 years is probable.

6 References

1. Spur, G., LAN application on shop floor. 18th CIRP Manufacturing Systems Seminar, Stuttgart, Germany, 5–6 June 1986
2. Warnecke, H. J., Computer networks for production planning and control. 18th CIRP Manufacturing Systems Seminar, Stuttgart, Germany, 5–6 June 1986
3. Gondert, S. J., Computer-Integrated Manufacturing – The Technological Imperative. Computervision Corp
4. Halevi, G., The Role of Computers in Manufacturing Processes. John Wiley & Sons, 1980
5. Shapiro, S. F., Artifical Intelligence key to computer integrated manufacturing. Computer Design, December 1984, pp. 50–59
6. Savage, E., and Mikurak M., Finite scheduling: staging a comeback? CIM Technology, Spring 1986, pp. 26–31
7. Halevi, G., CIM – the future technology. 10th World Computer Congress, IFIP Congress 1986, Dublin, Ireland, 1–5 September 1986
8. Mather, Hal., Logistics efficiency – product design and process requirements. CIM Review (USA), Vol. 2, No. 2, Winter 1986, pp. 16–20
9. Halevi, G., and Weill, R., On-line scheduling for flexible manufacturing systems. Annals of the CIRP, Vol. 33, No. 1, 1984, pp. 331–333

Part VI
Industrial Applications

Chapter 21
Multi-Product Batch Production on a Single Machine – A Problem Revisited*

Samuel Eilon

Professor Eilon graduated in Electrical Engineering at the Technion – Israel Institute of Technology. Haifa, and, after a spell in industry and in the Israeli armed forces, he came to Imperial College where he was awarded a Ph.D. in Mechanical Engineering and later the D.Sc. (Eng.) degree for his work in Production Management. He became Professor in 1963 and established the Department of Management Science, of which he was Head until 1987. He helped in the design and introduction of Operational Research and Management Studies as a Part II Tripos in Cambridge and was a visiting lecturer in the Engineering Department at Cambridge for five years. He has also held a Professorial Research Fellowship at Case Western Reserve University, Cleveland, Ohio, and a Visiting Fellowship at University College, Cambridge. He was a Founder Fellow of the Fellowship of Engineering. He has acted as a consultant to industry for some twenty-five years, was a consultant for the P-E Consulting Group, a director of the consulting company Spencer Stuart & Associates and of Campari International Ltd. He is now a director of ARC Ltd. In addition to his academic duties and industrial activities Professor Eilon is Chief Editor of *OMEGA*, the International Journal of Management Science, and editor of the *OMEGA* Management Science Series. Professor Eilon's personal research interests cover company performance analysis, production and inventory control, distribution systems, OR in banking, and management control. He is the author of twelve books and over two hundred and fifty papers and articles. Professor Eilon is a member of IFIP WG 5.7.

1 Introduction

Multi-product batch production is very common in industry. It stems from the fact that when rate of production exceeds the rate of demand, production of a given product has to stop at some stage to allow its inventory to decline and the production facilities are then used to produce other products. When each product requires several operations on different machines, as is often the case in machine shops in the engineering and electronics industries, then the plan becomes a complex queueing system, in which scheduling needs to take account of availability of facilities, capacity constraints, work-in-progress inventories, as well as due dates to satisfy demand.

---

* This paper was first published in *OMEGA*, Vol. 13, pp. 453 – 468, and acknowledgement is given to the Editor of *OMEGA* for permission to reprint the paper.

There is a vast literature on various aspects of production scheduling, from the determination of the production level in continuous production (the so-called "production smoothing problem", which is concerned with a single product subject to fluctuating demand) to the design of loading rules in job shops (where each job has its own processing specifications and its assigned due date for completion). Between these two extreme categories is the case of *multi-product batch scheduling* (or MPBS for short), concerned with intermittent production of several products, which share the same production facilities and for which demand is known. Unlike job shop production, in which each job is a single entity (often being "one-off"), MPBS requires the determination of the batch size for each product, either at the outset or for each of its operations.

The diversity of approaches to production planning and materials control is exemplified in a recent review, published in *Industrial Engineering* [1], in which the contrasting philosophies of several control systems in current use are described. These include MRP (materials requirements planning), which starts with the computation of total demand for each item based on a master production schedule, OPT (optimized production technology), which starts with a production network for each product (not dissimilar to a conventional bill of materials) and proceeds to load work on production facilities with the aim of giving priorities to bottlenecks and reducing work-in-progress inventories (the details of the algorithms used have not been published), and JIT (just-in-time, similar to the Japanese Kanban system), which aims to produce items only when needed by combining work-stations and reducing work-in-progress, often at the expense of increased idle-time on production facilities. There are numerous other production control systems, not mentioned in the *Industrial Engineering* review. They usually take the form of proprietary computer packages, which incorporate some of the features listed above.

Comparison of performance of these systems is difficult. There is not enough empirical evidence in the form of detailed case studies to allow comparisons of like-with-like, and it is too much to hope that a given plant could be subjected to alternative control systems in sequence to provide a consistent framework for measurement purposes. More often than not, the introduction of a new package is accompanied by a reorganization of the plant and a complete change in managerial attitudes, so that it is just impossible to ascertain what proportion of the ensuing dramatic improvement is due to the algorithmic process and ingenuity of the package itself and how much is due to the corresponding managerial reorientation. It is often asserted that the latter is far more significant than the former in transforming the production environment, though most practitioners in this field would probably agree that the two go hand-in-hand.

Be that as it may, it is useful to focus on some test problems as a basis for comparison of various algorithmic approaches. In this paper an old problem involving production in sequence of several products on a single machine is examined. This problem is encountered in the process industries, though it is rare in engineering machine shops. Nevertheless, the structure of the problem, coupled with the simplicity of its definition, serves to highlight some interesting issues relating to the development of a methodology for solution. Though the computer packages mentioned earlier (such as MRP, OPT, and JIT) or their derivatives have not been constructed with a single production facility in mind, it might be interesting if their exponents applied their approaches to the problem described below and compared their results with the method outlined in this paper.

2 A Test Problem

Multi-product batch scheduling (MPBS) has attracted the attention of many writers. Perhaps the most famous MPBS problem is that set by Bomberger in 1966 [2], for which numerous solutions have been proposed since. One review [3] cites over ten solutions, thereby demonstrating the many heuristics that can be designed with various degrees of sophistication and success. The Bomberger problem, though, has one serious failing, namely that the total production time (excluding setup) required to produce the specified product range amounts to only 22% of the demand time. This very high level of over-capacity, coupled with the fact that for three out of ten products the production rate exceeds the demand rate by a factor of at least 300 and that half the products involve relatively low setup cost (between $ 5 and $ 50), creates a lot of room for manoeuvre for scheduling purposes. Not only is there no need for overtime, but some of the products with low setup costs may be scheduled frequently to be produced in small batches, if need be, with little regard to the availability of capacity. In fact, with such a low level of capacity utilization, additional products (other than the ten listed) should be taken on.

The question of setup costs is central to the methodology adopted for the solution of an MPBS problem. The conventional approach is to seek to minimize the total variable costs per unit, i.e. to balance the setup cost per unit against the holding cost by the use of the well-known EBQ square-root formula (EBQ stands for the "economic batch quantity", sometimes called in the literature the EOQ, i.e. the "economic order quantity", or the "optimal lot size"; see Appendix A1). When goods are ordered from outside suppliers, the use of the EBQ for each product independently of the others seems logical enough. But in the manufacture of multi-products, which share the same production facilities, the application of the EBQ to each product is indefensible. When the stock for a given product 1 is exhausted and a fresh batch needs to be produced, the machines may be busy on another product 2, so that either production of product 2 has to be terminated prematurely (i.e. before its EBQ is completed) or product 1 will run out. To protect against such shortages, safety stocks are held, so that supply can be maintained while products wait for machines to become available, and these safety stocks often nullify the cost advantages claimed for the EBQ. MRP and other production and inventory control systems rely heavily on the application of the EBQ, which is blamed by some as the main cause for the high level of work-in-progress inventories found in industry.

Clearly, the smaller the setup cost, the smaller the EBQ and the cost per unit, and this is what development of flexible manufacturing systems has managed to achieve in recent years. Ideally, setup costs should be drastically reduced or even eliminated, so that the EBQ approach becomes irrelevant and the JIT concept could then be adopted in production control. Indeed, some control systems deliberately ignore the EBQ, implying that the manpower employed to set up machines is part of the total fixed overhead cost and is incurred irrespective of the number of setups involved. Such an assertion has a great deal of substance in the case of machine shops, and if applied to the Bomberger problem, would result in very low-cost solutions (admittedly, involving many setups, but the overcapacity can readily absorb the resultant added setup time). Among the control systems that dispense with the need to account for setup costs (though not setup time) are JIT and OPT.

However, there are circumstances when the setup costs may not be ignored. For example, in the process industries the setup costs may consist of cleaning materials and replacement of machine parts in preparation for a new product, as well as material lost during the running-in period to ascertain that the output will be of the right specifications and quality. Such costs are incurred whenever a changeover of product is called for, and in some cases they may be quite substantial. To ignore them and allow numerous setups to take place would result in low inventories but mounting total costs. At the same time, it is inadvisable – for the reasons argued above – to resort to the indiscriminate use of the EBQ to individual products. It is necessary, therefore, to apply a method that takes account of the EBQ as well as the fact that the products are dependent on each other through the production schedule that loads them on the same manufacturing facility.

The problem presented in Table 1 is based on one (with minor modifications) published several years [4] before Bomberger's. It consists of six products with relatively high setup costs and setup times (particularly for products 5 and 6). The fixed unit cost c (consisting of materials, labour, and processing costs, but excluding setup and holding costs) also varies and is particularly high for products 4 and 5. The holding costs h (per one unit of c per day) includes maintenance cost in store in addition to interest charges, and this is why in this case h is not the same for all products. Other asssumptions are:

1. Normal working time consists of 8 hours per day, 5 days per week, 250 days per year.

Table 1. Data for a six-product problem [4]

Product		1	2	3	4	5	6
Demand/day	a	20	24	30	36	40	50
Production rate/day	b	100	150	200	110	400	280
Setup cost	s	3000	1800	3600	1500	6000	30000
Setup time (days)	t_s	4.0	2.4	4.8	2.0	4.0	8.0
Holding cost/money unit/day ($\times 10^{-4}$)	h	12.20	20.40	11.35	11.04	7.65	14.82
Fixed unit cost	c	4.0	1.6	6.0	12.0	16.0	6.0

Note: A list of notations is given in Appendix A8

21. Multi-Product Batch Production on a Single Machine

2. Overtime may be worked up to 8 hours per day during weekdays, i.e. up to 5 overtime days per week at the cost of 20 per hour (or 160 per overtime day); no weekend working is allowed.

3 Initial EBQ Considerations

The data in Table 1 suggest that, in marked contrast to the Bomberger problem, capacity is inadequate to meet demand during normal working hours and overtime would have to be employed. This is found by computing γ (the ratio of demand rate to production rate) for each product, as shown in Table 2, where the sum of all the γ's amounts to about 1.116. This sum expresses the ratio of the amount of required production time (excluding setup time) to the available normal time, so that 11.6% of normal time (or just under one hour on the average per day) will have to be worked as overtime for production purposes, in addition to the total setup time, which will depend on the number of setups scheduled, hence involving a minimum cost of $0.116 \times 160 = 18.56$ per day for overtime.

If a batch of Q units of a given product is ordered from an outside supplier and is calculated to arrive then the previous batch is depleted after a cycle time T, as shown in Fig. 1(a), then the peak stock level is Q and the average $\frac{1}{2}Q$. In the case of production as in Fig. 1(b), however, demand is met during the pro-

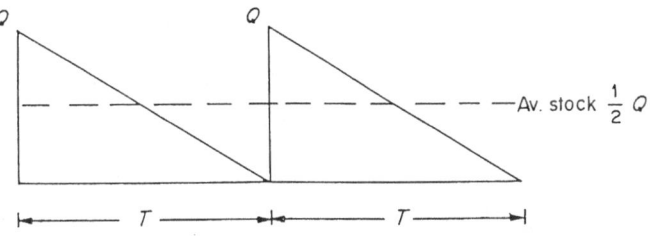

(a) Ordering from outside supplier

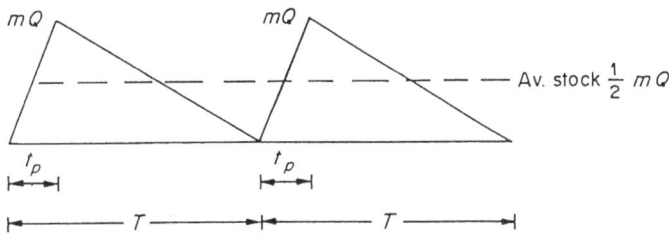

(b) An internal production facility

Fig. 1. The average stock level

Table 2. Some basic EBQ computations (ignoring overtime)

Product	1	2	3	4	5	6	Total
$\gamma = a/b$	0.2	0.16	0.15	0.327	0.1	0.179	1.1158
$K' = hc(1-\gamma)/(2a)\ (\times 10^{-4})$	0.9760	0.5712	0.9648	1.2378	1.3770	0.7304	
Notional optimal batch $Q'_0 = \sqrt{s/K'}$	5544	5614	6109	3481	6601	20266	
Individual cycle $T' = Q'_0/a$	277.2	233.9	203.6	96.7	165.0	405.3	
Production days $t_p = Q'_0/b$	55.44	37.43	30.55	31.65	16.50	72.38	243.94
Min. unit variable cost	1.0823	0.6413	1.1786	0.8618	1.8179	2.9606	
$y'_0 = 2s/Q'_0$ Min. variable/fixed unit cost	0.271	0.401	0.196	0.072	0.114	0.494	
y'_0/c Notional min. cost/day	21.645	15.390	35.358	31.026	72.716	148.031	324.166
$C'_0 = a/y'_0$							

21. Multi-Product Batch Production on a Single Machine

duction period t_p, so that the peak inventory is $t_p(b-a)$, where a and a are the daily rate of demand and production, respectively, ignoring overtime. But as the batch produced is $Q = b\, t_p$, it follows that the peak is $m\, Q$, where (ignoring overtime)

$$m = 1 - a/b = 1 - \gamma$$

so that the average stock level is $\frac{1}{2}\, m\, Q$, from which the average stock value is then derived. The EBQ formula is based on the criterion of minimizing the total cost per day (where the cost in this context is defined as the sum of setup cost and holding cost) or minimizing the total variable cost per unit, defined as

$$y = s/Q + K\, Q$$

where K is a holding cost factors. The resultant square-root formula for the EBQ is then easily found (see Appendix A1).

Applying this approach to the six-product problem in Table 1, we get the notional results in Table 2. They are notional because they do not take account of the use of overtime, so that the holding cost factor K', and the optimal batches Q'_0 (see Appendix A8 for notation) are computed on the assumption that capacity during normal hours is adequate to meet demand. From Q'_0 (which has been rounded off to integer values) the cycle time T' for each product is derived, giving a range of 96.7 days for product 4 to 405.3 for product 6, whereas the total number of production days required (excluding setup time) is approximately 244. These calculations clearly demonstrate the fundamental mismatch between the production cycle and the demand cycle, so that the indiscriminate application of the EBQ cannot work, unless the system is prepared to tolerate either high safety stocks or frequent run-outs (or both). Table 2 also reveals that

- compared with the basic fixed cost c the minimum variable cost y_0 varies greatly from product to product, adding over 40% to the basic cost in the case of products 2 and 6;

- the notional minimum cost per day also varies enormously from product to product, the major offenders being products 5 and 6, which are responsible between them for some 68% of the total.

The total notional minimum cost per day amounts in Table 2 to 324.17, which may be regarded as a crude lower bound. Such a bound is a useful comparator for proposed solutions, but the notional figure in Table 2 is deficient on two counts: first, it assumes adequate capacity and does not take account of the cost of overtime, and secondly it ignores the effect of overtime on the average stock values. Both factors are included in the computation of a better lower bound C_{B1} in the Appendix; the bound is shown to be a function of the cycle time T.

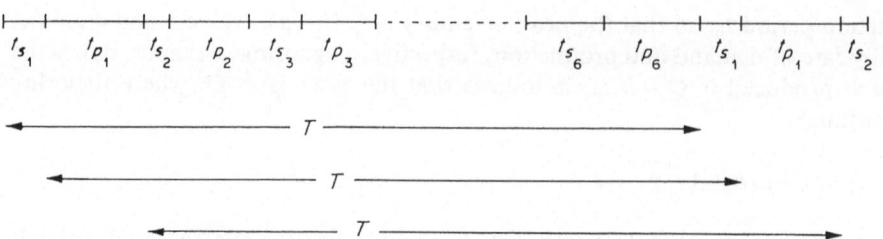

Fig. 2. The cycle time T for Model 1 when capacity matches demand

4 Solution Methods

To keep stock levels low, it is desirable to start production of a new batch at the same time as the previous batch of the same product is depleted, as shown in Fig. 1(b), and in principle this is attainable with the use of overtime. The resultant schedule would look something like that shown in Fig. 2, where setup and production of products follow in sequence to fill the cycle time. It is clear, though, that the batch quantities will have to deviate from the individual EBQs for the solution to become feasible. The following models are explored here:

Model 1: No batch splitting
Model 2: The cycle is divided into two subcycles with batch splitting
Model 3: The cycle is divided into four subcycles with batch splitting.

In principle, further alternatives are possible, but these three models with their variants should be adequate to illustrate the solution methodology.

4.1 Model 1

In this model it is assumed that the products are produced in sequence, as suggested in Fig. 2, each product produced once during the cycle in a sufficient quantity to meet the demand throughout the cycle. Since demand is assumed in this test problem to be constant, the pattern of the stock level for each product is as shown in Fig. 1(b).

It is further assumed that overtime is evenly spread throughout the cycle, and it is then convenient to distinguish between *calendar days* and *production days*. The cycle consists of T calendar days (in the absence of any designation it will be understood that a "day" is a "calendar day", so that there are five calendar days per week, the weekends being excluded), the batch size for product 1 to meet the demand for T days is

$$Q_1 = a_1 T$$

and the batches for the other products are similarly computed. The number of production days needed to produce product 1 is

Q_1/b_1 for production
plus 4.0 for setup.

With the use of overtime every calendar day is equivalent to λ production days, so that the time needed to produce product 1, expressed in calendar days is

$$\frac{1}{\lambda}(Q_1/b_1 + 4.0)$$

and the total time for all the six products is equal to the cycle time T. As shown by Equ. (16) in the Appendix, the value of λ can be easily ascertained for any given cycle T. For example, a cycle of 100 days leads to $\lambda = 1.368$, which means that a calendar day consists of 10.94 hours (of which 2.94 are overtime). λ declines as T increases, as shown in Fig. 3; in the hypothetical case of $\lambda = 1$ there is no need for overtime, and capacity during normal hours would match demand, but in our example $\lambda > 1$ irrespective of the cycle length.

If the total variable cost per day is chosen as the performance criterion, then the method developed in Appendix A2 yields the optimal cycle length, from which the batch size for each product is readily computed. The results are shown in Fig. 4, from which we find that the optimal cycle length is 260 days, the total cost being 406.25. Further details of this solution are summarized in Table 3, which also shows the individual EBQs and the bound C_{B1}. First we note that the optimal individual batches are somewhat lower than the notional values of Table 2. The cost per day corresponding to these optimal batches plus the cost of overtime yield the lower bound C_{B1} (further details on the computation of the lower bound are given in Appendix A3), which amounts to 364.27 for a cycle of $T = 260$, so that the solution of Model 1 is 11.5% above the bound. A closer scrutiny of the batch sizes and the cost per day for the individual products reveals that the major deviations are caused by products 4, 5 and 6: for products 4 and

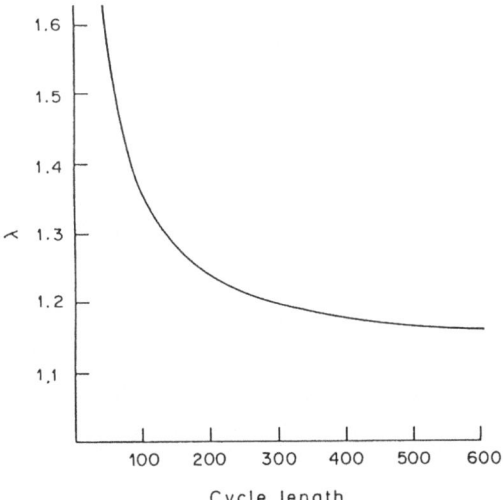

Fig. 3. The value of λ for Model 1

Table 3. Results for Model 1 for $T = 260$ Days

Product	1	2	3	4	5	6	Total
Holding cost factor K ($\times 10^{-4}$)	1.0188	0.5903	0.9946	1.3435	1.4038	0.7583	
Optimal individual batch Q_0	5426	5522	6016	3341	6538	19891	
Minimum cost/day C_0	22.12	15.65	35.90	32.33	73.42	150.82	330.23
Overtime cost							34.04
Total cost for the bound C_{B1}							364.27
Batch size Q for Model 1	5200	6240	7800	9360	10400	13000	
Cost/day C for Model 1	22.13	15.76	37.12	51.04	81.45	164.67	372.21
Overtime cost							34.04
Total cost for Model 1							406.25

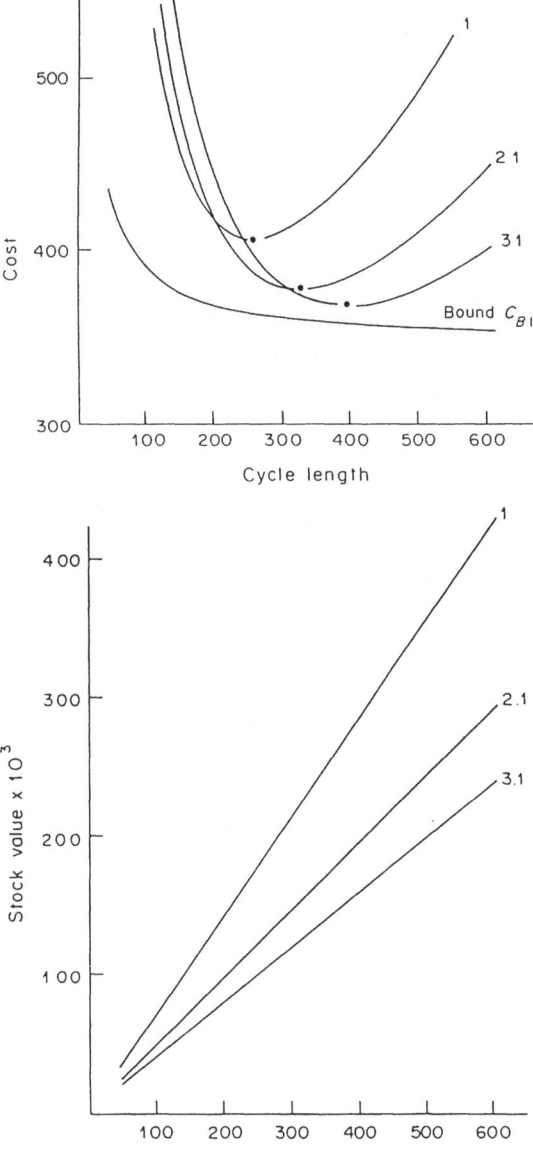

Fig. 4. Cost per day as a function of cycle length for three schedules

Fig. 5. The average stock value for three schedules

5 each batch is considerably larger than its corresponding EBQ (in the case of product 4 it is larger by almost a factor of 3), while for product 6 it is considerably smaller. It should be noted that the cost function is quite flat in the vicinity of the optimum. If, for example, $T = 250$ is chosen instead of the optimum value of 260, the effect on cost is rather marginal, as shown by Fig. 4 and the first two rows in Table 4, though the effect on the average stock value is more noticeable, since the average stock value increases linearly with the cycle length (see Fig. 5).

Table 4. Results for five models

Model	Cycle time	Cost/day	Av. stock value	Av. λ	λ for subcycles	Bound	Dev. from bound (%)
1	260 (Opt) 250	406.25 406.52	184,126 177,146	1.213 1.217		$C_{B1} = 364.27$ $C_{B1} = 364.98$	11.5 11.4
2.1	327 (Opt)	377.85	157,668	1.211	1.223, 1.199	$C_{B1} = 360.64$ $C_{B2} = 363.99$	4.8 3.8
	330	377.86	159,093	1.210	1.222, 1.198	$C_{B1} = 360.49$ $C_{B2} = 363.84$	4.8 3.9
2.2	327 (Opt)	377.93	157,735	1.211	1.218, 1.205	$C_{B1} = 360.64$ $C_{B2} = 363.99$	4.8 3.8
	330	377.94	159,162	1.210	1.217, 1.204	$C_{B1} = 360.49$ $C_{B2} = 363.84$	4.8 3.9
3.1	395 (Opt)	368.67	156,019	1.223	1.188, 1.281 1.143, 1.281	$C_{B1} = 358.18$ $C_{B3} = 366.18$	2.9 0.68
	390	368.70	154,072	1.225	1.189, 1.283 1.144, 1.283	$C_{B1} = 358.33$ $C_{B3} = 366.44$	2.9 0.62
3.2	392 (Opt)	371.03	156,624	1.224	1.429, 1.041 1.384, 1.041	$C_{B1} = 358.27$ $C_{B3} = 366.33$	3.6 1.3
	390	371.03	155,836	1.225	1.430, 1.042 1.385, 1.042	$C_{B1} = 358.33$ $C_{B3} = 366.44$	3.5 1.3

4.2 Model 2

In order to improve on the solution in Model 1 it is necessary to split the batches of products 4 and 5, i.e. to divide the cycle T into two subcycles, so that split batches of 4 and 5 are produced in each subcycle, while the other products are produced only once throughout the entire cycle. There are obviously many ways in which subcycles can be constructed with respect to their relative length as well as the distribution of products between them. We confine this investigation to the following (the symbol = separates the two subcycles):

Model	Sequence
2.1	$4-5-1-2 = 4-5-3-6$
2.2	$4-5-1-3 = 4-5-2-6$

In these schedules the subcycles are assumed for convenience to be of equal length ($T_1 = T_2 = T/2$) and the only difference between them is that products 2 and 3 are interchanged in the sequence. The first split batch of product 4 is aimed at satisfying the demand during the first subcycle, the second split batch is for the second subcycle and so are the split batches of product 5, while the batches of the other four products meet the demand during the whole cycle T.

Two possible lower bounds are computed for these models: the first (denoted as C_{B1}) assumes that each product is produced only once during the entire cycle. This bound is exactly the same as the one derived for Model 1. The second bound takes account of the added setup times required for products 4 and 5 to be produced twice each during the cycle (this bound is denoted as C_{B2}; details in Appendix A4).

The results for Models 2.1 and 2.2 are very close to each other for a very wide range of values of T (values from 50 to over 600 were explored), the results for 2.1 being very marginally better, as shown in Table 4. The optimal cycle here is $T = 327$, i.e. appreciably longer than the solution for Model 1 (in order to absorb the added setup costs), and again the cost function is very flat in the vicinity of the optimum. (If T deviates to the nearest multiple of 10, namely to 330 days, the results are hardly affected.) The performance of Model 2.1 is considerably better than that of Model 1, the cost being reduced to 4.8% above the bound C_{B1} and only 3.8% above C_{B2} (see also Fig. 4).

It is interesting to note that even when the subcycles are equal in length, as in Models 2.1 and 2.2, the λ values in the two halves may well be different, owing to the difference in the required number of production days. The results for the two λ values, compared with the average λ (which, incidentally, is not identical to the λ in Model 1, shown in Fig. 3) are plotted in Fig. 6 as a function of the cycle time T. A note as to when the λ values become identical is given in Appendix A5.

One consequence of the different λ values in the two subcycles is that if the split batches of product 4 are equal (as assumed in the computational procedure), the production time in calendar days for this product will not be the same in the two subcycles. This means that while the first split batch of product 4 aims at meeting demand for $T/2$ calendar days, the starting dates for producing the two

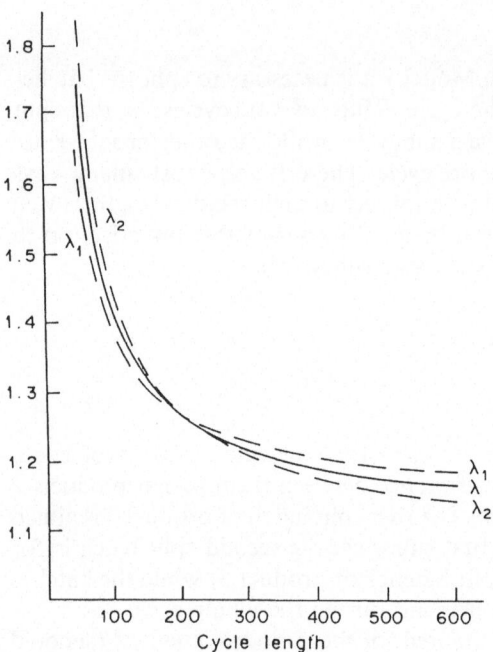

Fig. 6. λ values for the subcycles (broken lines) and the average λ for Model 2.1

split batches being $T/2$ days apart, the dates when production of these split batches terminates will not be $T/2$ days apart. The result of this so-called "subcycle mismatch" will then affect the level of stock availability of product 5. This effect is small, as shown in Appendix A6, and can be handled by "fine-tuning" of the split batches for products 4 and 5 or the level of localized overtime. This effect, though, is ignored in the results reported in Tables 4 – 6.

4.3 Model 3

The improved performance of Model 2, compared with that of Model 1, naturally raises the question as to whether further batch-splitting could be beneficial. In Model 3 we assume 4 subcycles to allow product 4 to be produced four times (once in each subcycle), three products to be produced twice, while the batches for products 1 and 6 remain unsplit. Two variants are explored:

Model	Sequence
3.1	$4-1 = 4-3-2-5 = 4-6 = 4-3-2-5$
3.2	$4-1-5 = 4-2-3 = 4-6-5 = 4-2-3$

The difference between the two models lies in the scheduling of product 5. In both cases the subcycles are equal in length (each being $T/4$). The computational procedure is similar to that used for Model 2.

The results are also included in Tables 4 and 5 and Fig. 4. Again, the cost function is insensitive to T in the vicinity of the optimum, and a deviation of T

21. Multi-Product Batch Production on a Single Machine

Table 5. Further details for the optimal solutions for three models

Product	1	2	3	4	5	6	Total
Model 1							
$T = 260$; $C_T = 406.25$; $\lambda = 1.213$							
Batch sizes	5200	6240	7800	9360	10400	13000	
Cost/day C	22.13	15.76	37.12	51.04	81.48	164.67	372.21
Setup cost/day							176.54
Overtime cost/day							34.04
Model 2.1							
$T = 330$; $C_T = 377.86$; $\lambda = 1.210$							
Batch sizes	6600	7920	9900	2×5940	2×6600	16500	
Cost/day C	22.56	16.69	40.40	37.80	73.42	153.34	344.20
Setup cost/day							161.82
Overtime cost/day							33.66
Model 3.1							
$T = 390$; $C_T = 368.70$; $\lambda = 1.225$							
Batch sizes	7800	2×4680	2×5850	4×3510	2×7800	19500	
Cost/day C	23.52	15.92	36.05	32.41	74.78	150.09	332.77
Setup cost/day							158.46
Overtime cost/day							35.93

to the nearest multiple of 10 hardly affects the cost. We find that the performance of Model 3.1 is better than that of 3.2 and both are better than that of Model 2. In the case of Model 3.1, the cost is just 2.9% above the bound C_{B1} and only 0.7% above C_{B3}. The value of the bound C_{B3} is computed in a similar fashion to that related to Model 2, except that the added setup times for all the split batches are taken into account (details and a comparison of the three bounds are given in Appendix A4). The average stock value is considerably lower in Model 3.1 compared with that of Models 1 and 2, as shown in Fig. 5.

It may be possible to effect a further improvement by constructing even more complicated schedules, but it is questionable whether such exercises would be worthwhile, in view of the very satisfactory results from Models 2.1 and 3.1.

The effect of subcycle mismatch, alluded to earlier in the discussion of Model 2, is exacerbated if the schedule involves more batch splitting. Although overstocking and run-outs can be overcome by fine tuning, as shown in the Appendix for Model 2.1, implementation may become cumbersome for more complex schedules. The effect of fine tuning on the cost function is ignored in the results in Tables 4 and 5.

5 Concentrated Production

All the results cited above are based on the assumption that a constant level of overtime is maintained throughout the cycle in Model 1 or each subcycle in the other models. Since the data allow overtime to be used up to 8 hours per day, it may be possible to affect the average stock level by postponing production for as long as possible, as illustrated in Fig. 7: the production period for each product is split into two parts, the first using only normal time and no overtime, while the second uses the maximum overtime allowable. Although the batch size is exactly the same as that shown in Fig. 1(b), and the peak inventory level is also the same, concentrated production results in a lower average stock level than when the production rate is held constant, as in Fig. 1(b).

Models 1 and 2.1 were run for this mode of production and the results, summarized in Table 6, show an improvement in the cost level of just over 1%. For

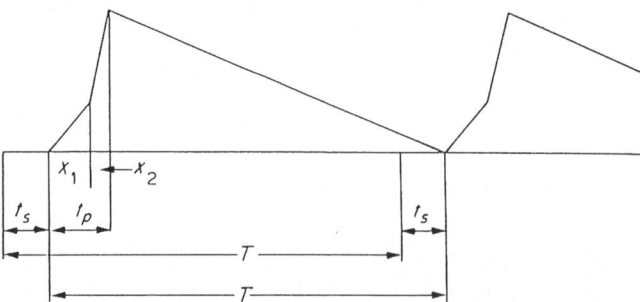

Fig. 7. Normal production during x_1 and concentrated overtime during x_2

Table 6. The effect of concentrated production

Model	Production mode	Cycle time	Cost/day	Av. stock value	λ values	Av. λ
1	EVEN	260	406.25	184,126	1.213	1.213
	X	260	401.11	179,565	1.213	1.213
2.1	EVEN	330	377.86	159,093	1.222, 1.198	1.210
	X	320	373.52	150,622	1.225, 1.202	1.213

X denotes concentrated production during the latter part of the production period

Model 1 the optimal cycle time (to the nearest multiple of 10) and the value of λ are unaffected. For Model 2.1 the optimal cycle time falls slightly (from 330 to 320 days) while the value of λ rises slightly. As expected, this mode of concentrated production helps to lower the inventory level.

6 Effect of Setup Costs

As stated earlier, the level of setup costs may have a profound effect on the daily cost function. To explore this effect, Model 1 was run with eight different levels of total setup costs, and the results are summarized in Table 7. The first case A relates to the orginal data in Table 1, in case B the setup cost for product 6 is halved, while the other setups are unchanged, and so on. The final case H involves no setup costs as all. In all these cases the setup times remain as in Table 1.

The results in Table 7 and Fig. 8 (where the cost function for 5 of the 8 cases is plotted against the cycle time T) show that the effect is indeed dramatic, with respect to both the optimal cycle time and the corresponding optimal cost. More detailed results for case B (where the setup cost for product 6 is halved) are given

Table 7. The effect of setup costs in Model 1

Case	Setup Costs $\times 10^3$							Cycle time	Cost/ day	Av. stock value	λ	Note
	s_1	s_2	s_3	s_4	s_5	s_6	Σs					
A	3	1.8	3.6	1.5	6	30	45.9	260	406.25	184,126	1.213	Original data
B	3	1.8	3.6	1.5	6	15	30.9	220	343.33	156,199	1.230	s_6 is halved
C	1.5	0.9	1.8	0.75	3	15	22.95	190	304.31	135,244	1.248	All setups are halved
D	1.5	1.5	1.5	1.5	1.5	2.5	10	140	225.41	100,285	1.296	
E	1	0.5	1	0.5	1	1	5	110	184.97	79,273	1.345	
F	0.75	0.45	0.9	0.375	1.5	7.5	11.475	140	235.94	100,285	1.296	Setups reduced by 75%
G	3	1	3	1	6	26	40	240	382.84	170,164	1.221	
H	0	0	0	0	0	0	0	70	130.60	51,157	1.476	No setup costs

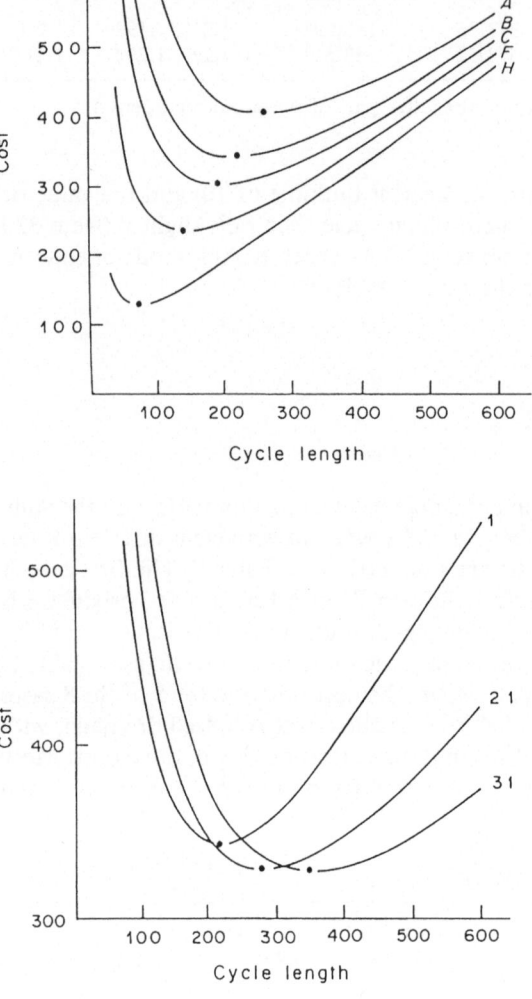

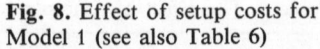

Fig. 8. Effect of setup costs for Model 1 (see also Table 6)

Fig. 9. Three schedules when the setup cost of product 6 is halved

for the three basic models in Fig. 9, from which we conclude that there is hardly any improvement in cost for Model 3.1 compared with Model 2.1 and that the improvement of the latter over Model 1 is far smaller than in Fig. 4 (which is based on the original setup cost data). It is also interesting to note that Table 7 and Fig. 10 (which relate to Model 1) suggest that it is the total setup cost which is the decisive factor, rather than the way the total is split among the products (Fig. 10 also shows the optimal cycle). The conclusion from these results is that efforts to reduce setup costs are potentially very rewarding and are likely to be sig-

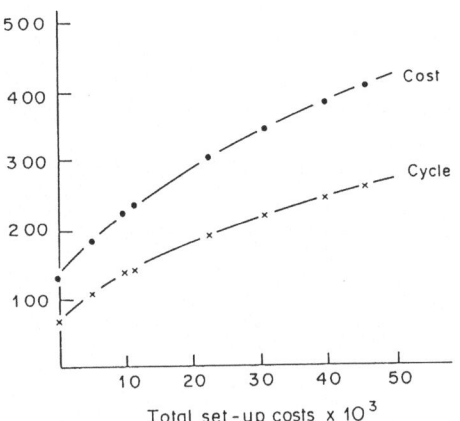

Fig. 10. Effect of the total setup costs on the optimal cost per day and the optimal cycle length for Model 1

nificantly more effective than the promise of increasingly more complex algorithms.

7 Conclusion

The methodology described in this paper starts with a simple solution for Model 1, based on a cycle in which each product is produced once in a quantity calculated to meet the demand during the cycle. The optimal cycle length is the one for which the total variable cost per day (including the cost of setups, stock holding, and overtime) is minimized. Although this solution inevitably causes most (or even all) products to deviate from their individual EBQs, the result compares reasonably well with the computed lower bound, which is clearly unachievable. This solution can be greatly improved with batch splitting, as indicated by the results for Models 2 and 3 in Table 4 and in Fig. 4, but the improvement by moving from Model 2 to 3 is less spectacular, and for practical scheduling purposes production controllers may well prefer the solution for the former to the more complicated cycle required by the latter. In either case, the results are sufficiently close to the lower bounds to allow us to question the need to consider more complicated schedules. Another indication of the quality of the solutions is given by the ratio of cost due to setups to the total cost (excluding overtime). In the case of the EBQ for a single product the setup cost per day is half the total cost. The results in Table 5 show that this ratio is about 0.47 for the three models. The results also reveal that the cost functions are very flat in the vicinity of the optimal cycle time, so that deviations of $\pm 10\%$ or even more from that optimum point have a relatively small effect on the cost.

Several variants of Model 2 can be constructed, but the results for the two examples 2.1 and 2.2 in Table 4 show that the former is only marginally superior to the latter. It is interesting to note that the range of the values for the two sub-cycles in the two models (given in Table 4) is almost the same. A comparison of Model 3.1 with 3.2 shows that the former yields a lower total cost and that its λ values are more evenly distributed, whereas the relatively high λ in the first subcycle of 3.2 has an adverse effect on the average stock value and hence on the total cost. Also, the need for fine tuning is reduced when the λ values are more evenly distributed.

It is clear that delaying production for as long as possible helps to reduce inventories and costs, and the results in Table 6 confirm that conclusion. However, the greatest potential benefit by far is obtainable by reducing the total setup cost, as shown in Figs. 9 and 10.

8 References

1. Swann, D., Jacobs, F. R., Fox, K. A., and Schonberger, R. J., New approaches to material flow and inventory control. Industrial Engineering, Vol. 16, No. 10, 1984, pp. 25–63
2. Bomberger, E. E., A dynamic programming approach to a lot size scheduling problem. Management Science, Vol. 12, No. 11, 1966, 00. 778–784
3. Eilon, S., Production scheduling. In Operational Research '78 (Proceedings of the Eighth IFORS Conference, Toronto, 1978), K. B. Haley (ed.). North-Holland, 1979, pp. 237–266
4. Eilon, S., Production Planning and Control. Macmillan, New York, 1963

9 Appendix

A1 The EBQ for a single product

If overtime is used to the extent that the ratio of the average number of hours worked per day to normal hours is λ and if the normal production rate is b units per day, then λ causes the production rate per calendar day to increase to λb. The peak stock in Fig. 1(b) then becomes

$$t_p (\lambda b - a)$$

Since the batch size $Q = t_p \lambda b$ it follows that the average stock level is $mQ/2$ where

$$m = 1 - \gamma/\lambda \qquad (1)$$

$$\gamma = a/b \qquad (2)$$

In the special case of no overtime, $\lambda = 1$ and this corresponds to the notional results in Table 2. The average stock value is

$$V = \tfrac{1}{2} mQc \qquad (3)$$

and the holding cost per day is Vh, so that when the setup cost is added, the total cost per day is as given in the main text. By substituting

$$T = Q/a \qquad (4)$$

$$K = hcm/(2a) \qquad (5)$$

it is simple to show that

$$C = s/T + Vh$$
$$= a(s/Q + KQ)$$
$$= ay \qquad (6)$$

where y is the variable unit cost, namely

$$y = s/Q + KQ \qquad (7)$$

from which the classical EBQ formula to minimize y is derived as

$$Q_0 = (s/K)^{1/2} \qquad (8)$$

at which the minimum value for y is shown to be

$$y_0 = 2s/Q_0 = 2KQ_0 \qquad (9)$$

corresponding to the minimum cost per day of

$$C_0 = 2sa/Q_0 = 2KaQ_0 \qquad (10)$$

i.e. the setup cost per day constitutes half the total cost.

A2 Optimizing the multi-product schedule [4] – Model 1

Assuming that the products are produced once each during the cycle T, as prescribed in Model 1 and shown in Fig. 2, it follows that

$$T = Q_i/a_i \qquad (11)$$

hence

$$Q_i = a_i Q_1 \qquad (12)$$

where a_i/a_1, i.e. the ratio of demand of product i to that of product 1. As it is meaningless here to consider unit cost, we need to resort to another criterion, such as the ratio of profit to the cost of production. If the selling price of product i ist p_i, the total unit cost of which is $c_i + y_i$, then the profit is $(p_i - c_i - y_i)Q_i$, so that the ratio becomes

$$\frac{\Sigma(p_i - c_i - y_i)Q_i}{\Sigma(c_i + y_i)Q_i} = \frac{\Sigma p_i Q_i}{\Sigma(c_i + y_i)Q_i} - 1 = \frac{\Sigma Q_1 p_i a_i}{\Sigma(c_i + y_i)a_i Q_1} - 1$$

Q_1 cancels out and since a_i, p_i, and c_i are independent of the batch size Q_i it follows that the above ratio is maximized when $\Sigma a_i y_i$ is minimized. The optimal value of Q_1 is then derived from

$$\frac{d\Sigma a_i y_i}{dQ_1} = 0$$

and it is easy to show that the solution is [4]

$$Q_1 = \left(\frac{\Sigma s_i}{\Sigma K_i a_i^2}\right)^{1/2} \qquad (13)$$

from which the other Q_i and the optimal cycle time are readily found by Eqs. (11) and (12). Minimization of the total cost per day yields the same result, since

$$C = \Sigma C_i = \Sigma a_i y_i = a_1 \Sigma a_i y_i \tag{14}$$

Applying this result to the data in Table 1 we find that $\Sigma s_i = 45900$, whereas the value of $\Sigma K_i a_i^2$ depends on λ, which can be shown to depend on the cycle time T: since the total production days are $\Sigma(t_{pi} + t_{si})$ and this total needs to be divided by λ to obtain calendar days, it follows from Fig. 2 that

$$\Sigma t_{pi} + \Sigma t_{si} = \lambda T \tag{15}$$

Substitute $t_{pi} = Q_i/b_i = Ta_i/b_i = T\gamma_i$, and $\Sigma \gamma_i = 1.1158$, also $\Sigma t_{si} = 25.2$ (from Table 1) and we obtain

$$\lambda = 1.1158 + 25.2/T \tag{16}$$

The solution may be found by the following simple procedure:

1. Assume T
2. Derive λ from (16)
3. Calculate K_i by (5)
4. Find Q_1 by (13)
5. Find $T = Q_1/a_1$

then go back to step 2 and iterate again. The value of T converges rapidly, as shown in the following table (for practical purposes T is confined to integer values):

Initial assumption:	$T = 100$	$C = 535.88$
Iteration 1	244	371.91
2	247	371.86
3	247	371.86

The solution converges to $T = 247$. It should be noted that the solution in (13) is derived on the assumption of constant values of K, whereas in our case K is a function of T, so that the true optimum of C occurs at $T = 249$, but this hardly affects the cost (which is 371.85).

Notice that this result takes no account of the cost of overtime, which is excluded from the cost function C in Eq. (14), and hence the solution underestimates the value of the optimal cycle. The cost of overtime is derived by taking the average amount of overtime per day $(\lambda - 1)$ times the cost per day (160). The total cost would then be

$$C_T = C + 160 \, (\lambda - 1) \tag{17}$$

where C is found from (14) or from

$$C = \Sigma(s/T + Vh) \tag{18}$$

A simple solution procedure may easily be programmed on a microcomputer (or a suitable desk calculator) as follows:

1. Assume T
2. Derive the corresponding batch sizes
3. Find the value of λ and hence the average stock value for each product
4. Find the total cost C_T from (17) and (18)
5. Repeat by scanning other values of T to find where the minimum of C_T occurs.

Results for Model 1 are given in Tables 3–5 and in Fig. 4.

A3 Computation of the bound C_{B1}

As stated in the main text, the total notional cost per day given in Table 2 can be greatly improved upon for the purpose of providing a lower bound. First, the cost of overtime can be accounted for by adding $160(\lambda - 1)$. Secondly, the effect of overtime on the rate of production, and hence on the average stock level, is dealt with by including λ in the computation of m, as in Eq. (1), and as shown in the previous section we can derive λ for any given value of T. The bound is therefore obtained by

$$C_{B1} = \Sigma C_0 + 160(\lambda - 1) \tag{19}$$

The computation procedure is as follows:

1. Assume a given cycle T
2. Derive λ from Eq. (16)
3. Find m and hence K for each product by Eq. (1) and (5)
4. Compute Q_0 by Eq. (8)
5. Find C_0 for each product by Eq. (10)
6. Compute the bound C_{B1} by Eq. (19).

For $T = 260$ (which is the optimum cycle length for Model 1) the results are shown in Table 3. The result for the bound C_{B1} as a function of the cycle time is shown in Fig. 4. Note that the bound computed here is unachievable in practice, since it is based on the assumption that all the products are produced at their individual EBQ values and, as already stated in the main text, such a solution is infeasible.

A4 Computation of the bounds C_{B2} and C_{B3}

The bound C_{B1} is a useful yardstick for the results of all the models. However, in the case of Model 2, products 4 and 5 are produced twice each, so that additional setup time is involved, increasing the total setup time for the whole cycle from 25.2 days (in Model 1) to 31.2 days. A modified bound C_{B2} may then be computed by the use of Eq. (19), except that now λ is determined by

$$\lambda = 1.1158 + 31.2/T \tag{20}$$

to replace Eq. (16), which applied to Model 1. Similarly, for Model 3 the value of λ is found from

$$\lambda = 1.1158 + 42.4/T \tag{21}$$

To summarize, we have three bounds with which to make comparisons:

C_{B1} for Model 1, but useful for the other models as well, based on (16)
C_{B2} for Model 2, based on (20)
C_{B3} for Model 3, based on (21)

A comparison of the numerical values of the three bounds is given below:

Cycle T	C_{B1}	C_{B2}	C_{B3}
250	364.98	369.36	377.50
275	363.30	367.29	374.71
300	361.90	365.56	372.38
325	360.71	364.10	370.39
350	359.61	362.85	368.70
375	358.81	361.75	367.23
400	358.04	360.80	365.94

A5 λ values for the subcycles in Model 2.1

The production sequence for Model 2.1 calls for four batches in each subcycle:

$$4-5-1-2 = 4-5-3-6$$

The number of production days needed for setups for the two subcycles is 12.4 and 18.8, respectively. The total number of production days for setups and production for the first subcycle is

$$t_{p4} + t_{p5} + t_{p1} + t_{p2} + 12.4$$

The ratio of this sum to the subcycle length is λ_1. If the batch Q_4 is intended to meet demand during the subcycle (the length of which is taken in this model as $T/2$), i.e. $Q_4 = a_4 T/2$, then

$$t_{p4} = Q_4/b_4 = (a_4/b_4) T/2 = T\gamma_4/2$$

and a similar expression applies to product 5, whereas for product 1 the batch is expected to meet the demand for the whole cycle, i.e.

$$t_{p1} = T\gamma_1$$

Thus, the ratio λ_1 is obtained from

$$\tfrac{1}{2} T\lambda_1 = (\tfrac{1}{2}\gamma_4 + \tfrac{1}{2}\gamma_5 + \gamma_1 + \gamma_2) T + 12.4$$

Substituting the values of γ from Table 2, we get

$$\lambda_1 T = 1.1473\, T + 24.8 \tag{22}$$

and similary for λ_2 for the second subcycle

$$\lambda_2 T = 1.0844\, T + 37.6 \tag{23}$$

From the last two equations it is clear that $\lambda_1 = \lambda_2$ at $T = 203.5$ and that $\lambda_1 > \lambda_2$ when $T > 203.5$ (see also Fig. 6).

A6 Effect of subcycles on cost in Model 2.1

If we take the optimal cycle length of $T = 330$ then the λ values for the subcycles are derived from the last two equations as

$$\lambda_1 = 1.222, \quad \lambda_2 = 1.198$$

and these values may then be used to convert production days into calendar days when the detailed schedule is constructed, as shown in the following table. For example, product 4 requires 2 days for setup and 54 for production, i.e. a total of 56 production days, which are equivalent to 45.8 calendar days. If the calendar starts at 0 at the beginning of the cycle, then production terminates on day 45.8. Stock is available from day 1.6 (when the setup time of $2/\lambda_1$ expires and production starts) until day 166.7. Similarly, the batch for product 5 meets the demand from day 49.1 to 214.1, while the batch for product 1 lasts for a whole cycle, namely from day 65.9 to the same date in the next cycle, and so on.

Product	Batch size	Production and setup time		Calendar time		Demand calendar	
		Production days	Calendar days	Start	Finish	Start	Finish
First subcycle							
4	5940	2+54	45.8	0	45.8	1.6	166.6
5	6600	4+16.5	16.8	45.8	62.6	49.1	214.1
1	6600	4+66	57.3	62.6	119.9	65.9	65.9X
2	7920	2.4+52.8	45.2	119.9	165.1	121.9	121.9X
Second subcycle							
4	5940	2+54	46.7	165	211.7	166.7	1.7X
5	6600	4+16.5	17.1	211.7	228.8	215.0	50.0X
3	9900	4.8+49.5	45.3	228.8	274.1	232.8	232.8X
6	16500	8+58.9	55.9	274.1	330.0	280.8	280.8X

Note: X signifies the calendar date in the next cycle.

As we can see from this schedule, the only problem that arises concerns product 5: the first split batch lasts to day 214.1 while stock from the second split batch is available only from day 215.0, i.e. run-out occurs for one day. Similarly, the second split batch lasts until day 50, whereas stock is already available from the first split batch on day 49.1. This minor mismatch, caused by the different λ values, can easily be overcome, if needs be, by adjusting the sizes of the split batches, or by changing the overtime rate. Such fine tuning hardly affects the schedule or the total cost function. In the case of more complex models, however, where fine tuning is required to deal with several products, the final schedule may be deprived of its initial simplicity, though the effect on cost is likely to remain marginal.

A7 Concentrated production

As suggested in Fig. 7, the maximum allowable overtime may be concentrated at the end of the production period, leaving the rest of the time for normal working without overtime. Thus, the production period would consist of

x_1 calendar days at the rate b

x_2 calendar days at the rate $2b$ (using 8 hours overtime per day)

If λ is the average for the whole cycle, then the total production days required for a given product is t_p equivalent to t_p/λ calendar per days, leading

$$\lambda (x_1 + x_2) = t_p \qquad (24)$$

The batch produced is $Q = bt_p$, which is also obtained from Fig. 7 by $Q = bx_1 + 2bx_2$, and it therefore follows that

$$x_1 + 2x_2 = t_p \qquad (25)$$

For any given batch Q (for which t_p is known) the last two equations yield the values of x_1 and x_2 and the resultant average stock level (which is lower than that in Fig. 1(b)) can be computed.

A8 Notation

a = demand rate per calendar day (excluding weekends)
b = production rate for 8 hours day
c = fixed costs per unit (materials, labour, and processing)
h = holding cost per money unit of c per day
m = a factor affecting the peak inventory = $1 - \gamma/\lambda$
p = selling price per unit
s = setup cost for a batch
t_p = production period expressed in production days = Q/b
t_s = setup time for a batch
y = variable cost per unit = $s/Q + KQ$
y_0 = minimum value of y when the EBQ is produced

C = cost per day, excluding overtime = $s/T + Vh$
C_0 = minimum value of C when the EBQ is produced, excluding overtime
C_B = bound for the total cost C_T
C_T = total cost per day, including overtime
K = holding cost factor = $hcm/(2a)$
Q = batch size
Q_0 = the EBQ to minimize the variable cost per unit of an individual product
T = cycle time in calendar days
V = average value of the stock

$\alpha_i = a_i/a_1$
$\gamma = a/b$

Chapter 22
Production Control in Small Companies

Kai Mertins

Dr. Ing. Kai Mertins has a B.S. (Ing.) in Electrical Control Theory from Hamburg and an M.S. (Dipl.-Ing.) in Economics and Electrical Engineering from Technical University, Berlin. He has a Ph.D. in Production Control of Computer-Aided Manufacturing Systems. At present he is Head of Department for Manufacturing Systems, Fraunhofer-Institut für Produktionsanlagen und Konstruktionstechnik, Berlin. Dr. Mertins has experience in Factory Control systems, complex manufacturing systems, simulation of large manufacturing systems, and CIM. He has specialized in the machine tool industry and small and medium batch production. He has managed international multi-partner projects for innovation of manufacturing systems.

1 Introduction

Flexible automation of manufacturing systems is characterized by computer integration in subsystems such as processing, transport, storing, handling, and control. The structure of production is influenced by the assembly of specifications represented by the variety of workpieces, orders, and operating materials. Each one has a different influence on the execution of different tasks in production control. Modular program systems for production control, which are conventionally task-oriented, consider stochastic influences and human behaviour insufficiently, so that in small lots down to a lot size of one the goal of corresponding planning and production sequences may not be achieved.

The manufacturing system presented here has a competence-oriented design. Conditions of manufacturing are created which allow the preparation and modification of the production process on different factory levels by relying on the direct users' experience and creativity.

Competence-oriented production control considers the actual working process and resource requirements by a disposition closely attached to the manufacturing process. The dispositive and operative tasks of such a production control are carried out at a computer-integrated shop floor control station.

Production systems such as flexible manufacturing cells, flexible manufacturing lines, and flexible manufacturing networks are used in a growing number of small companies. As these systems are capital-intensive, high capacity utilization is needed, and this leads to new requirements for production planning and control. At the same time the development of the market directly leads to greater demands such as a faster delivery process and a shortening of flow time. The growing complexity of products is correlated with a growing complexity of informational relations in the structure of the production process. New perspectives are given on the other hand to production planning and control by technical developments. Together with growing efficiency, computer systems are becoming smaller and cheaper. Thus the opportunity is available to expand computerized control of algorithm-described processes and to reinforce human intelligence by intelligent machines.

2 Problems in Small Companies

Implementing a new technology is a greater burden for small enterprises than it is for large companies, which have better resources through a larger number of employees and a broader basis of know-how. Solutions for shop floor control have been worked out already by research institutes. One special problem is given by the necessity to adapt these solutions to the needs of small companies.

The implementation of technical and organizational novelties in small and medium-sized companies is considerably different from their implementation in large ones for the following reasons.

- Working methods, software, and production techniques must be newly conceived for the specific needs of small companies if they are to be economically applied.
- The dimensions of resources, structure, and organization of the process are smaller.
- Financial resources are limited and/or expenditure is more tightly restricted.
- Because of the smaller number of employees there is not much positive allowance for the deployment of staff to the new technology in the company.
- For many tasks there is a lack of special knowledge. Mostly only those skills are available which are required for the work practices already operating in the small enterprise.
- Because of the small number of emloyees the manufacturing design has to accept lower standards of specialization.
- Personal relationships are of high importance.

These difficulties have great impact, especially on optimizing the flow of materials and production control in small and medium-sized firms, because of the large step from theoretical solution to practical realization. The use of computers to optimize these tasks is state-of-the-art in large companies. In small and medium-sized companies, suitable developments have been realized only partially.

In small and medium-sized companies, shop floor tasks which are closely related to production have mostly involved a high degree of human decision-making and execution. For example, the operator decides the composition of lots; he transport the raw materials and tools. The foreman decides the order sequence and the machine loading. The positive allowance for decisions of the employee in this form of staff-oriented organization is large. With relatively little effort a structure of action can be assigned to the staff which is in good correlation to its qualifications.

This flexibility can be maintained after the implementation of computer-aided disposition and control devices if the criteria for software design are adapted to these needs and an economically acceptable solution is found.

3 Production Control as a Strategic Decision-Making System

The concept of strategic production control is now set against the conventional production control like MRP-systems (Fig. 1). The strategic planning is process-oriented and gives much consideration to the enterprise's goals and objectives at the different decision levels. It lays down infrastructures that are expected to remain unchanged over longer periods of time and limits to some extent the scope of goals and actions of a decision maker, thereby guaranteeing sucessful reaction to actual conditions. This way the experience of the operators is used to rule out the variations which characterize single-part and small-lot production.

In accordance with the horizons of planning, a hierarchical assortment of goals can be defined (Fig. 2). In the long-run the delivery readiness is to be ensured through planning and adjustment of marketability and production factors. Middle-term goals are the minimization of inventory through control of processing time and the utilization of capacities to maximum extent. Short-term goals

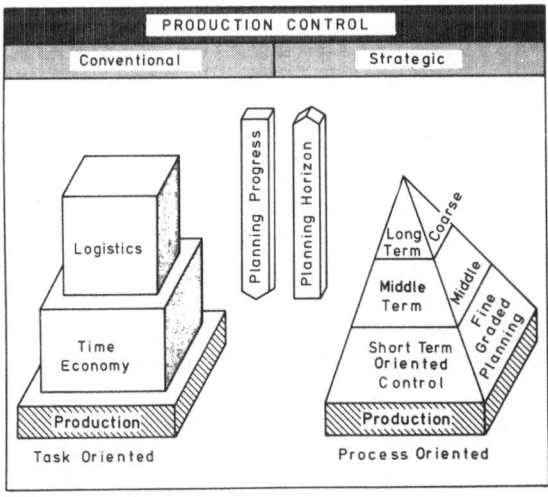

Fig. 1. Production control concepts

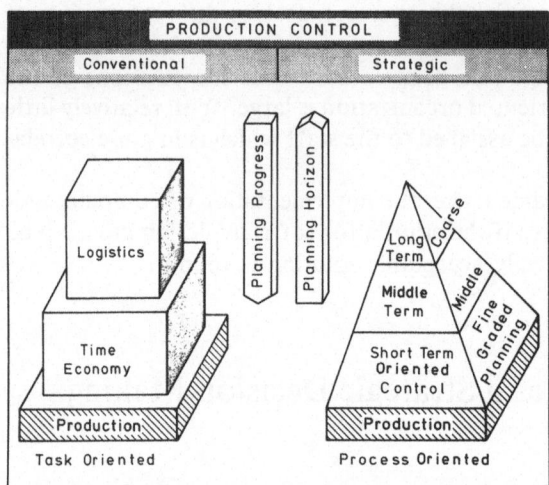

Fig. 2. Control hierarchy

are to ensure that the production can be executed and that it is carried out. According to the work progress, different actions at the different levels are to be untertaken. The planning object of the coarse planning can be the plant as a whole, of the medium planning the department, and of the fine planning each single machine.

The classification of the actions into different levels leads to the fact, that the amount of information to be processed at the long- and middle-term level is small compared with conventional production control. The information can therefore be processed more often. An information pyramid arises out of such a concept. A decisive advantage of this concept is that through progressive planning the planning data are collected simultaneously with the production process. That results in complete compliance of the data with the real-time data in the short-term control.

In order to process the multitude of information at each level and between levels an electronic data processor (EDP) is necessary. The realization of an EDP concept for strategic production control has to consider the following:

- The "planability" of discontinuous processes depends on the ease with which they can be surveyed. The amount of information should therefore not be greater than absolutely necessary.
- The quality of information must be as high as possible; that is, the most important and the desired facts must be represented in a relevant way.
- The production process in single-part and small-lot production cannot be represented fully by algorithms. Moreover, a larger extent of human intervention in decision-making can lead to more flexibility in execution.
- The possibility of frequent data manipulations by means of the computer, for example simulation of decision results, leads to more flexibility in planning.
- The responsibility of the workers for the results of their planning must be clearly retraceble.
- The origin of data and their relation to one another must be determinable.
- Software ergonomy is to be kept in mind, so that the principles of flexible, differential, dynamic, and participative work design are considered.

Such a computer-aided production control system based on hierarchical decision levels uses both human experience and creative initiatives as well as the enormous data-processing capabilities of EDPs.

4 Interactive Shop Floor Control

The shop floor control system is the most important module for small companies. Optimal capacity utilization, keeping the delivery dates in view, is the goal. The term "shop floor" hints at the economic application field of the control concept: the multi-level manufacturing of different products in small- and medium-size batches. Production is characterized by frequent changes of operation sequences. Deviations and variations from the planned schedule and the final determination of planning data should not be seen as disorders, but as normal behaviour of the system. In the control system presented here, an effort has primarily been made to make the situation on the shop floor as transparent as possible in its influences. Online with the production process, with the help of human competence, optimal production can be achieved. With the help of simulation, the system behaviour in the case of release of an order or an operation sequence is examined and optimized.

The tasks of dispositive shop floor control can be classified into five modules:

- Data administration
- Placement of orders
- Release of orders
- Distribution of orders
- Control of production.

The data administration manipulates orders, capacity, and personal data and takes care of the data exchange with the operative shop floor control. The placement of orders transfers the customer orders into shop floor orders. In the module, before release of orders an initial check on the availability of materials, jigs and fixtures, tools, and the necessary NC programs is carried out. Only when all availability conditions are fulfilled, will the order be released. This is connected with a reservation of the required materials. If a shop floor order is to be speeded up, this is possible by the input of an external priority.

The control module distribution of orders consists of sequence selection, printing of workpapers, and check of availability. Before an order is finished at a capacity unit, the next order from the queue will be selected online. In the case of this selection, the disponent can consider a number of factors:

- External priority of the queued orders
- Delay of orders
- The following order with the least setup time
- Human factors.

Depending on the extent of waiting orders, a decision can be made directly or with the help of priority calculations by means of a computer. The preparation

Representation / Object	List	Matrix	Bar Chart	Block Chart	Gantt Chart	Flow Chart	Pie Chart	Function Chart
Order Sequence	●	○	○	◐	●	○	○	○
Order Interaction	○	●	○	○	●	○	○	○
Capacity Loading	○	●	●	●	◐	◐	◐	●
Availability	●	●	○	○	○	◐	○	○
Queues of jobs	●	◐	◐	◐	◐	◐	○	○
State of Machining	●	◐	○	○	○	●	○	○
Disturbances	◐	●	○	○	○	●	○	○
Trends	○	○	●	●	○	○	●	●

● Good way of Representation ◐ Medium way of Representation ○ Small way of Representation

Fig. 3. Clearness of representation

of orders for material removal, work orders, wage calculations, material transport, and acknowledgements should be made as late as possible, in order to avoid changes on forms printed already. The long-term goal is tokenless production.

As a basis for decisions about the functions of shop floor control, information is to be prepared and represented. The statement contained in a piece of information has to be represented in a way best suited for a specific problem. The information system has to represent the actual data, planned data, and the combination of both in an easily identifiable way and also in their dynamic changes. In the case of computer-aided systems, clarity and comprehensibility can be enhanced through the use of colour graphics.

The possible ways of representation, namely lists, matrices, diagrams and graphics, are suitable for shop floor control to different extents (Fig. 3). For the representation of the capacity utilization of a machine tool, for example, a column diagram is very useful. For a larger shop floor area, it is not possible to represent all information in such a way, so in that case a matrix representation will suit much better. The list form has an advantage when there is a dependence upon the sequence of information. In such a case a precondition is that the lists can be obtained already sorted out. Flow charts can be used to indicate disturbances in the case of extensively branched plants. Pie charts and probability curves can be used to represent general developments and trends.

It is important to give a lot of significance to the interaction between the user and the computer. Based on the example of short-term balancing of the capacity, the interactive process is explained as follows. The capacity balancing requires a lot of information that is to be made available simultaneously:

− The workload of each machine group in the shop floor area, for example for the next 20 working days
− The workload peaks of a machine group for the next 20 days
− The list of waiting shop floor orders of a machine group together with the organizational and technological data.

In order to represent the necessary combination of information and its manipulation, a disposition work-station was developed (Fig. 4). On three

Fig. 4. Prototype shop floor control station

displays the disponent is informed about the actual situation by means of matrices, graphics, and lists. An alphanumeric keyboard, function keys, and a joystick for cursor movement are provided. Only two measures are available for short-term capacity balancing. These are temporary dislocation of an order on the same machine or technical dislocation to a deviation machine. On the basis of the process knowledge and the experience of the disponent, online replanning of orders is carried out with the help of the joystick and function keys. The results of planning are determined by simulation and displayed as work-load

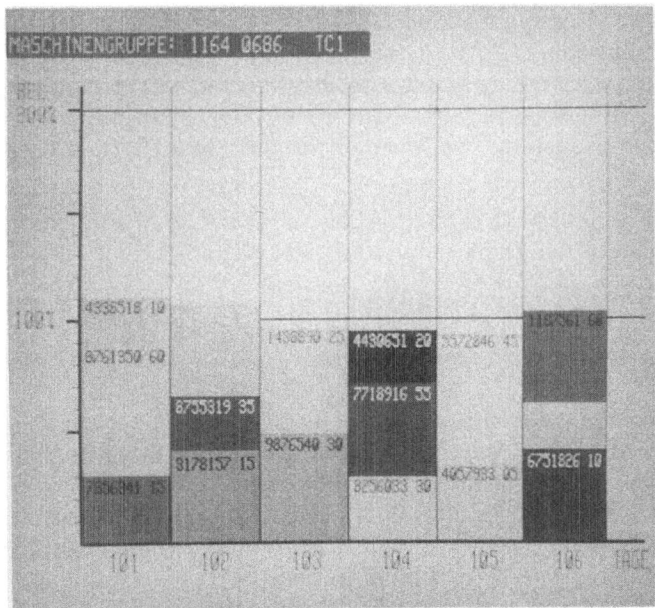

Fig. 5. Workload peaks of capacity groups

peaks with reference to that particular machine group and as percent declaration with respect to the whole shop floor area (Fig. 5). These results are shown on the display terminals. Process overlapping, either with reference to the orders or with reference to the capacities, is reported as error. In the event of a replanning because of delays in delivery or capacity bottlenecks, the process network will be postponed either automatically or in dialogue.

The disponent can either accept his decision in dialogue or reject it and start all over again. The program can offer further support for structuring the work:

- Indication of the capacity units that are over- or underloaded
- Display of time-critical operations.

Based on these few data the disponent can estimate quite easily whether, for example, the postponement of an order has any effect on other operations or orders, and hence avoid unnecessary test planning using his experience.

5 Realization

In realization of the production control system in a small company we chose a 16-bit microcomputer with the operation system UNIX. Furthermore, an easy transfer of the installed software to a broad spectrum of computers could be expected because of the strong efforts made by manufacturers of microcomputers in standardization of the UNIX operating system.

For the following reasons the computer language 'C' was choosen for the software realization of the production control system:

- It is highly effective concerning the modification of the system and the efficiency of the machine code.
- It allows structured programming and has modern mechanisms of monitoring.
- The use of the systems' functions implies no difficulties, since UNIX in most parts is written in 'C'.

The following devices have been used to achieve clear design of the screen, high informational value of the graphics, and efficient planning:

- Framing connected data with semi-graphic symbols for easier location of the information
- Video techniques such as reversing colours, blinking, and increasing intensity of important data
- Colour graphic screen for presentation of the diagrams
- Immediate monitoring of the influence of decisions by continuous actualization and presentation of data at the appropriate screen
- Automatically printed documentation of critical decisions which must be written down for the further use; thus the concentration of the user on the essentials of his planning activity is supported
- The use of a cross-hair cursor and colour bar to select the jobs which are to be moved and the days for scheduling helps to speed up handling.

6 Conclusion

In the last years standardized programs which always try to define exactly the production process have been used to solve production control tasks. Numerous studies in factories and reports of experiences are showing that the organization and implementation of production control by current computer programs do not meet increasing needs, especially in the case of small companies. A expensive capacity-oriented computed schedule may turn out to be useless after the first problem. The consequences are long flow times, low utilization of capacity, and delays in delivery. The available production capacity can be better used if modifications of plans can be compensated immediately by counteractions. This may be guaranteed if also the dispositive measures are taken by employees locally concerned with the process and if the planning results are checked by simulation.

Possibilities of design of dispositive control systems for small and medium-sized companies in competence-oriented interactive shop floor control systems have been presented. The information necessary for the implementation of shop floor control tasks has been investigated by analysis and evaluation of small and medium-sized machine tool companies in Berlin (West). The shop floor control system has been realized as a prototype in the Fraunhofer-Institut für Produktionsanlagen und Konstruktionstechnik. In several small enterprises it has now passed to the phase of final specification or implementation.

7 References

1. Mertins, K., Steuerung rechnergeführter Fertigungssysteme. Carl Hanser Verlag, München, Wien, 1985
2. Seliger, G., Wirtschaftliche Planung automatisierter Fertigungssysteme. Carl Hanser Verlag, München, Wien, 1983
3. Spur, G. and Mertins, K., Dialog oriented shop floor control by decentralized computer networks. In Proceedings of COMPCONTROL 83, 6th International Conference an Application of Computer Technology. Czechoslovak Scientific and Technical Society, Bratislava, 1983
4. Spur, G., Mertins, K. et al., Planung, Einführung und Steuerung eines Fertigungssystems der mittleren Technologie. KfK-PFT 75, Karlsruhe, 1983

Chapter 23
Production Control in the Car Industry

Wolfgang D. Thurow

Dr. techn. Wolfgang D. Thurow has his education from Technical University, Graz, Austria in commercial engineering and mechanical engineering. At present he is Manager of material planning with BMW AG, Munich, West Germany. He has previously been Assistant Professor with TU, Graz, Austria in industrial science, Manager of production with Hilcona Ges.m.b.H., Neusiedl/See, Austria and Manager of logistics with BMW Motoren Ges. m. b. H., Steyr, Austria. His publications cover Marketing, Logistics, Production Control, Just-in-Time, and Computer-Integrated Manufacturing. Dr. Thurow is Chairman of the Association "GF+M" Ges. für Fertigungssteuerung und Materialwirtschaft, Stuttgart, West Germany.

1 Introduction

The necessity of cutting down on development time, and the demand for market-specific variants, both derived from international competition, are now forcing us to adopt new ideas with regard to the creation of factory layouts, processes, and control systems.

Furthermore, the factory layout in the motor industry is characterized by relatively high manufacturing penetration and high capital intensity. This capital intensity necessitates greater utilization of daily operating times and usage of production equipment over several product life cycles. This means that production equipment should be characterized by both successor and variant flexibility.

Whereas an increase in the number of models and variants results in increased horizontal complexity, the manufacturing penetration has an influence on complexity in the vertical direction.

The complexity of horizontal factory layouts affects product quality and reliability and superimposes on the temporal dependencies resulting from the vertical layout. Business management can cope with these multidimensional dependencies only with the help of integrated control systems.

2 Integrated Systems of Production Control

The state-of-the-art in the field of production and especially the achievements in the field of communication technology make it possible to step into the areas of total integration of all company processes by means of computers. Computer-Integrated Manufacturing (CIM), however, is not a product, but decisive target planning in order to cope with a changed setting. The highly complex factory layouts and the implementation of DP-assisted systems in the motor industry demand the adoption of the CIM concept according to three basic parameters:

1. All integration plans must be in line with the strategic aims of the company.
2. All departments concerned must agree on a common approach.
3. There must be comprehensive planning and localized realization.

In the motor industry, the CIM concepts are based on a joint database and determined by three basic factors (Fig. 1):

1. The product relations
2. The materials flow
3. The order flow.

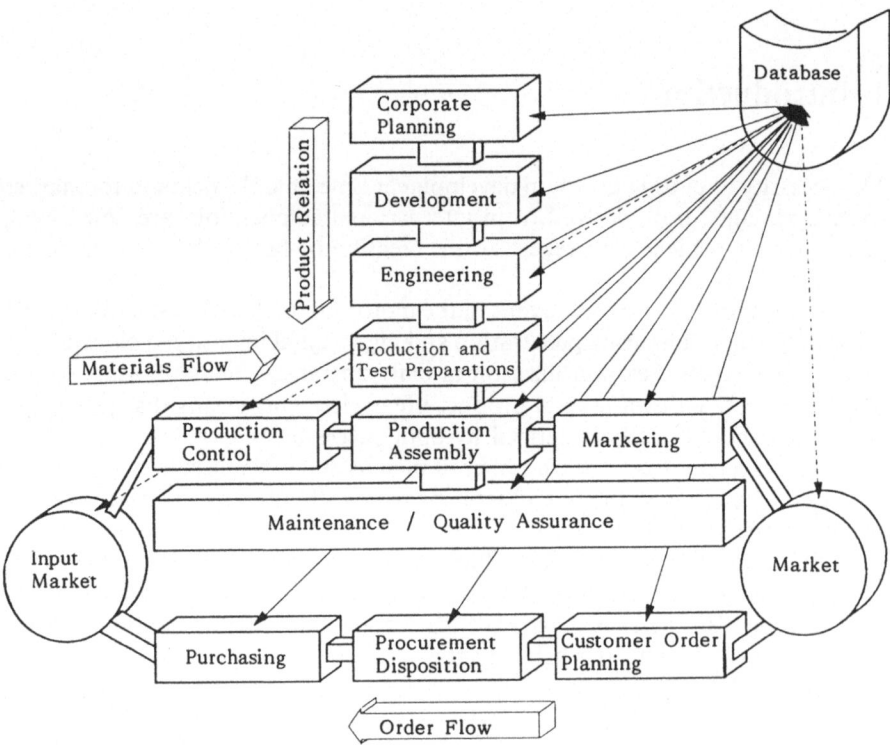

Fig. 1. CIM concept

Only the total integration of all information flows pertaining to these factors will guarantee that the wants of the market are satisfied by highly automated serial production.

For a product to meet the requirements of the customer, the individual orders must be considered and all current information saved over the entire life cycle of the product. The shortening of product life cycles called for by the market, in conjunction with frequent changes in the engineering design of the product, determines the type of relationship to be formed between the CIM components – CAD, CAE, CAM, CAQ – and existing logistic systems. Only after these integrating steps have been taken will the basic data have the quality standard required for the production control systems. Only his stage is reached, it will be possible to react correctly and with a high degree of reliability to short-term and highly complex changes.

3 Just-In-Time Production and Procurement

The basis of all JIT attempts is a comprehensive way of thinking which should start at product development and pass via the process engineers to the operations people. Only this approach will ensure meaningful utilization of the computer instruments of the just-in-time process. The necessary prerequisites of the just-in-time environment – e.g. top product quality, process stability, increased participation, and data networks – must coincide with the JIT processes.

In phase I, using BMW AG as an example, the basic systems for materials planning were extended to facilitate the complete explosion of vehicle-related customer orders on a daily basis. Exploding the complete parts spectrum of the company takes about 50 minutes. Here, the central module of the materials requirements planning system is a specific bill of materials which is based on the principle of the parts grouping.

In the course of actual realization, the requirements information determined by the materials requirements planning system is made available to the supplier on a continuous basis in the form of a so-called finite call-off for the following 15 days (Fig. 2). The daily finite call-off represents an improvement of call-of information in the short term. This can never be achieved with the monthly call-off method which has been used so far in the same period of time. The advantages for both parties are evident. The supplier can adapt his production to customer demand much more accurately than before and thus supplies his parts only in the required quantities right on the scheduled delivery date. This means that both the supplier and the motor manufacturer will have less stock in hand and consequently less capital tied up for material and transport containers.

Just-in-time production and supply requires a rapid exchange of information. Remote data transmission is the ideal solution. Formal data transfer is regulated by the standards of the German Society of Motor Manufacturers (VDA). The finite call-off can either be automatically transmitted to the supplier's computer immediately, or the supplier can transmit the information at the time of his

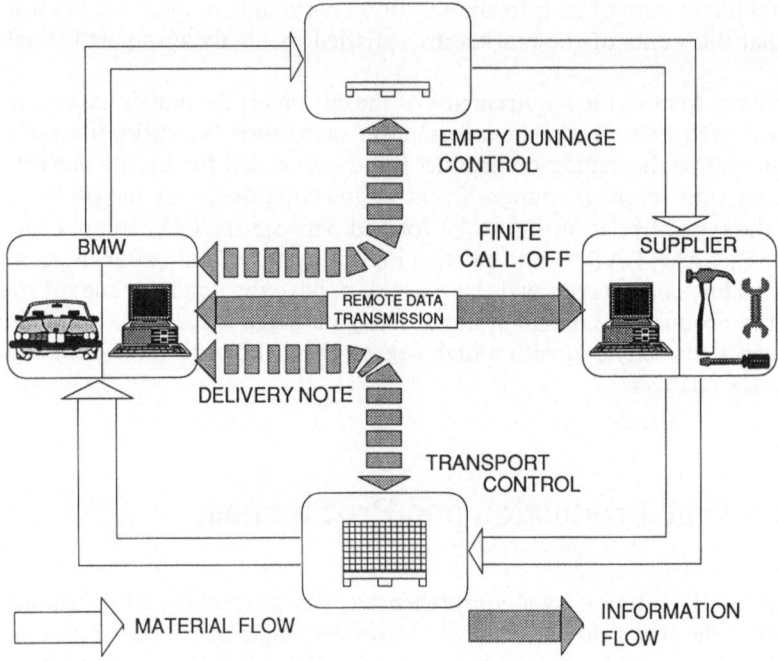

Fig. 2. Just-in-time BMW AG

choice. Delivery call-offs for the medium and long terms are only transmitted when required, i.e. especially in the case of changes.

Within the framework of direct control, the procedure for in-house parts is identical. Here, the requirements ascertained are used directly for the order control of the various production areas. The ensuing synchronous customer-order-related production of single parts by the motor manufacturer and the supplier enables the following:

– Reduced stocks and storage areas
– Reduced administrative expenses
– Flexible reactions to market fluctuation

Requirements which go beyond the three-week period are determined in the traditional way. These requirements are based on the motor manufacturer's production programme, which usually contains the requirements for at least 4 months, summarized in monthly values. This information is handed on to the suppliers for the purpose of capacity allocation (manpower, machine, materials) at the various suppliers and their sub-suppliers.

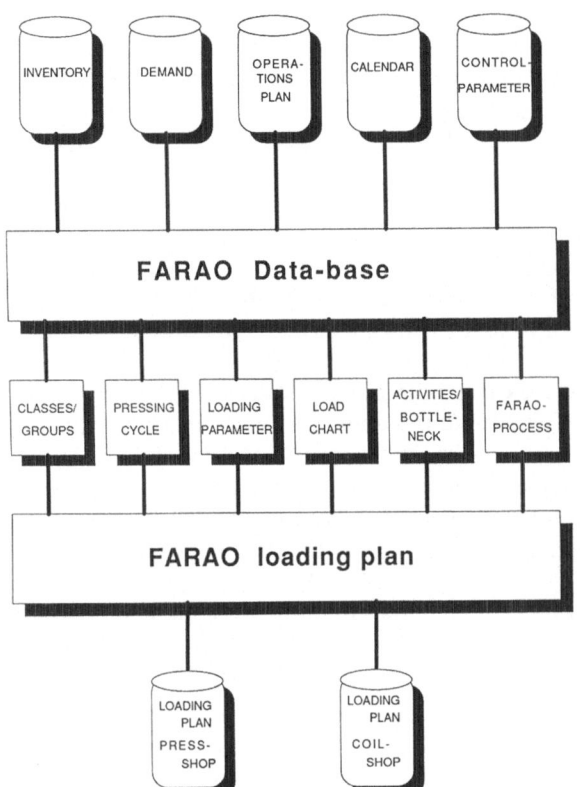

Fig. 3. FARAO − production control in the press shop of BMW-AG

4 FARAO − Flexible Pressing Cycles for Sequence and Order Optimization

In a large press shop, manufacturing means the coiling and pressing of sheet metal parts on presses or press lines in a certain sequence and in certain cycles. By resetting these presses, numerous parts can be produced on the same press or press line. FARAO calculates the press occupation by part numbers and ensures maximum utilization of existing press capacities (Fig. 3).

The pressing cycle for each production line is select on the basis of minimum costs, and the labour requirements are levelled out and minimized over the period of computation. In addition, the limitations imposed by simultaneous resetting activities in the press shop are taken into account.

The capacity-loading problem in a large press shop is rather complicated (approximately 1000 times). In order to simplify the matter, FARAO divides the part number spectrum into classes. These classes are then subdivided into groups. In this way it is possible to delimit the immense variety of loading possibilities for all part numbers, so that an optimum solution can be found within a reasonable computation period.

Part numbers from different classes can only influence each other as far as press utilization is concerned through the work-force or resetting activities. It is therefore possible to assign each class to its own flexible pressing cycle which is selected on the basis of storage and setup costs. Over and above that, FARAO establishes the classes and the groups within these classes with are to be occupied by certain part numbers for an optimum utilization of existing press capacities.

All values which appear in FARAO are parameters, in so far as this is useful and can be adapted to the prevailing condition at any time.

Press occupation can be calculated for several days, several months, or even one year. The pressing cycles of the different classes, groups, and part numbers for press shop utilization, the capacity of working days, the number of available workers and setup teams can all be selected interactively.

The coil-loading plan is derived from the press-loading plan. The items derived from both loading plans determine the appropriate production sequence. The orders are processed by the controllers and released for production. In addition, the raw materials handling and the allocation of production facilities for production are carried out. Once the release has been issued, production order papers are drawn up and orders are transmitted to the data collection level and the central order file.

The data collection level registers and follows up the logistical and technical status of the orders and the production facilities. Part of these data are passed on to other systems for information and appropriate action.

5 Future Trends

In the factory of the future it will be necessary to integrate existing and future partial solutions in an overall concept such as CIM. At present, step-by-step integration selected on the basis of priorities has pushed an important production control subsystem into the foreground. We have called this area of responsibility NPL – New Product Logistics.

The transition from preproduction to serial production is often characterized by incomplete computer information in the production control system (Fig. 4, graph 2). In order to cope with product launches during shorter development times, all planning and operation departments must be provided with the necessary information in time, even if this information has not yet been released by the engineering department. This is particularly important if the information is already available on computer in the development area (Fig. 4, graph 3). Greater integration, as in the case of CIM concepts, means the product descriptive data can be collected already in the preliminary design phases. With the NPL system it is possible to provide the production areas with the necessary information in time to secure the scheduled production start date. Taking the launch program as a basis means that all preproduction activities can be initiated, controlled, and monitored by the materials planning functions of the NPL system.

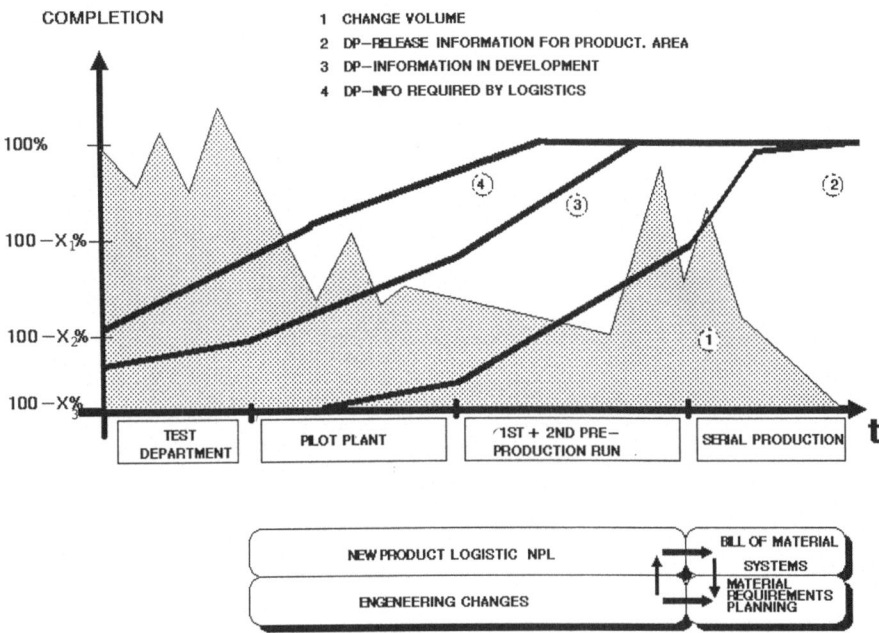

Fig. 4. Product development phases

Finally, the objectives of production control in the car industry — if CIM concepts are put into practice — should be emphasized again. The main objectives are:

- Improving market-related flexibility
- Reducing customer order and materials flow-related lead times
- Securing serial production supply
- Securing product launches
- Increasing productivity.

Due to the high degree of complexity in the production areas, the dependencies in the production network, and the increasing engineering change frequency, the systems design must be comprehensive. This means that all participants must be incorporated into the planning and implementation processes.

6 References

1. VDI-Gesellschaft, Strukturen der Automatisierung für die Fabrik mit Zukunft. VDI-Verlag, Düsseldorf, 1985
2. IPA-IAO, Produktionsplanung, Produktionssteuerung in der CIM-Realisierung. Springer-Verlag, Stuttgart, 1986
3. Weck, M. and Dern, U., Integrated manufacturing and assembly. In Proc. Int. Conf. on Robotics and Factories of the Future, Charlotte, N.C. USA. Springer Verlag, 1984

Chapter 24
Production Control in the Aircraft Industry

Bernd Hirsch and Gustav Humbert

Professor Bernd Hirsch has university education in mechanical engineering and production engineering from the Rheno-Westphalian Technical University, Aachen (RWTU), West Germany (1965). He has a degreee from 1969 on automatic determination of machining data, considering particularly the computer-aided programming of NC-lathes. His professional experience covers a mechanical apprenticeship at Daimler-Benz AG (1957–1959), part-time assistant in the Dept. of Industrial Business Administration, RWTU, Aachen (1960–1963), and part-time assistant and after passing the final examination scientific assistant to Professor Dr.-Ing. Dr. h.c. H. Opitz at the Machine Tool Laboratory, RWTU, Aachen (1963–1969). He has carried out various aerospace production development and management functions at VFW-Fokker's Bremen headquarters (1969–1979) and been Section Director for corporate production planning and control. During 1979–1981 he was team-leader of a Brazilian/German university project at the Universidade Federal de Santa Catarina, Fliranopolis S.C., Brazil. From 1982 to 1983 he was Head of the Central Production Development Dept. of MBB's transport aircraft division in Hamburg and Bremen. Professor Hirsch is at present Director of the Bremer Institut für Betriebstechnik und angewandte Arbeitswissenschaft (BIBA).

Dr.-Ing. Gustav Humbert studied production engineering at the University of Hannover, West Germany. He has been assistant at Institut für Umformtechnik und Umformmaschinen at the University of Hannover, Visiting Professor at the Department of Mechanical Engineering, McGill University, Montreal, Canada, and since 1980 in the aircraft industry with Messerschmitt-Bölkow-Blohm GmbH, Transport Aircraft Group where he is at present Director of data processing systems production.

1 Introduction

Owing to the extremely high development costs and the limited national markets, new aircraft programmes in Europe today are now only executed in multinational cooperation. Usually, a work-sharing scheme is devised corresponding to the financial participation of the individual partners, with the result that design and production of the aircraft are spread over several countries. This is accentuated by subcontracting components to the industry of customer countries (offset transactions).

Another feature of the aircraft industry comprises the high demands on the quality of the components. In each case the quality is subject to constant documentation; for "traceable parts" it is even recorded per part and per production progress stage.

In addition to these organizational necessities there are the problems of the great variety of parts and the customer-related solutions in the aircraft industry: for the construction of a passenger aircraft of the Airbus family approximately 300000 different parts must be controlled and, in particular for Payload Systems, there are specific and sometimes short-term solutions for nearly every airline. This presents problems since these payload details often have repercussions on the aircraft structure.

All of these points illustrate the fact that Production Control in the aircraft industry has to meet special requirements.

The following is a description of the state of the art, taking as an example the sequence of Production Control in the Airbus Programme at the Transport Aircraft Group of Messerschmitt-Bölkow-Blohm GmbH (MBB-UT).

2 Principles of Cooperation

To operate an international work-sharing scheme, many common decisions have to be taken in Engineering and Production by the participants in order to ensure compatibility of the large number of interfaces. An important principle in this respect is the maxim that only fully equipped components ready for installation are delivered for final assembly. In other words, every component is fully equipped with electric and hydraulic systems and has been function-tested, i.e. ready to fly. The following is an example of the work-sharing scheme practised in the Airbus programme (see Fig. 1):

– Cockpit, fuselage – with section, fuselage connection, pylon, final assembly: Aérospatiale, France
– Wing structure: British Aerospace, UK
– Horizontal stabilizer: CASA, Spain
– Forward fuselage section, centre fuselage section, fuselage tail, vertical stabilizer, wing equipment, payload systems: Messerschmitt-Bölkow-Blohm, West Germany
– Engines are vendor items.

Here the rule applies that the responsibility of the relevant partner fully covers the major components he delivers for final assembly, and he also has the logistic responsibility for items produced by subcontractors in his workshare.

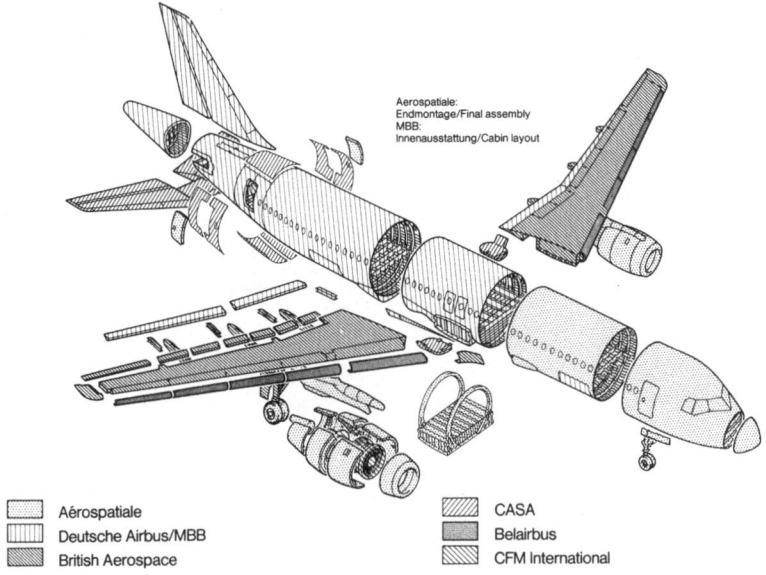

Fig. 1. International worksharing in the Airbus program (A 320)

3 Production Schedule Sequence

3.1 Schedule Planning of Major Components of Each Partner

Once the production rates have been fixed by the management committees responsible at Airbus Industrie and the partners, the detail schedule planning is drawn up among the planning departments of all those involved. Based on the earliest/latest delivery dates to customers, the capacity requirements of the partners, and the production lead time, a schedule is produced for the delivery of each major component to the final assembly line and of the finished aircraft to the customers (Fig. 2). This alignment process occurs only in individual critical cases in joint meetings, usually by data exchange among the partners on computer terminals.

The earliest/latest delivery dates for major components are converted into detailed internal schedules for each partner.

3.2 Detail Schedule Planning of Major Components of One Partner

The main task of Production Control in the construction of large aircraft today does *not* consist in bringing together the major components of each partner, because as a rule there are less than 10 such components. The principal tasks rather involve disposition, tracing, and allocation of the many thousands of

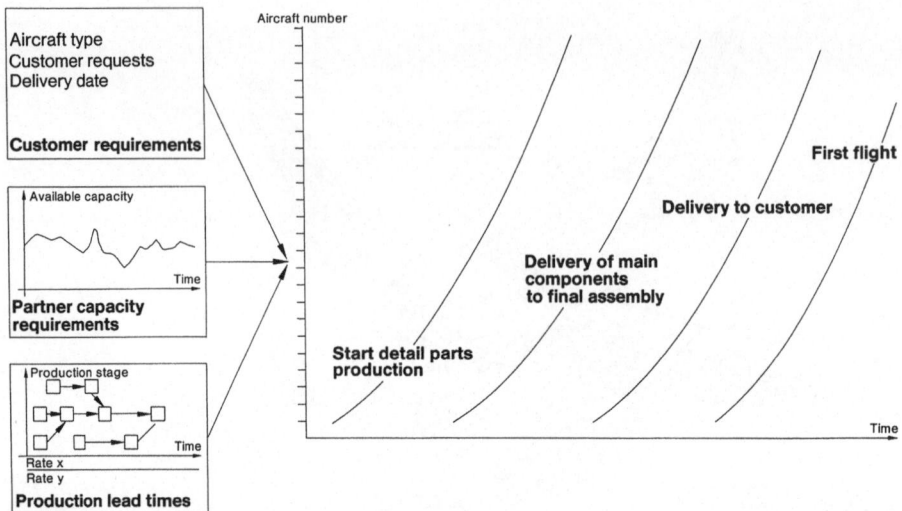

Fig. 2. Principles of scheduling for major components

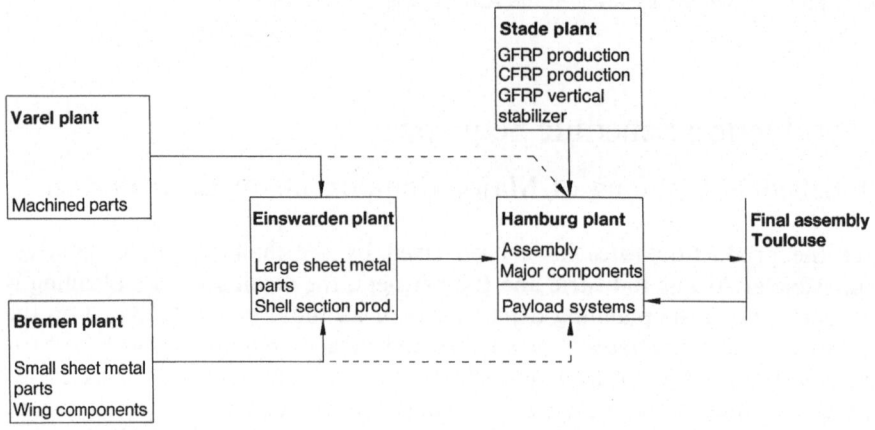

Fig. 3. General structure of Manufacturing at MBB Transport Aircraft Group

detail parts and components that make up the assemblies ready to be delivered. This calls for a procedure organization which, similar to the other branches of machine engineering, fulfils the requirement of the highest degree of flexibility (with regard to plan changes), retaining at the same time maximum adherence to deadlines, minimum production lead times and minimum stocks.

Manufacturing at the MBB Transport Group is spread over five works specializing in certain fields of production technology (Fig. 3): one plant specializes in machining, another in small sheet-metal components, one in fuselage shell sections, another in plastics, and an assembly works for the major components to be delivered to final assembly. This complementary manufacturing structure, chosen for economic reasons, entails a high degree of logistic dependency among the works and requires constant ordering and information systems. Thus there is also a typical distribution of roles. There is always a manufacturing plant and a user plant.

The user plant is in principle responsible for the disposition of the parts to be ordered in the manufacturing plant. This relative requirements control produces a just-in-time effect in the control between the plants. It is, however, the duty of the manufacturing plant to optimize the order control within the spectrum of ordered components (i.e. components with fixed quantities and deadlines), according to the specific criteria of the Manufacturing Department involved (capacity usage, setup times, waste optimization, etc.).

In order to give the works a general plant and thus the necessary foresight, the earliest/latest dates agreed with Airbus Industrie are submitted to them by the central Schedule Planning Department of the Group.

Detail disposition then follows, as mentioned above, from the user works of the detail parts assemblies with a firm order to the appropriate manufacturing works.

Further details are given on procedures after the following brief outline of the system.

Procedure until Disposition

The Design Office prepares the geometrical and parts list data per CAD in a system. With this data the Production Planning Department defines the necessary tooling (design of fixtures and NC programming) and also the process sheets. The Materials Budgeting Department plans the medium-term and long-term requirements for vendor items using the materials data from the parts lists. The means and procedures are thus described for the Manufacturing Department. In parallel to production planning, the Quality Assurance Department draws up test schedules to test the quantitative properties of the parts, specified in the process sheets. The quality inspections are defined and documented in the process sheets as a workstep.

The questions "how many?" and "when?" are then tackled by the Disposition Department by adding quantity and deadline data to the process sheet. This now becomes a manufacturing order.

Disposition

The current proximate disposition with a horizon of several months, or in the case of vendor items sometimes several years, is effected by each plant for its component spectrum allocated by Central Manufacturing, on the basis of the parts list data per aircraft type from the Design Office. This occurs periodically and by the corresponding computer programmes for each amendment to the delivery schedules. In addition to the Ident. No. each proprietary component has a procurement code providing information on the manufacturing and user works. This ensures that the responsibility for disposition and manufacturing can be seen from the identification mark on each component.

Detail disposition of the orders is effected periodically, approximately 30 days prior to the scheduled order start, using a net requirements system. It is worth noting that the control programmes previously check whether all the process sheets and tooling/NC data are available by the start of the order. Negative reports to this check are output by the system and processed accordingly.

For the assembly plant delivering the major components to final assembly, all requirements with deadlines, quantities, and manufacturer are stored in this net requirements system. These are taken from the principal production plan, from the list of structures and production lead times. The requirement calculation results in orders to the manufacturing plants for the necessary assemblies, which are transferred by computer to the net requirements system of the relevant plant. This net requirements system identifies in the same way the components to be ordered for the completion of this assembly, e.g. in the case of plants producing detail parts, and informs the net requirements system of the plants. This breakdown of requirements over various stages is effected by computer in batch run for all five works, so that actual, homogeneous information on the ordering status is available in all Disposition Departments.

In this way, each net requirement system of a plant is aware of the requirements for detail parts or assemblies in the disposition period. These gross requirements are then compared by computer with the physical stocks in the parts stores and with the orders already in the Manufacturing Department (or with the orders already placed). If the requirements can be met from stocks or by an order being processed, the order is reserved. Only when the requirements cannot be met in this way does a net requirement arise which automatically leads to an ordering proposal for the disposition clerk. After specific inquiries and amendments, if necessary, the proposal is approved and thus the manufacturing order is released (or the order to the manufacturing plant).

If the order is an assembly order requiring several detail parts (proprietary or vendor items) the order is not offered for release as long as the starting data has

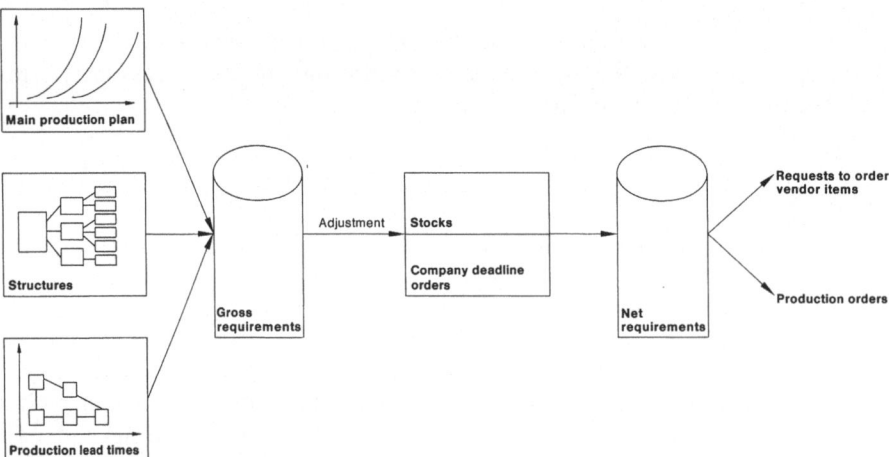

Fig. 4. Disposition procedures for proprietary and vendor items

not been exceeded. Otherwise the missing parts are placed in a missing parts file with a reference to the relevant assembly order and the file is processed by trouble-shooting. The time of these materials availability checks prior to the commencement of order and their frequency can be regulated and is at present approximately 30 days before scheduled start. Release of order is approximately five days prior to start of manufacture. The aim is to provide an online printout at the control station immediately prior to start of manufacture.

In the case of vendor items the disposition procedure is similar. However, the disposition periods vary according to the material lead times stored in the system. The result of disposition is the order application to the central Purchasing Department which collates the requirements of various plants if necessary and forwards them to the supplier. Supervision of the supplier's deadlines is in the hands of the central Purchasing Department which has supervision systems at its disposal, until the goods are received.

The disposition procedure is shown as a diagram in Fig. 4.

Order Control

The control of the orders released by the Disposition Department is performed by control stations and job distribution offices located within the Manufacturing Department to be as close as possible to the scene of operations.

The manufacturing orders which only have a start and close date in the disposition file are transferred immediately after release to a production lead time scheduling and capacity comparison program (in the case of alternative manufacturing possibilities), to set deadlines for the worksteps.

Only this detail disposition of schedule and capacity for the manufacturing procedure makes it possible for Production Control to distribute the work using job distribution lists per capacity unit. The control stations or job distribution offices have the neccesary links to data processing, to have this information of the control system available in dialogue, or to enter comments and feedback into the system online. It is worth noting that the start and close dates are no longer printed on the orders, since these data are usually out of date a few days after printout. The actual deadlines are linked in the Production Control System. At present there is a daily recalculation of amendments/new entries for detail parts production.

An organization data acquisition system is installed in three plants and is being introduced in the other plants, which acquires the operator's workstep completed report (by bar code entry) – apart from data on premium wage calculation – and processes this data online in the control system. Special processing of these reports by Production Control is therefore no longer necessary, which in addition to a reduction in expenditure, produces a considerable increase in transparency with regard to the manufacturing status.

Only with this dialogue-oriented application with possibilities of online amendment is Production Control in a position to accelerate the orders through the organization in a flexible manner taking into account the latest requirements. Flexibility of the staff and of the systems are particularly needed at this interface

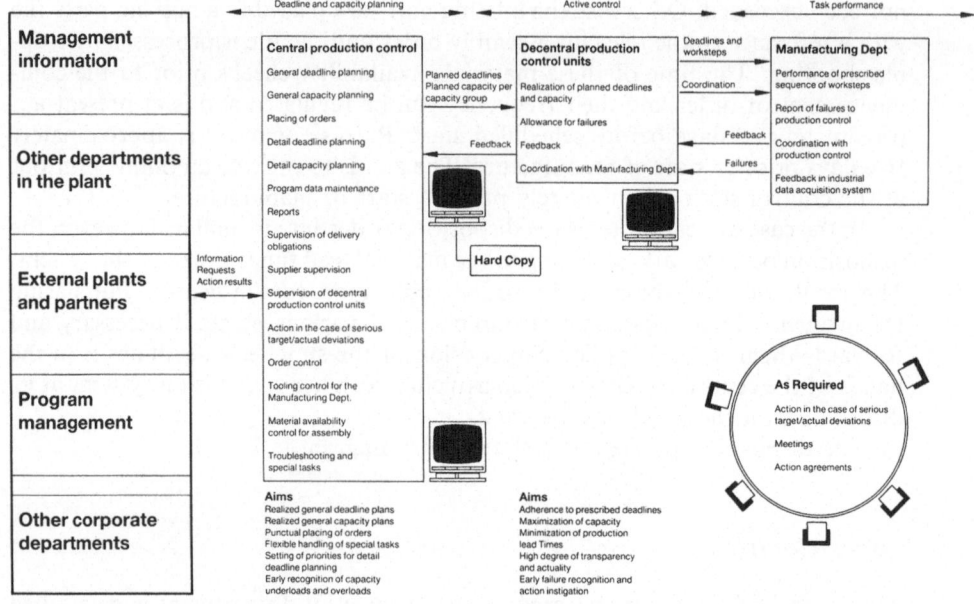

Fig. 5. Order control flow chart

to the manufacturing stage since for reasons which cannot be foreseen (missing parts, absence of staff, machine failure, etc.) plans must be repeatedly changed in spite of all the precheck systems. Therefore, the presence of a Production Control disposition clerk in the control station is still required even with all the computer systems, although the systems must be developed more and more in the direction of aids to decision-making. A start has been made on developments within the MBB Transport Aircraft Group, particularly in the field of specific assembly control systems.

The order control procedure is shown as a diagram in Fig. 5.

Special Problems of Order Control

In the past few years several CIAM stations (*C*omputer *I*ntegrated and *A*utomated *M*anufacturing) have been set up in the works specializing in manufacturing technology: one for the production of small pressed metal sheets, one for the production of shell sections, and one for the production of plastic panels. Further works are in the planning stage. These are highly automated production lines using the latest technology, linked together logistically with transport systems and connected to the relevant parts and materials stores. The necessary information flow is integrated by a computer control system. These control systems for specific manufacturing areas within the plant usually have separate detail scheduling modules which for technological reasons must sort again the order sequence given to the system by production lead time scheduling. To achieve greater productivity of the manufacturing installations, the CIAM

24. Production Control in the Aircraft Industry

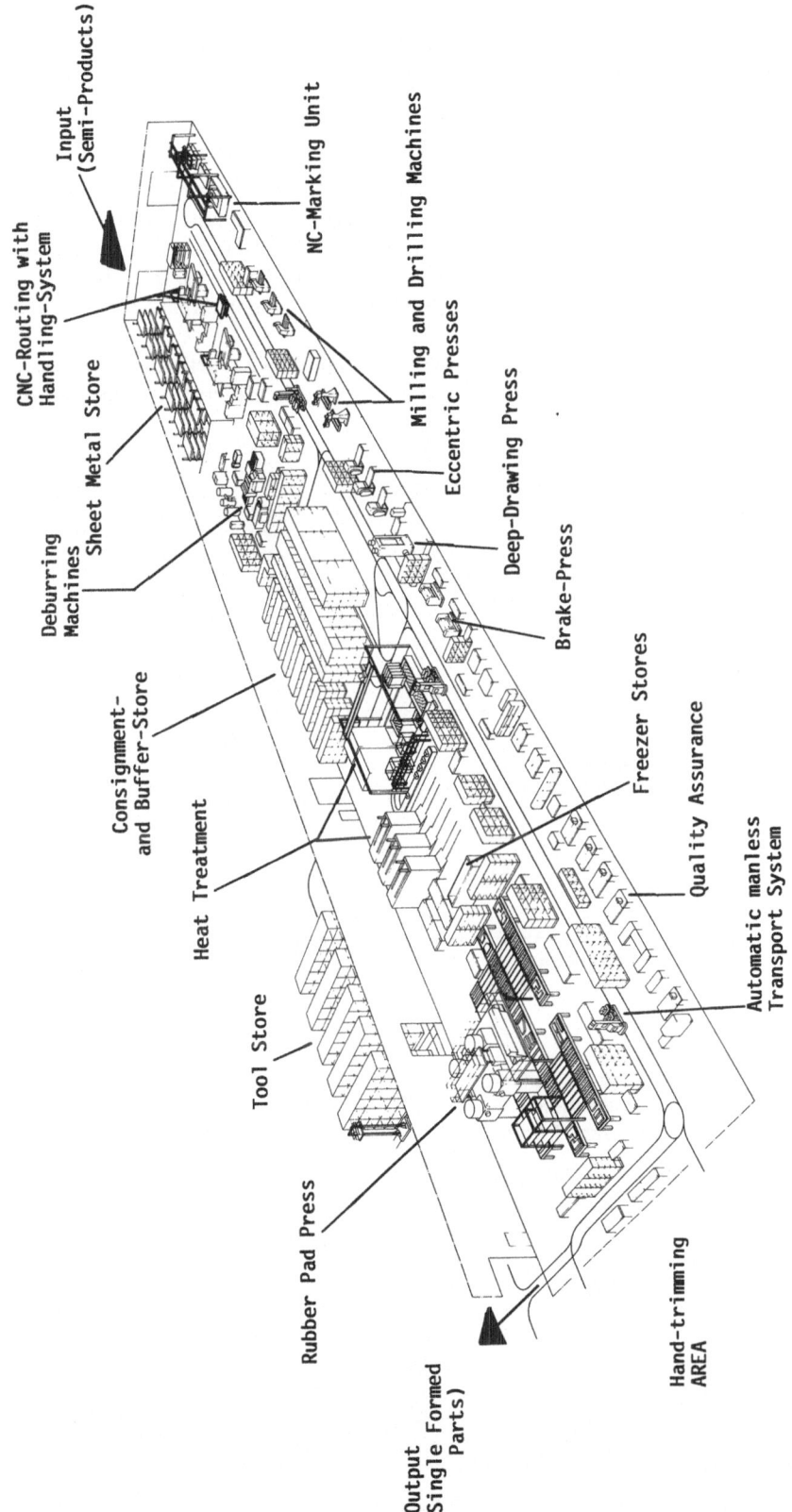

g. 6. View of the CIAM Forming Centre for the production of pressed sheet metal components

centres often work on batches/menus. These consist of several similar orders, deviating from the classic individual processing of a manufacturing order. Examples of the formation of such batches/menus are the nesting of components for the optimization of setup time and wastage, combination of orders for heat treatment cycles, and the loading of press pallets.

This procedure inevitably disrupts the order sequence prescribed by the superior conventional control system, which assumes individual processing of the orders. This could happen several times when passing through a CIAM centre. Therefore, a special detail disposition module must be employed taking into account both scheduling and technological parameters. An example of such a manufacturing setup is the CIAM Forming Centre for the production of pressed metal sheets (Fig. 6).

In the CIAM systems, capacity allocation planning is particularly complex because of the many formation criteria and can only be effectively optimized by simulation.

Company CSS	**Central** **Control System**	- Design, preparation of parts lists - Preparation of process sheets, net requirements system - Stocks, availability control, ordering
Plant PCS	**Plant** **Control System**	- Acceptance of orders after release in the planning system - General deadline planning of worksteps - Order tracing in other divisions - Feedback processing - Stocktaking
Divisions DCS	**Division** **Control System**	- Acceptance of orders from plant control system with earliest/latest deadlines - Autonomous workstep control in CIAM areas - Transfer of "intelligent" activities to subsystems batch formation, sequence formation - Availibility control - Coordination/syncronization of installation administration systems - Feedback processing from installation administration systems - Feedback of relevant data to PCS
Installation IAS	**Installation** **Administration System**	- Acceptance of activities - Queue formation - Activity selection and transfer to installation control - Acknoledgement processing and status monitoring - Complete local operation
ICS	**Data Allocation** **Installation** **Control System**	- Synchronization and coordination of sub-controls - No administration functions only "non intelligent" control

Fig. 7. Function hierarchy of the computer systems for Production Planning and Control at MBB Transport Aircraft Group

4 Structuring the Computer Aid Systems

Owing to the indispensable application of numerous computer programmes to process large quantities of data sufficiently quickly and economically, there is a structuring of the systems according to function, in levels which are being created at present at the MBB Transport Aircraft Group. The architecture is such that all systems for design, job planning, materials planning, and net requirements (including the appropriate pre-check systems) run on the host of the Group (principle of centralized planning systems).

Once the order is released, the active manufacturing order is transferred to control systems of the individual manufacturing plants and processed according to priorities set by the plant, i.e. allocated a production lead time. If necessary, the order is submitted to a control system of a CIAM area and followed through production by the internal industrial data acquisition system (principle of decentralized control systems).

Order completed reports for the entire order and stock levels are reported online by the plant to the host and can be called up by all the other plants. This basic structure is shown in Fig. 7.

5 Conclusion

The main aspects of Production Control in the MBB Transport Aircraft Group plants are net requirements definition, online Production Control systems including industrial data acquisition, and the structuring of the hierarchy of the computer systems involved.

Effective Production Control also requires a high level of work planning and flexibility of production facilities.

Chapter 25
Job Shop Production Control

Oddmund Oterhals

Oddmund Oterhals holds an M.Sc. in General Science from the Norwegian Institute of Technology, Trondheim, Norway, 1974. He has a Ph.D. in Production Engineering from the same University, 1980. He was at Production Engineering Laboratory NTH-SINTEF during 1975–1980, Head of Production Control Section from 1977. He has been Head of Planning Department in Shipyard Industry 1980–1984, and is now Company partner and Senior consultant of TRIANGEL Management Service. Oddmund Oterhals has experience of system development and several industrial implementations of production control systems.

1 The Problem

A job shop is commonly known as a workshop with a functional layout of production equipment, where work-stations are organized around particular types of equipment or operations.

The characteristics of job shop production are that

- production demand is converted into a set of production orders, each consisting of one or more identical parts
- each production order is processed by a predefined routing through a set of work-stations in the workshop.

The challenge of planning in these environments is to make an optimal utilization of workshop stations, while throughput time and resource consumption are kept to a minimum, and deliveries are kept within due dates. In other words: a rather complex problem with conflicting goals.

This problem has for decades been a subject of theoretical studies. The problem can be treated by analytical methods within operational analysis, or simulated on computer, as an optimizing problem, defining

- n orders with defined process operations through
- m workstations,

with

- Throughput time
- Resource utilization
- Resource consumption
- Finish dates

as examples of optimum criteria. The parameters than can be manipulated are

- lot sizes
- setup/change-over times
- process times
- priority rules, etc.

Placed within corporate management the problem is even more complex. This chapter deals with experiences and problems that occur when computerized techniques for solving the job shop production control problem are implemented in an industrial organization.

2 A Job Shop Production Concept

According to Fig. 1, the production flow is divided into a part production phase and a product assembly phase. This production concept is common within mechanical and electronic industry. We find storage of raw materials, components ready for assembly, and in most cases finished, and even wrapped, goods.

As long as the customer's required delivery time is less than the total throughput time, we are forced to prepare components ready for assembly on the basis of forecasted demand or exceeded re-order point on component stock.

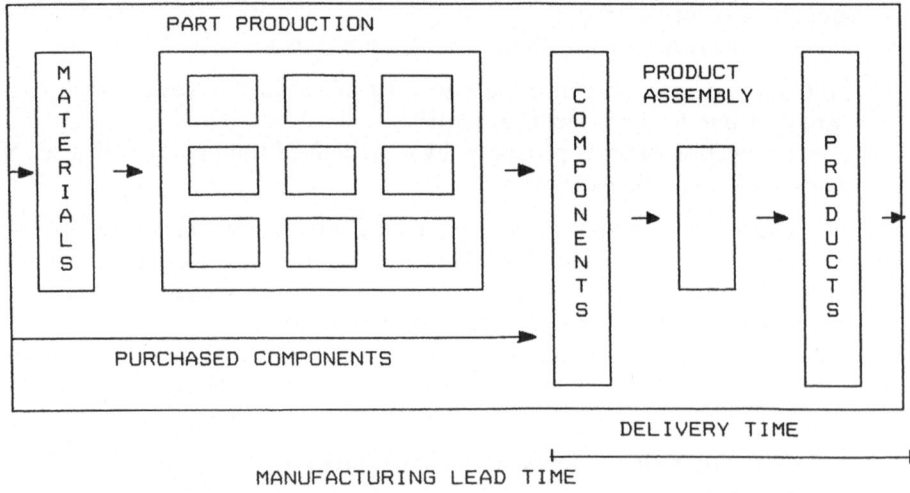

Fig. 1. Conceptual layout of job shop production

In these environments the production planning problem is close to the original job shop problem. The fundamental "n-m-problem" is, however, embedded in a complex management system, where unpredictable events, lack of communication between company departments, and conflicting goals tend to cover the original problem.

3 Applied Solutions

Online computer systems for job shop production control have been available for more than 10 years. The main features of available systems are

- online communication with central computer from screen terminals placed in the different departments involved
- Menu-driven system modules with user guidance through predefined user routines, or
- Available tools for programming of user transactions against the common data base.

Software packages dedicated to production control are today available for most of the mini- and microcomputers on the market, comprising modules for

- Maintenance of the database
  (parts, products, bill of materials, resources, vendors, customers, calendar, etc.)
- Master scheduling
- Materials requirements planning
- Capacity requirements planning
- Sales order control, invoicing
- Purchasing
- Production planning, dispatching
- Shop floor control
- Inventory control
- Price- and cost-calculation.

Most of these system modules just serve the purpose of keeping track of information. The planning functions can be more or less sophisticated, but they seldom search for the optimal production program regarding given criteria for optimizing. Lot sizes and lead time elements are in most cases predefined parameters, with no optimizing of work-station utility or manufacturing lead times. The information system is in this case a tool for supervising production progress and stock levels. On the other hand, this kind of system is open, in the sense that a planner can manipulate lot sizes, priorities, dispatch dates, etc., and the database is the communication medium between the system users. Another big advantage is the efficiency in calculation and reporting of progress towards plan, resources load, and resource consumption.

We can reach the preliminary conclusion that applied computer systems for job shop control suffer a lack of features for production optimization. However, they contain the basis from which an optimal production program can be derived. A step in this direction is the realizing of the MRP II concept, as described in Ref. [1].

3.1 Data Structures

The first step in computerizing production control is to formalize logical data structures, and establish the fundamental files of records in the computerized database. Fig. 2 describes possible content of some typical data records, illustrated by symbols that indicate fundamental logical links.

To establish the company-specific database, with parts, product compositions, and routings, is the first, and a major, task in the process of realizing a computer-aided production management system.

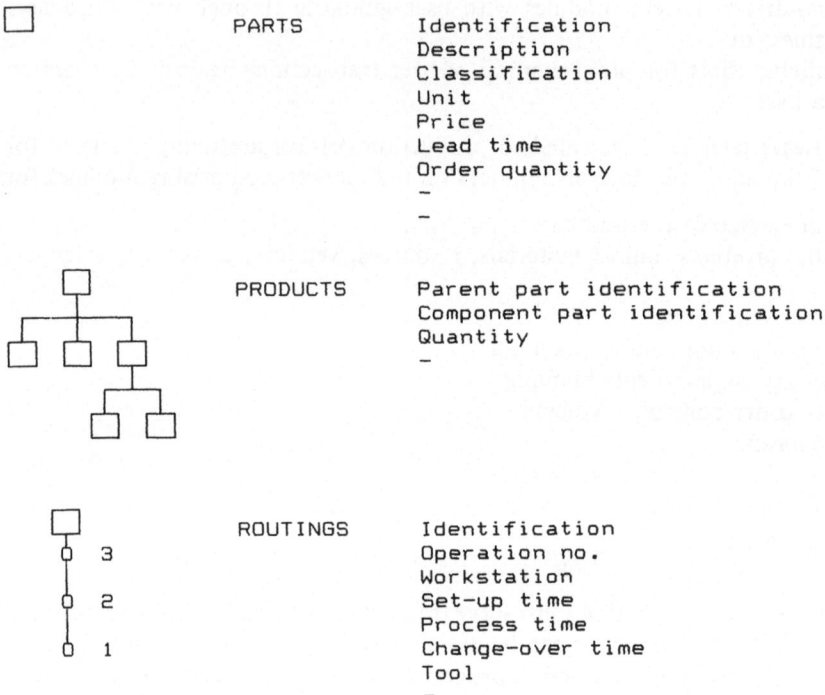

```
         PARTS        Identification
                      Description
                      Classification
                      Unit
                      Price
                      Lead time
                      Order quantity
                      -
                      -

         PRODUCTS     Parent part identification
                      Component part identification
                      Quantity
                      -

         ROUTINGS     Identification
                      Operation no.
                      Workstation
                      Set-up time
                      Process time
                      Change-over time
                      Tool
                      -
```

Fig. 2. Typical data records and data structures within job shop production control

3.2 System Functions

Only system functions that involve calculations, and resulting plans, are mentioned here.

Price calculation of products is in principle performed as an aggregation of costs related to parts and operations. The calculation algorithm can be simple or sophisticated, according to the number of price items that have been established.

The basis of requirements planning is an entered master schedule or a delivery plan that is exploded by means of the bill of materials and process routing registers. Required components on lower levels and required production capacity are calculated. On this level, capacity planning is performed towards groups of work-stations or production departments.

Purchase orders and production orders can be lot sized and dispatched automatically if sufficient decision rules are implemented. In most cases, however, group-to-order and dispatching are performed manually by the planner.

Detail planning of part production is based on a set of production orders with quantities and due dates given from requirements planning. The time schedule for each order is calculated from the routing data for the actual part. Different scheduling techniques exist, based on forward or backward scheduling towards a production calendar. Normally, it is the planner's responsibility to adjust the production programme to fit within capacity limits, but some computer systems also offer capacity levelling.

Available computer systems aim at finding a possible solution that fits within capacity constraints and order due dates. Upper-level objectives like minimum

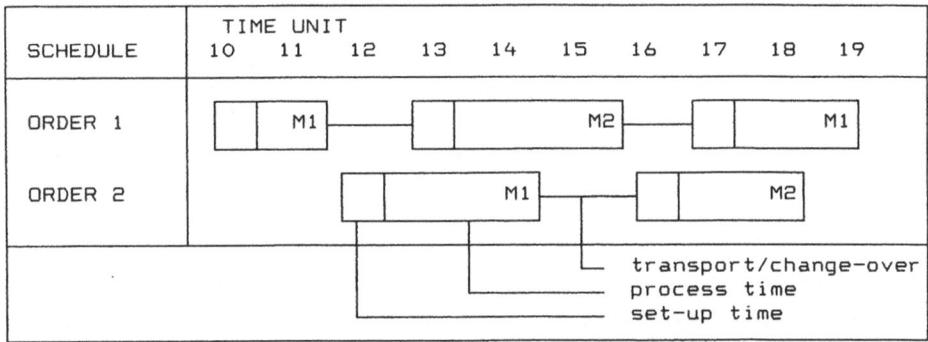

A

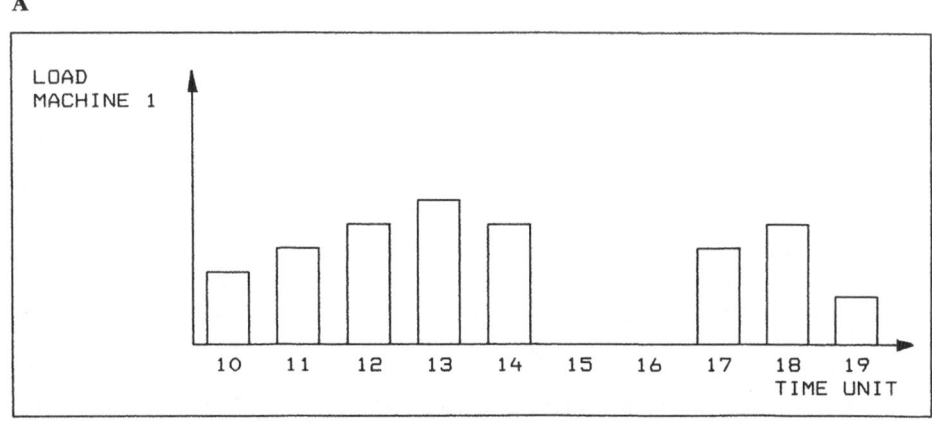

B

Fig. 3A,B. The fundamental plan documents. A Progress plan for production orders; B Planned load for work-station

manufacturing lead-time or maximum total utilization of equipment are not brought into consideration. Neither are new production principles like OPT or Kanban included in existing computer-aided production planning tools.

The result of planning is normally documented by the two principal plan documents shown as bar graphs in Fig. 3.

4 Experiences

Positive experiences from implementation of this kind of computer-aided production management are summarized as follows:

- The computer system is a flexible tool for planning and control, and not a black box that produces rigid plans and reports. The production manager is able to adjust plans according to events as they arrive, and has got an efficient tool for general information processing.
- The system allows continuous watch of work in progress, resource load, and stock levels.
- Reported feedback from production is collected in the database, and is the source of cost calculations and accounting after order completion.
- Access to up-to-date production data from user terminals is a major improvement in distribution of management information and communication between involved personnel.
- Historical data collected in the database is a basis for different kinds of analyses and a platform for further development of planning functions.

However, companies that have implemented computer-aided production management systems have met several complications and unforeseen problems during the implementation project:

- The need of a parallel *education* of involved personnel, and *training* of system users, have been underestimated. The consequence is bad utilization of system benefits.
- User interface and working routines must be designed according to company-specific needs, and on the other hand users must show the necessary *discipline* in preparation and registration of data. Correct data entry from production is perhaps the weakest part of this kind of management system.
- Work in connection with database building and design of new routines has to be carried out by key personnel in the production organization. In a common conflict between daily trouble-shooting and long-term project work, the latter is often sacrificed. This is a matter of *project management,* which is another weak point.
- Formalization of information means coding, classification, and formal descriptions. Changed notation of parts, operations, etc. also means change of technical specifications and drawings. This is another major task that is often neglected when personnel are assigned to the project.

5 New Requirements

Existing computer-aided production management systems work best under the following conditions:

- There is a frozen planning horizon of, for example, 4 weeks, within which the master schedule cannot be altered.
- The bill of materials for each product should be well defined with a minimum of component replacement over time.
- Each part or product should have one single routing.
- Components and subassemblies are well-defined milestones in the production line.

These conditions are ideal and difficult to satisfy in reality. We build a rigid, and maybe artificial management structure when we apply the computer solution to our production problem. Most likely we obtain suboptimizing of production objectives, and we become more eager to satisfy system requirements than economical production. Some new tendencies tend to disturb our ideal conditions even more:

- Increasing degree of customizing means an increasing number of product variants. The bill of materials is no longer well defined.
- Delivery time is becoming the most important competitive factor. There are no longer 4 weeks available for master scheduling, requirements planning, order dispatch, etc.
- Reliability of delivery promises is also becoming a competitive factor. Required service level is increased.
- The amount of goods on stock or in production have to be reduced, because this means reduced need of expensive capital and shorter throughput time.

These elements point towards another kind of management system. We have to consider the total production system, including plant layout, human organization and management system, as a total unity. In addition, each company has to reconsider product design, because standard elements that can be configured to a variety of end-products reduce the need for stored components.

6 A Response to New Requirements

The database established during what we can call the first generation of computer-aided production management systems is a platform when we start responding to these new requirements. Further development of a complete production system requires analysis based on collected historical production data. The main elements of a suggested *change process* are described below:

Product-oriented layout – reduction of stock locations. The traditional workshop for job shop production has a functional layout. This is in many ways incompatible with new requirements:

- Splitting of the process routing through different, geographically separated work-stations implies complicated routings, long transport times, and complicated queue situations. The total manufacturing lead time is far longer than the theoretical minimum.
- This splitting after operation type gives many operations and a complicated planning task.
- This kind of production system is not well suited to quick response to customer demands.

The ideal answer is to divide the production system into product-oriented lines that are specialized for the routing of a product family. This line should have a total throughput time not longer than the customer delivery time, if possible. The planning problem is reduced and the number of stock locations is reduced. Figure 4 illustrates this production concept.

MA-oriented organization. In traditional line organizations the production management functions are assigned to different departments, as illustrated in Fig. 5.

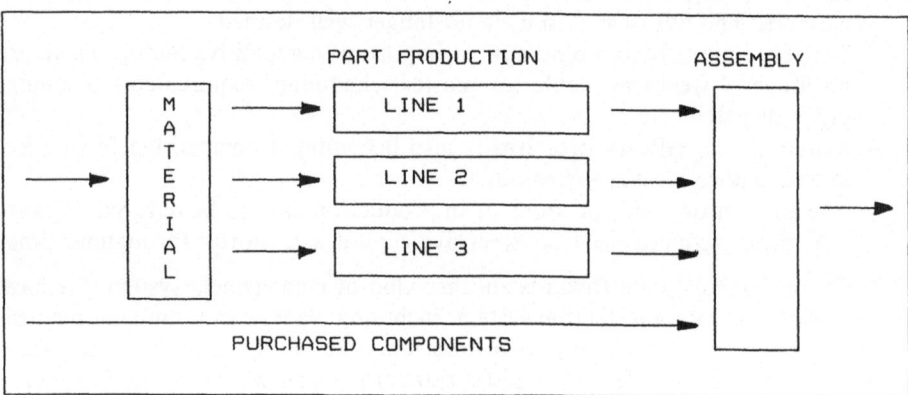

Fig. 4. Conceptual layout for product-oriented production

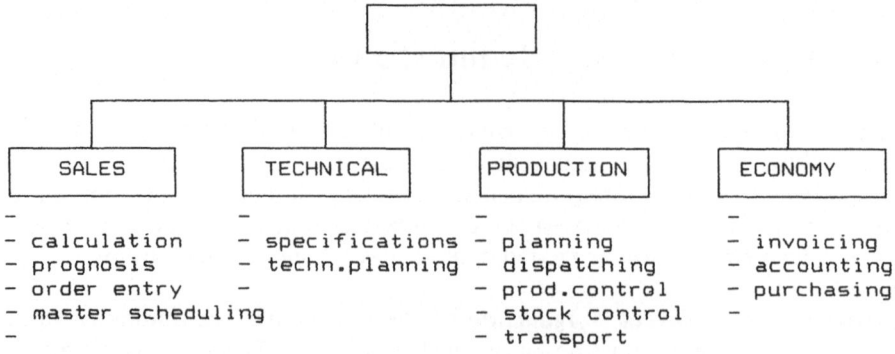

Fig. 5. Traditional organization of production management functions

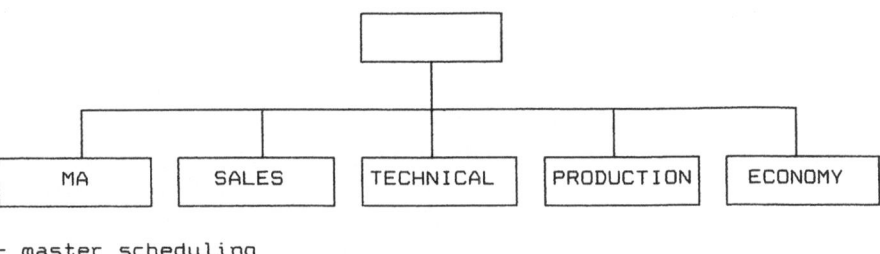

- master scheduling
- purchasing
- stock control
- transport

Fig. 6. MA-oriented organization

The traditional organization has required a horizontal integration of management functions through different departments, with complicated communication routines. This organization form is not the most efficient for customized production.

One way of handling this is to reorganize towards an MA-oriented organization — that is, an organization that separates all functions connected to order execution and materials administration in a separate organizational unit. This department performs all planning, purchasing, logistic functions, and customer order control. Figure 6 shows a possible organization chart.

Focus at A-items and bottlenecks in production. The idea is to allocate different planning and control effort to different parts and products. All stored items are classified according to average value on stock, and divided into A-, B-, and C-category. Management routines are made different for each category.

New management principles also point to the fact that there will always be bottlenecks that determine the capacity of the total production system. Planning should first of all take care of an optimal utilization of these work-stations.

Education and training. Practising new philosophies within production management is a matter of understanding underlying mechanisms. Together with the introduction of computer-aided management systems, all involved personnel have to be taught about production planning and control, the economic impact, objectives, and of course the possibilities of computerization. This teaching is also a central part of the necessary motivating work that has to be carried out by managers to make the total organization adapt to the change process.

7 Conclusions

Computer-aided systems for production management have been available the last 10 years, and have been implemented in many companies with varying success.

The most successful installations are found in companies that have taken care of the following:

- Take care of motivation and supply new competence. This leads to an acceptance of new technology and new working routines.
- A change of production layout and reorganization of management functions can reduce the complexity of production management, and should be a predecessor to implementation of a computer-aided management system.
- Establish an overall management concept and identify subsets that can be computerized. Perform the development project by controllable steps.
- Relate production management to economical objectives. Change the routines that have most economic impact first.

Companies that have been through a first generation of computer systems for production management have struggled with coding and classification. The achieved database is, however, the key to further development of a profitable production system, and the experiences are a platform for future development according to new requirements.

New technologies and market demands have to be considered when future management techniques are developed. Key factors are

- Standardizing on component level
- Product-oriented flow
- Flexible, adjustable equipment
- "No storage"
- Focus on A-items and bottlenecks
- Education.

Finding optimal management techniques for job shop production is more challenging than ever.

8 References

1. Wight, O. W., Manufacturing Resource Planning: MRP II. Oliver Wight Limited Publications, Vermont, USA, 1984
2. Burbidge, J. L., The Principles of Production Control. Macdonald & Evans, Plymouth, England, 1978
3. Aaram, J., New Principles in Production Management (in Norwegian). Production Engineering Laboratory, NTH-SINTEF, Trondheim, Norway, 1986

Chapter 26
Production Control in the Electromechanical Industry

Siegfried Augustin

Dr. Siegfried Augustin studied at the Montan-Universität in Leoben, Austria and received his M.S. in 1970. From 1970 to 1973 he worked with a consultant firm in West Germany and wrote his doctoral thesis on production planning and control. After receiving his Ph.D. in 1973 he joined Siemens AG, Corporate Research and Technology, and was engaged in the field of production management, information systems, and logistics. He managed projects in basic research as well as in practical application. Since 1985 he has been manager for R & D planning in the Department of Technology Planning and Cooperation (Siemens), and, since 1986, also lecturer for production control at the University of Karlsruhe. Dr. Augustin is a member of IFIP WG 5.7.

1 Today's Situation

Today's situation in the electromechanical industry is characterized by a thorough-going change of structures − in the markets as well as in products and production. As is also the case in other branches, the time of quiet and protected markets has gone. A lot of competitors, especially coming from the "triad" − as Japan, the USA, and Western Europe are often called − are struggling for market shares. The market itself is becoming more and more customer-oriented; this means that the products, their functions, quality, and availability are being defined by the market, no longer by the producer. So delivery time, service level, and quality are the most important weapons in this struggle now. As they are all relevant for production, production control has to contribute to sharpen them; in particular, high flexibility and quick reaction can be influenced by production control, which is being challenged first and foremost in this situation.

The change of structures of electromechanical products is dominated by "electronization", i.e. the increasing employment of microelectronics (Fig. 1) [1].

The penetration of electronic components into electromechanical products can be regarded as representative of the effect of new technologies on products. The first consequence of these structural changes is a quite new relation between the efforts in manufacturing and development. This is confirmed by practical experiences with telecommunication products (Fig. 2). In this case the effort for development (in hours) has increased to up to 1000%, whereas the effort for

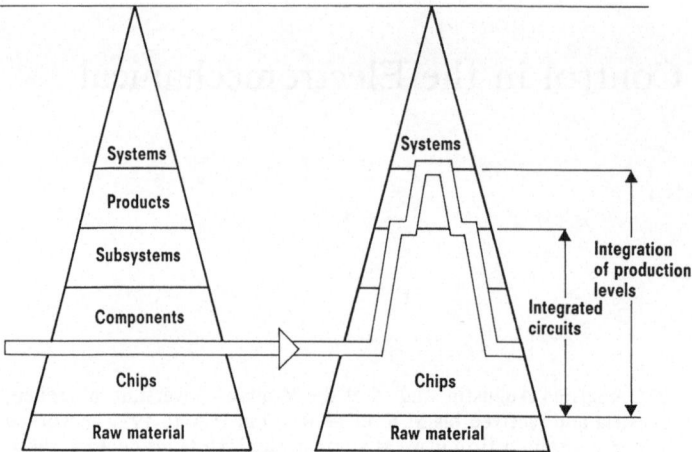

Fig. 1. Effect of electronization on the structure of products

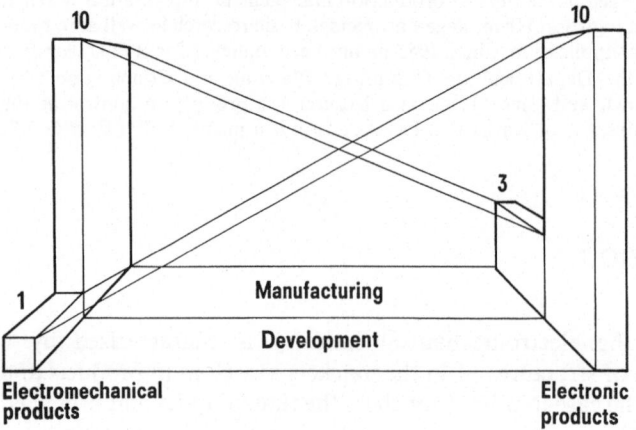

Fig. 2. Effects of electronization on the relation between manufacturing and development for telecommunication products

manufacturing (prefabrication and assembly) has decreased to a third. But also the cost structure of the products has changed, as the experience with the same products shows (Fig. 3).

All these effects initiated by the product innovation are overlapped by the influence of new technologies on production processes. The application of microelectronics in machine tools has made possible an exact steering of processes; as such processes are full reproducible they can be automated. Up to the last years especially, processes in serial production have been automated which have been characterized by a high degree of Taylorism. However, these processes were rather inflexible. Only by advanced technologies for setting-up could manufacturing processes be made more and more flexible, so that flexible manufacturing cells and flexible manufacturing systems (FMS) could be realized. They represent the first step of the quick evolution of automation.

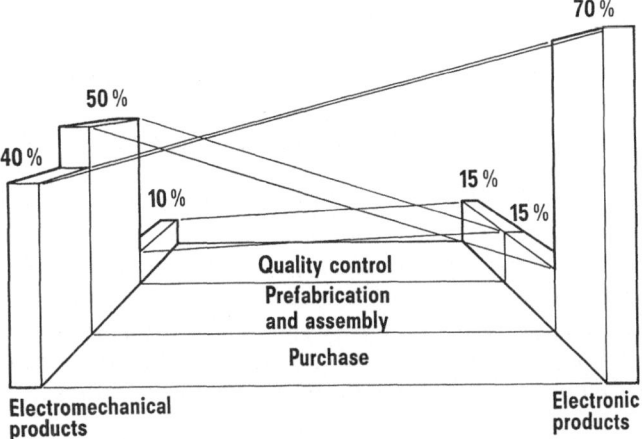

Fig. 3. Effect of electronization on the cost structure for telecommunication products

One of the effects of automation — besides the main advantages such as high flexibility, better productivity, better quality of products etc. — is the simplification of production control systems. Batch sizing and sequencing to minimize setup times are unnecessary; hence the influence of setup times need not be considered.

In addition to these influences the new technologies in information and communication processing also have to be taken into account. They play, of course, an important role in the realization of CAM systems, but are also relevant for production management and therefore for production control. How these technologies can help to master the problems of the electromechanical industry will depend on the individual goals of the production management systems.

Last but not least, one phenomenon should be mentioned which seems to be typical for "big" production management systems. In the course of time many systems have become so complicated that it is impossible for the user to recognize any relation or causality between input and output of the system. So the users are often unable to strive to achieve defined goals by decided actions. On the contrary, they have lost their trust in the whole system. Attempts to solve these problems by adding new EDP functions are being recognized as wrong. New ways have to be found to attain the new goals of the production of electromechanical products and to master the situation, which can be summarized as follows:

- Increasing importance of customer-orientation
- Decreasing number of production levels
- Decreasing importance of prefabrication and increasing importance of purchasing and assembly
- Higher degree of automation in manufacturing and transport
- New priorities for production management goals.

2 Logistic Strategies

The requirement of high service level and short delivery time can be fulfilled in principle in two ways: by high inventories or by short lead times in the production areas. Because of the extended capital tie-up (cost of storage, stock in transit, pipeline stock) the first way to high flexibility is undiscussible today. This should be illustrated by an example from practice. The delivery time of a certain product family of contactors was about four weeks on an average. Suddenly the sales department required a delivery time of one day, otherwise the contactors would be no longer competitive on the market. For economic reasons it would have been impossible to produce all variants on stock and to hold them available. So the problem had to be solved by realizing logistic principles as there was a continuous flow of materials and information along the whole pipeline of orders. As such problems are typical for the electromechanical industry, this kind of solution can be regarded as representative.

To make such goals feasible it is necessary to understand the area "production" no longer as the sum of prefabrication and assembly, but as the chain or pipeline of all areas through which orders have to pass. That means that all functions and activities from order clearing to shipping have to be subordinate to these goals and to contribute to their achievement. Today, the main criterion for the efficiency of an industrial company is the flow rate within the above-mentioned pipeline. Other criteria useful for this purpose are the stock turnover and the difference between delivery time and lead time, which describes the coverage. The flow rate is the relation between operation time and the whole lead time. Despite the high degree of computer aid in the area of production management and control the flow rate of orders in the manufacturing area as well as in the "administrative" areas before and after is only about 10%. Today's production systems have to make feasible, however, flow rates of at least 90%, turnovers of more than 20, and lead times which are almost coincident with the delivery times of the products.

In general, the attainment of these goals, which corresponds with the realization of a just-in-time (JIT) principle, is not possible by merely redesigning the existing production control system; on the contrary it is necessary to call all hands and to use all new technologies, methods, and cognitions. As the aspect system approach [2] has been shown as practicable and successful for designing production control systems, the four aspects

- Goals
- Functions/methods
- Responsibility/organization
- Information equipment

are used as the thread of the following section in which the state-of-the-art of the present activities is given.

3 Production Control on the Way to JIT and CIM

3.1 Goals

Only quantified goals will be reachable. So it is necessary to derive the goals of production control from the concrete overall goals of the company – for example, return of investment, market shares, revenue. Here the working out of a hierarchy of goals and means has proved to be a useful method (Fig. 4). By such a procedure the contribution of each area of the pipeline will be quantified. Based on these results a rough feasibility study can be carried out already. In most cases the precondition for the achievement of goals is a new production layout with some of the new technologies mentioned above. Only the cases which cannot be cushioned by the layout must be regulated by a production control system. That is a fundamental difference to the production control philosophy of the last decades.

Before the goals of production control are fixed definitely, care has to be taken that they are not contradictory to each other. The experiences from practice show that this problem is one of the big barriers in the way of a JIT system. Sometimes a third of the whole effort in a design project for production control must be used for harmonizing the goals. Typical contradictions are

- High capacity loading vs. avoidance of floor stock
- Minimization of ordering costs vs. deep stock level
- Possibility of alteration during final assembly vs. no alterations after order release.

As mentioned already the priorities of the "classical" goals of production control have changed. In particular, the importance of high capacity loading has decreased in favour of short lead times and low inventories.

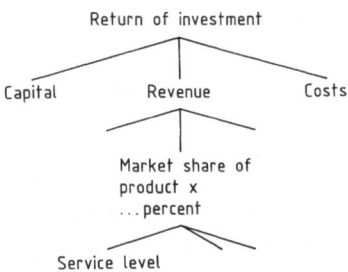

Fig. 4. Hierarchy of goals and means: return of investment

3.2 Functions/Methods

The necessary functions of a production control system have to be derived from its concrete goals and subgoals. These functions first and foremost have to guarantee the continuous flow of orders (material and information). So two groups of functions can be distinguished:

- Functions for guaranteeing the 100% availability of all production resources (materials, capacity, tools, information)
- Functions for synchronizing all production processes by situation-oriented balancing of offer and demand of capacity.

For executing these functions different methods are available. Which of them will be elected depends on

- The structure of the product
- The degree of customer-orientation
- The type of production (mass production, serial production, small batch production, etc.)
- The system of manufacturing (job shop production, group production, assembly line production, etc.).

There exist a lot of articles in which the most important methods of production control used in the electromechanical industry are described in detail [3–7]. It should be mentioned that is has proved very useful to design a production system with the help of simulation: in many cases the simulation model can be used – in a modified version – for production control.

3.3 Responsibility/Organization

The execution of functions to attain the goals needs capacity, know-how, and responsibility. Therefore, it is an important task in designing a production control system to ascertain the appropriate organizational structure of this cybernetic system, i.e. to define organizational units with clear responsibility and to attach responsibility-holders to these units. Furthermore, the rules of interaction between the units have to be fixed.

In designing the organization it has to be considered that there is a strong reciprocation between the required know-how and the capacity for decision on the one hand and the qualifications of staff and the performance of information equipment on the other hand. According to the individual situation the organizational parameters such as managing scope, controlling range, number of hierarchical levels, degree of decentralization, allocation of functions to controllers, and degree of formalization can be laid down [8]. The better the qualification of staff, the more functions can be delegated and executed autonomously, and the lower will be the costs of controlling [9].

In the electromechanical industry many functions of production control are decentralized now. At the same time production as well as organization have become product-oriented in a relatively large measure. The consequence of this trend is the dividing of "big" productions into small units [10]. As practical experiences show, the importance of designing and realizing the appropriate organization is often underestimated. But especially for the realization of JIT and CIM systems which need a very high degree of integration it has to be ensured that the logistic pipeline won't be interrupted by organizational borders.

3.4 Information Equipment

According to the logistic principles by which production control today is influenced, the information system has to provide the right information at the right point of time at the right place. The information requirement – quality, quantity, actuality, availability date – is clearly defined by the functions and methods of the special production control system as well as by its responsibility-holders. The choice of information equipment and communication equipment and of the functions which will be definitely formalized, i.e. the information processes which will be automated, is an economic consideration [11].

One other main consideration in planning and realizing the information layout in the electromechanical industry is the slogan "More personal responsibility again!". That means that the big central EDP systems of the 1960s and 1970s (which have the image of "big brothers") have to be decomposed more and more. It is a prerequisite for a flexible production system, of course, to have a flexible information system. However, this presumes highly qualified staff – and highly qualified managers! [12].

Based on an integration concept, functions and decisions can be decentralized with the help of personal computers or work-stations which enable the user to control the production processes in an interactive way with a high measure of comfort (graphic screens, simulation tools, expert systems, etc.). According to the approach of small units, there often exist different "tailored" production control system for each unit.

For the design of production control systems in the future it should be taken into account that this will be done on the basis of an overall integration concept for all information processes in the production area as well as in the whole company. Such a concept has been worked out by the electromechanical industry and has proved already to be a useful approach [13].

4 References

1. Eidenmüller, B., Auswirkungen des technologischen Wandels auf die Produktion. In Proceedings of 2nd CIM Conference of Betriebswissenschaftliches Institut der ETH Zürich, 8th April 1986, Rüschlikon
2. Augustin, S. and Hübner H., Designing computer supported production management systems using the aspect-system-approach. In Proceedings of the IFIP WG 5.7 Working Conference on Production Management Systems, Vienna, 1983, H. Hübner (ed.). North-Holland, Amsterdam, 1984
3. Takeda, K., Analysis of Just-in-Time production management systems. In Proceedings of CAPE '86, 20–23 May 1986, Copenhagen, K. Bø, E. A. Warman, and L. Estensen (eds.). North-Holland, Amsterdam, 1986
4. Wiendahl, H. P., Integration von Prozeß- und Auftragssteuerung. In Proceedings of 1st Euromanagement Congress, 24–25 September 1986, Stuttgart: "Zukunftsorientiertes Produktionsmanagement in Europa". Springer, Berlin, 1986
5. Beier, H. H., AZR – A new approach to shop-floor management for automated and non-automated factories. In Proceedings of the IFIP WG 5.7 Working Conference on

Decentralized Production Management Systems, Munich, 1985. S. Augustin, R. Gündling, and J. Ohanian (eds.). North-Holland, Amsterdam, 1985
6. Gündling, R., Entwicklung allgemeingültiger Bausteine für Design-Projekte von Montagemanagementsystemen auf der Basis des Aspekt-System-Ansatzes für die Gestaltung von Produktionsmanagementsystemen. Dissertation, Montanuniversität Leoben, Austria, 1985
7. Inoue, I., Yamada Y., and Adachi, T., A tools-system in decentralized production management systems. In Proceedings of the IFIP WG 5.7 Working Conference (details in Ref. [5])
8. Oberhofer, A. F., Corporate management and organization. In Proceedings of APMS '85, 28 – 31 August 1985, Budapest, E. Szelke and J. Browne (eds.). North-Holland, Amsterdam, 1985
9. Kneip, K., Management by corporate identity. W&V (Werben und Verkaufen), 1978, Nos. 47 – 52
10. Tress, D. Kleine Einheiten in der Produktion. zfo (Zeitschrift für Organisation), 1986, No. 3
11. Augustin, S. and Gündling R., Computers – a drug for PM-problems? Planning and realizing the information layout. In Proceedings of APMS '82, Bordeaux; G. Doumeingts, and W. Carter (eds). North-Holland, Amsterdam, 1984
12. Eidenmüller, B., Human aspects in the factory of future. In Proceedings of APMS '85. (details in Ref. [8])
13. Augustin, S., Beier, H. H. et al., Atlas der innerbetrieblichen Informationsverarbeitung. ZVEI (Zentralverband der elektrotechnischen Industrie e.V., Frankfurt/M.) (ed.). Sachon, Mindelheim, 1985

Chapter 27
Production Control in the Electronics Industry
Ichiro Inoue

Ichiro Inoue graduated in physics from Kyoto University, Japan in 1969 and studied at the graduate schools of the University of Hawaii and MIT, USA from 1969 to 1971 and at the Max-Planck Institute, West Germany from 1972 to 1973. He joined the Computer Systems Research Laboratory, Central Research Laboratories, NEC Corporation in 1974. He has been engaged in the research and development of system sciences and technologies, on complex system evaluation, system life cycles and, as a practical application field, on production systems. He is a member of IFIP WG 5.7, of the Information Processing Society of Japan, of the Japan Society for Simulation Technology, of the Japan Society of Operations Research, and of the Japan Industrial Management Association.

1 Electronics Industry Overview

Electronics industry products are segregated into three categories:
- Home use products: VTR, television, tape recorder, stereo set, home appliances, etc.
- Industrial use products: communication equipment appliances, computer, electrical measurement apparatus, office machines, etc.
- Electronic parts and components: general parts, active element parts (semiconductors, LSI, electronic tubes, etc.), etc.

The electronics industry is a newly evolved industry. Especially after the oil crisis in 1973, it started to play an important role in changing today's industrial structure in the social environment in which we are encouraged to save resources and save energy. Developing a stronger electronics industry, therefore, is a very important policy, commonly seen among industrial countries. In these countries, electronics industries have been making great efforts in R&D and have actively made heavy capital investment in support of the concept.

Electronics has reached the top level in most advanced technologies, and has become a leading industry in every industrial nation's economy. (Figure 1 shows an example of manufacturing industry in Japan. It can be seen that the relative position of electronics has been markedly advanced in recent years.) Further-

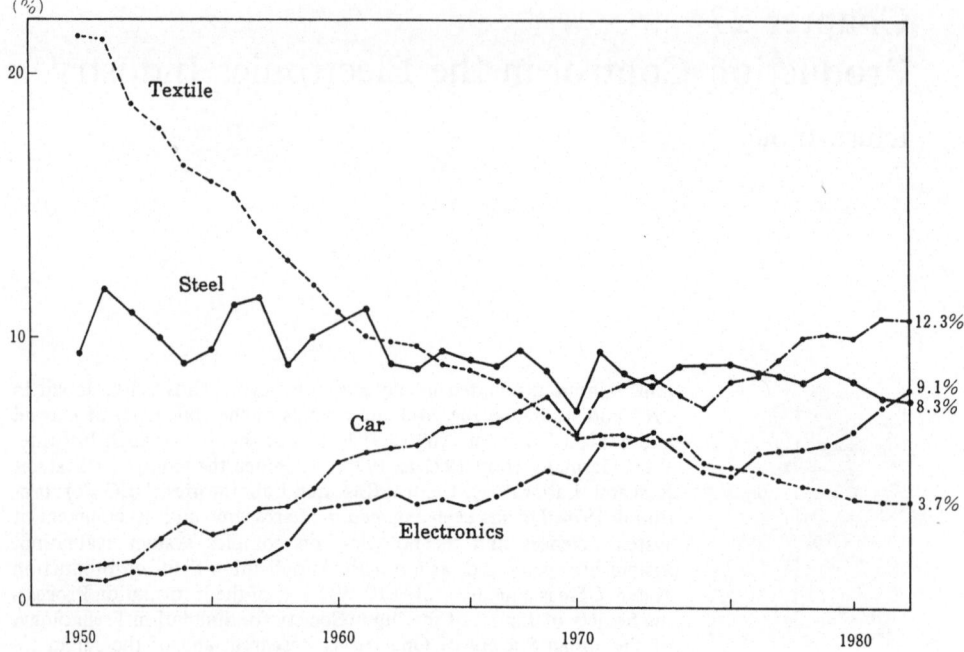
Fig. 1. Electronics industry position in Japanese manufacturing industry [EIA (1)]

more, as today's information-oriented society advances, the electronics industry has a greater effect and plays a greater role in society as a high-technology industry.

Three major electronics industry conscious areas are North America, Western Europe, and Japan. The total production in these three areas reached about US$ 3200 billion in 1984 (about US$ 2770 billion in 1983) and is still growing at a high rate.

2 Electronics Industry Environment

The main electronics industry features are as follows:

- Competition among enterprises is fierce. Competition includes technology development, quality, cost, delivery time, etc.
- Product life is short.
- Technological evolution is rapid.
- Customer's needs are diversified.

In order to cope with this severe environment, the following problems are expected to be solved in the production system domain.

- Quick adaptation of newly evolving technologies
- Flexibility enhancement towards frequently occurring technical changes, such as design changes, production engineering changes
- Continuous cost reduction
- Lead time minimization
- More multiple-products and smaller-quantity production.

Achieving this goal is not so easy. The phenomena commonly seen in production systems in the electronics industry can be listed as follows.

- Rush orders are frequent.
- Load imbalance among workshops is usual and this causes lower machine utilization efficiency and, at the same time, large work-in-process inventory.
- Production lead time cannot be minimized.

Home-use electronics product manufacturing is taken as an example. Even the manufacturing is usually multi-products-small/medium-quantity production and organized mixed production line. Short product life cycle and variety production bring about frequent line re-organization. TV manufacture, for example, often necessitates line re-organization and manpower layout changes, once or more every week, while taking into consideration a better combination between weekly job orders and worker skills for higher productivity. Usually such line re-organization requires much time to find one feasible solution and break down its solution into a concrete daily operational plan. Therefore, limited time does not allow sufficient optimization operation. Once practical manufacturing starts, factory line foremen and production management/control sectors are requested to take flexible action towards counteracting various fluctuations and to control the line output.

Another example is a case involving LSI manufacturing and PCB (printed circuit board) assembly. During the manufacturing, automated and semi-automated machines are used comparatively intensively. Despite the appearance of the manufacturing site, the manufacturing system usually still requires manual operation in machine-to-machine transfer, testing, etc. The very small quantity items in multi-product-small-quantity production, newly started-up production, and test/sample products inevitably are manually manufactured. Even in such a manufacturing system, consisting of a large number of automated/semi-automated machines, quite a few manual operations remain, and production flow, as a whole, is not as smooth as is desirable. The flow optimization and production management/control is not considered to be sufficient.

3 Production System Improvement Approaches

Aiming at high-productivity, high-flexibility production, top-down approaches and bottom-up approaches are vigorously taken in the electronics industry. The top-down approach involves

- Introduction of automated hardware, such as NC machines, robotics, FMS, etc.
- Computerization of production planning and control, EDP systematization of shop floor data collection, etc.

etc. The bottom-up approach involves

- Small-group activity, such as quality control (QC) circle, zero defect (ZD) circle, etc.
- Daily improvement proposal system

etc. In both approaches, production management is a key factor in achieving success.

4 Production Management Method

The electronics industry is a newly evolving, rapidly growing industrial area. Production management, under severe industrial circumstances, requires a new production management philosophy. Production management should achieve (1) quick start-up of production, (2) flexibility to adapt to widespread and frequent changes, (3) multi-products-small-quantity production, (4) continuous cost reduction, and (5) constantly positive self-improvement. In order to establish a new production management philosophy, the electronics industry has been taking a positive attitude and vigorous steps to adopt various production management methods and to improve them.

Production management in the electronics industry can be classified into the following four categories:

- Traditional
- Materials requirements planning (MRP) oriented
- Just-in-time (JIT) oriented
- Autonomous decentralized and integrated.

Traditional method. This includes "Seiban"" (product-oriented production), parts-oriented production methods, and other traditional methods. In the "Seiban" method, necessary materials and parts are arranged and acquired for each ordered product, and the product's identification number is assigned. (The name "Seiban" originates from its literal meaning of product number.)

When the specific materials parts are constantly and frequently used, it was considered to be more efficient to maintain a certain stock level for these items, in order to save on materials/parts handling time and labour. Production management on the basis of these methods is inclined towards "order launching and expediting". These methods are considered to be unsophisticated and out-of-date, but more than half of the world's small/medium sized companies are still using these methods. (In Japan, more than 95% of manufacturing companies are small/medium sized.)

MRP. As more multi-product-small-quantity production was continued, the "Seiban" method and inventory stock method brought about management time and labour losses. To solve these difficulties, the MRP method was invented in the USA in the late 1960s and attempts have been made to put it into practical operation.

The MRP method is, in principle, based on the planning-oriented centralized-control philosophy. It requires extensive computer power to realize its functions. The electronics industry, where product life cycle is very short and where product design and production schedule are frequently changed, has been trying to introduce it to its production management. However, no report has yet indicated that it has reached full operation. Some MRP sub-functions, such as the B/M processor, are built into the system and are contributing to production management quality enhancement. MRP II and MRP III, which are revised and expanded versions of MRP, were proposed in recent years, and their application to industries was started.

Kanban method and JIT method. The Kanban method or Toyota production method, which brought about a great success in the car industry, had quite a large impact on the electronics industry. In particular, home-use product manufacturing is similar in terms of multi-products-small-quantity production, big demand fluctuation, etc. Therefore, great efforts have been made in adopting the Kanban method to the electronics industry with much success. The Kanban method is, in principle, based on Just-in-Time (JIT) philosophy and has been modified and even improved in the electronics industry. The action plate method, signal method, used by Hitachi Co., and the non-stop production method, used by NEC Corp., are some examples on this line.

Autonomous decentralized and integrated production management. This is a newly proposed style. In production systems in the electronic industry, frequent daily changes are a routine occurrence and continuous self-improvement and self-growing for the future environmental change are fundamental. The origin for the innovative vitality is expected to be obtained from all the related sectors and management hierarchy level, from workers and supervisors to managers. Therefore, it is very important in the electronics industry for management to activate QC circles, ZD circles, etc. and to enhance management quality for effective utilization of improvement proposals. For these purposes, autonomous decentralization for the circles and organizational sectors and proper information integrity for necessary information acquisition are fundamental requirements.

In order to enhance production management quality and efficiency, combination of the above-listed methods is a good possibility and support tools are indispensable and important factors. Simulation tools as system evaluation support have to be further developed and advanced. Database and information networking as an infrastructure have to be effectively constructed to ensure the active communication among autonomous decentralized organizational sectors. Making most use of these technologies and tools, production management is expected to play a big role in the electronics industry.

5 References

1. Denshi Kohgyo Nenkan '86 (Electronics Industry Annual Journal), 1986
2. Rolstadas, A., Trends in production management systems. In Advances in Production Management Systems 85, E. Szelke et al. (eds.). North-Holland, Amsterdam, 1986
3. Porter, R. W., Rx For Sick MRP Systems. In Proc. APICS '80, 1980
4. Chushokigyo Jigyodan, Research Report on MRP in small/medium sized companies. Vol. 86, No. 6, 1985 (in Japanese)
5. Takeda, S., Shimoyashiro, S. et al., A decentralized control method for online shop floor expediting systems. In G. Doumeingts et al. (eds.). Advances in Production Management Systems. North-Holland, Amsterdam, 1984
6. Inoue, I., Yamada, Y., and Adachi, T., A tools-system in decentralized production management systems. Computers in Industry, Vol. 6, No. 6, 1985
7. IFIP WG5.7 working conference on decentralized production management systems. Computers in Industry, Vol. 6, No. 6, 1985

Appendix
A Drafted PM Glossary

John L. Burbidge*

1 Introduction

Most dictionaries and glossaries of terms today are not structured. They contain many different terms, arranged in alphabetical sequence, with little or no attempt to show how the terms are related.

A new IFIP "Glossary of Terms used in Production Control" has now been produced, which is structured. It is based on a classification of Production Management, of which Production Control is a part, and the terms and definitions it contains were chosen to fit the selected classes and subclasses.

The objective of the Glossary is to provide definitions for the principal special terms needed to carry on an intelligent discussion about Production Control. As Production Control is just one of a number of closely related subsystems of Production Management, it would be impossible to carry on such a discussion using only those terms which are special to Production Control. For this reason the IFIP glossary has been based on a broad classification of the field of Production Management, and a deeper classification of the subfield of Production Control.

2 The Primary Classification

The primary classification of Production Management divides the subject into eight main functions. These are tabulated in Table 1.

These Management functions are sets of closely related management tasks which require similar skills for their efficient performance. They are subsystems of the total Production Management System. All sciences depend for successful development on the adoption of an orderly and systematic classification framework. The division of Production Management into functions is, to give one example, equivalent to the division of the Science of Medicine into the

---

* Professor John L. Burbidge is also author of Chapter 5. His biography and photo appear on p. 71

Table 1. The management functions

	Function	Type of task	Controls
1.	Product Design	Plans final form of product	Quality Control
2.	Production Planning	Plans how product is to be made	Process Control
3.	Production Control	Plans materials supply and processing activities	Progressing; Loading and Inventory Control
4.	Purchasing	Finds sources and negotiates contracts	Purchase progressing
5.	Marketing	Develops markets; sales and distribution	Sales control
6.	Finance	Plans investment, profit and cash flow	Budgetary control, Standard costing
7.	Personnel	Plans conditions, employment and training	Merit rating, attendance control
8.	Secretarial	Plans communications and data processing	Data control

human subsystems named the Skeletal, Circulatory, Lymphatic, Respiratory, Nervous, Digestive, Urinary, Reproductive, Muscular, and Endocrine systems.

The functions provide a universal classification of production, but their significance varies with the type of Industry. Product design for example is of major importance in the aero-engine industry. It is still necessary but much less important in the cement industry. The main importance of the functions is in Training, Research, and System design. The functions do not provide a good basis for organization. In all the functions the higher-level decisions will normally be made at a higher level in the organization structure than the others, and it will often be advantageous to divide some of the functions into parts which are allocated to organizational units specializing in different families of products.

The processes of Management are Planning, or deciding what to do in the future; Direction, or the process of management by which plans are caused to be implemented; and Control, or the process of management which constrains events to follow plans. The methods of Direction tend to be the same for all functions, but each function has its own planning and control methods, as is indicated by Table 1.

3 The Relationship Between the Functions

The main relationship between the Production Management functions is that concerned with data flow. Fig. 1 illustrates the fact that each function originates some forms of data and processes some of it into higher forms. Each function sends the data it originates or develops to one or more of the other functions.

Appendix. A Drafted PM Glossary

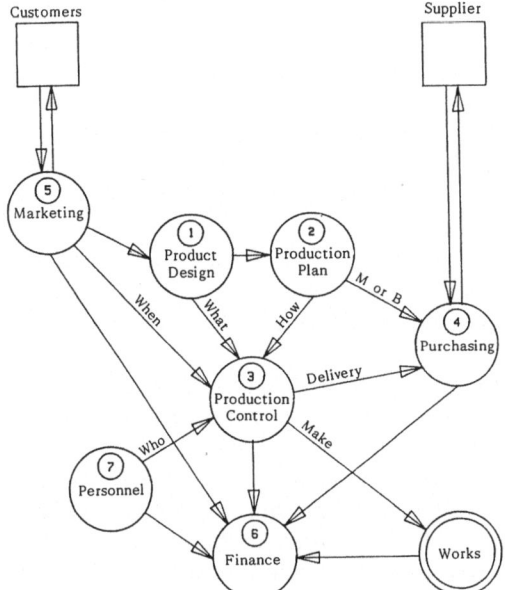

Fig. 1. Data flow

It will be seen in Fig. 1. that Production Control receives the principal data it requires from Marketing, Product Design, Production Planning, and Personnel. It in turn is a main supplier of data to Purchasing, the Works, and Finance. As far as the functions other than Production Control are concerned, the IFIP glossary deals mainly with this inter-functional data flow.

Another important functional relationship arises because the parameter changes made for the regulation of particular function tasks affect not only the output variables special to that function, but also the values of other output variables in other functions. Care has been taken to include the terms needed for studies of this variable connectance.

4 The Classification of Production Control

The classification of Production Control in the Glossary is more detailed than that for the other functions. It is based on a division of the function into the subfunctions of Programming, Ordering, and Dispatching, and into three feedback control systems, known as Progressing, Loading, and Inventory Control. Each of these classes is divided and subdivided at at least two further levels. Nearly half the terms in the glossary are included in this portion, dealing with Production Control.

5 The Choice of Terms and Definitions

As it was not the intention to create a new language, care was taken to choose terms for each concept which are in general use.

Nineteen different dictionaries and glossaries were used for comparative studies of the definitions of these terms. Some were defined in only two or three sources; other in as many as seven. It was found that these definitions differed not only in wording, but also in many cases in the general range of concepts with they covered. To simplify these studies they were finally restricted to a comparison of the definitions contained in British Standards and in the APICS dictionary (USA), which were seen as the most authoritative sources for Production Control terms in English.

The terms defined in the IFIP glossary can all be found in existing sources, or in recent literature. They are used in the general sense accepted by the majority of these sources. In cases where there are wide differences between the meanings accepted by different sources, it has been necessary to choose between them.

Two thirds of the terms in the IFIP glossary do not appear in BS or APICS. Their definitions are therefore special to the glossary. The definitions for most of the remaining terms were also written to suit the adopted subject classification. In only a small number of cases has it been possible to adopt British Standard or APICS definitions.

6 Conclusion

The new IFIP glossary of Production Control terms has been prepared for the IFIP Working Group WG 5.7, as a source for the terms needed to undertake an intelligent discussion on Production Control. It differs from all existing glossaries in that it is based on a classification of Production Management: that it contains a wide range of closely related terms from other management functions; and that the terms are presented in both the classified form and alphabetically.

Address List of Authors

Chapter 1
Asbjørn Rolstadås
Production Engineering Laboratory NTH-SINTEF, Richard Birkelandsv. 2B, N-7034 Trondheim-NTH, Norway

Chapter 2
John Harhen
Digital Equipment Corporation, 290 Donald Lynch Boulevard, DLB5-2/B3, Marlboro, MA 01752, USA

Chapter 3
Hajime Yamashina
Kyoto University, Faculty of Engineering, Dept. of Mechanical Engineering, Kyoto, Japan

Chapter 4
Oded Cohen
Avraham Goldratt Institute, 7 Parklands Drive, Finchley, London N3 3HA

Chapter 5, Appendix
John L. Burbidge
Wild Goose Leys, Abbots Ripton, Huntingdon, Cambs., UK

Chapters 6, 20
Gideon Halevi
19 Nehardea st., Tel-Aviv 64235, Israel

Chapter 7
Peter Falster
Technical University of Denmark, Building 325, DK-2800 Lyngby, Denmark

Chapter 8
Jim Browne
Dept. of Industrial Engineering, University College, Galway, Ireland

Chapter 9
Wing S. Chow
Sunderesh Heragu
Dept. of Mechanical and Industrial Engineering, The University of Manitoba, Winnipeg, Manitoba R3T 2N2, Canada

Chapters 9, 10
Andrew Kusiak
Dept. of Industrial and Management Engineering, College of Eng., The University of Iowa, Iowa City, Iowa 52242, USA

Chapter 11
J. C. Wortmann
Dept. of Industrial Engineering, Eindhoven University of Technology, P.O. Box 513, Eindhoven, The Netherlands

Chapter 12
Eero Eloranta
Helsinki University of Technology, Laboratory of Information Processing Science, Otakaari 1A, SF-02150 Espoo 15, Finland

Chapter 13
Guy Doumeingts
GRAI, Laboratoire d'Automatique de L'Université de Bordeaux 1, 351, Cours de la Libération, F-33405 Talence Cedex, France

Chapter 14
Jarle Aaram
Production Engineering Laboratory NTH-SINTEF, Richard Birkelandsv. 2B, N-7034 Trondheim-NTH, Norway

Chapters 15, 16
Harinder Jagdev
Computation Department, UMIST, P.O. Box 88, Manchester M1 7JA, UK

Chapter 17
John King
Imperial College of Science and Technology, Department of Management Science, Exhibition Road, London SW7 2BX, UK

Chapter 18
Kathryn E. Stecke
Associate Professor of Operation Management, The University of Michigan, School of Business Administration, Ann Arbor, Michigan 48109–1234, USA

Chapter 19
Birger Rapp
Docentbakken 15, S-10405 Stockholm, Sweden

Chapter 21
Samuel Eilon
Imperial College of Science and Technology, Department of Management Science, Exhibition Road, London SW7 2BX, UK

Chapter 22
Kai Mertins
Fraunhofer-Institut für Produktionsanlagen und Konstruktionstechn. – IPK, Kleinststraße 23–26, D-1000 Berlin 30, Federal Republic of Germany

Chapter 23
Wolfgang Thurow
Bayrische Motorenwerke AG, LZ-2, Petuelring 130, BMW-Haus, Postfach 400240, D-8000 München 40, Federal Republic of Germany

Chapter 24
Bernd Hirsch
Produktionssystematik, Universität Bremen, Bibliothekstraße, D-2800 Bremen 33, Federal Republic of Germany

Gustav Humbert
MBB GmbH, TF 4, Postfach 950109, D-2103 Hamburg 95, Federal Republic of Germany

Chapter 25
Oddmund Oterhals
Production Engineering Laboratory NTH-SINTEF, Richard Birkelandsv. 2B, N-7034 Trondheim-NTH, Norway

Chapter 26
Sigfried Augustin
Siemens AG ZTP FEP KF, Otto-Hahn-Ring 6, D-8000 München 83, Federal Republic of Germany

Chapter 27
Ichiro Inoue
NEC Corporation, C&C Systems Research Labs., 1-1, Miyazaki 4-chome, Miyamae-ku, Kawasaki, Kanagawa 213, Japan